The Rise of the New Network Industries

I0474312

Cutting through the confusion around the nature and implications of digitalization, this book explores the rise of the new digital networks, how they affect traditional infrastructure, and how they will eventually need to be regulated. The authors examine how digitalization affects infrastructure in telecommunications, transport, and energy, and how digital platforms establish themselves as a new network on top of and in addition to traditional ones.

Complex concepts are introduced through short and colorful stories about the founders of the most popular platforms (Google, Facebook, Skype, Uber, etc.) and how they grew to positions of power, drawing parallels with century-old traditional network industries' monopoly power (AT&T, General Electric, etc.). The authors argue that these digital platforms strongly interfere with traditional infrastructures that are heavily regulated and provide essential services for society–meaning that digital platforms should be considered as a new and much more powerful type of infrastructure and will require regulation accordingly.

A global audience of policy makers, public authorities, consultants, lawyers, students, and academics, as well as anyone with an interest in these digital platforms, will find this book enlightening and essential reading.

Juan Montero is Professor of Regulation at the Law School of UNED University, Madrid, Spain; part-time Professor at the European University Institute in Florence, Italy, in the Florence School of Regulation; and of counsel in MLAB.

Matthias Finger is Professor Emeritus from Ecole Polytechnique Fédérale in Lausanne, Switzerland (EPFL); part-time Professor at the European University Institute in Florence, Italy, where he directs the Florence School of Regulation's Transport Area; and Professor at the Faculty of Management at Istanbul Technical University (ITÜ).

The Rise of the New Network Industries

Regulating Digital Platforms

Juan Montero and Matthias Finger

Routledge
Taylor & Francis Group

NEW YORK AND LONDON

First published 2021
by Routledge
52 Vanderbilt Avenue, New York, NY 10017

and by Routledge
2 Park Square, Milton Park, Abingdon, Oxon OX14 4RN

Routledge is an imprint of the Taylor & Francis Group, an informa business

Library of Congress Cataloging-in-Publication Data
Names: Finger, Matthias, author. | Montero, Juan, author.
Title: The rise of the new network industries : regulating
digital platforms / Matthias Finger and Juan Montero.
Description: New York : Routledge, 2021. |
Includes bibliographical references and index.
Identifiers: LCCN 2020049495 (print) | LCCN 2020049496 (ebook)
Subjects: LCSH: Internet governance. |
Internet–Economic aspects. | Internet industry.
Classification: LCC TK5105.8854 .F54 2021 (print) |
LCC TK5105.8854 (ebook) | DDC 384.3/34–dc23
LC record available at https://lccn.loc.gov/2020049495
LC ebook record available at https://lccn.loc.gov/2020049496

ISBN: 978-0-367-69304-6 (hbk)
ISBN: 978-0-367-69305-3 (pbk)
ISBN: 978-1-003-14132-7 (ebk)

Typeset in Sabon
by Newgen Publishing UK

Contents

Introduction

The beautiful hills of Fiesole, above the Italian city of Florence, house the Florence School of Regulation, in the same villa where Boccaccio wrote the Decamerone seven centuries ago, where the ideas of a few Renaissance men transformed the world. Leaders in the telecoms, transport, and energy industries from all over the world regularly retreat to Fiesole with regulators, academics, and other stakeholders to discuss the regulation of the network industries. This is where we work.

Over the past years, digitalization has increasingly monopolized the debates in Fiesole. Excitement about the opportunities of the new digital technologies was mixed with apprehension about disruption by the new digital players. As professors at the Florence School of Regulation, we have witnessed the fear of the largest telecoms, transport, and electricity companies to be disrupted – or what we will later call "platformed" – by the new digital platforms.

This book builds on the debates at the Florence School of Regulation, as well as on the most solid currently available academic thinking on platforms in multi-sided markets, disruption, and, in particular, network theory. A favorite among academic economists is network effects, which are the effects that an additional user of a product has on the value of the product for the other users.

The traditional network industries are defined by network effects. Telephony provides the classic example: the value of a telephone network increases as the number of connected users increases. These are known as direct network effects. A network connecting just one user has a value of zero, while a network connecting all potential users reaches its maximum value. The same applies to all physical infrastructures: the more users share the infrastructure, the lower the cost of providing the service to each user, whether it is an electricity network, a railroad, or a postal network.

The first part of the book describes how digital platforms are also defined by the same network effects, even though this may not appear obvious from the outset. Platforms rely not only on the direct network effects identified in traditional network industries, but also on so-called "indirect network effects": the value of a product increases for one user group when a new user of a different user group joins the network. You must have two or more user groups or sides to achieve indirect network effects, which is why these are also called two-sided or multi-sided markets. The classical example is advertising: an increase in the number of Facebook users increases the value of the product for advertisers.

Furthermore, digital platforms rely on what we call "algorithmic network effects." Data play a fundamental role, as the interaction of the different sides in the multi-sided market is governed by algorithms that are fed with data. In turn, these algorithms improve as they are fed with more data to the point that they become predictive: based on the analysis of massive data from the past, algorithms are able to identify the most efficient complementarities between the different groups and users intermediated by the platform: In Uber's words: "Our network becomes smarter with every trip."

Analyzing digital platforms as network industries is not a mere academic exercise. In the following chapters, we will demonstrate how fortunes have been built by the small cluster of founders and investors in Silicon Valley that, very early and more deeply than others, understood the power of network effects in the internet era. Monetizing the network effects created by the internet was the explicit winning proposition of Thiel's PayPal, Zuckerberg's Facebook, and Kalanick's Uber. It was also the reason why venture capital firms such as Sequoia, Draper, Fisher, Jurvetson, and Benchmark Capital invested in them: "We had an internal thesis that other industries might benefit from a network layer on top of them."[1] Creating a network on top of things and monopolizing value created by such a network is the winning strategy behind the digital platforms.

The first observation to be made is that platforms can grow network effect much faster if their algorithms coordinate the interaction of previously uncoordinated and fragmented assets. Platforms do not have the ambition of producing and owning the assets they coordinate. Their strategy is to identify the potential complementarities between assets, make these complementarities real, and extract the value created by such complementarities. In other words, they create a more efficient system (network) out of a formerly inefficiently organized and fragmented pool of assets. These new network industries, as we will also call these new digital platforms, can only be properly regulated if it is fully understood that platforms mediate between third parties to create value in the form of new network effects. They do not need to own the assets they coordinate. They are not the providers of the intermediated services. They coordinate. They are intermediaries.

The second observation is that this strategy will usually be disruptive for the traditional players, including the players in the traditional network industries (telecoms, transport, energy), in two very different ways. Platforms can substitute traditional players. Substitution can take place, as platforms create a digital substitute for a traditional physical service (like email for letter mail services). This is not very common, however. Substitution mostly takes place as platforms identify idle assets that can be coordinated in a new and creative form that substitutes a traditional product or service. This is the case of user-generated-content empowered by YouTube and Facebook substituting media, drivers empowered by Uber substituting taxis and so on. Traditional players are totally or partially substituted because they cannot compete with the platform and they become redundant. Of course, this is the worst nightmare for the CEO of a traditional company. Speak to a postal operator in order to understand what digitalization does.

A more subtle effect can be identified when the platform does not substitute the traditional players, but "platformizes" them. Digital platforms identify the specific assets, products, and services produced by the traditional players, they detect new complementarities and new forms to coordinate them, which creates more value than the value previously created by the traditional system coordinator. Traditional players are not made redundant. They are always necessary to produce and maintain the underlying infrastructures and services. Fiber optic cables will always transport data, trains will always transport people, and electricity grids will always transport electricity. However, platforms displace traditional network industries players in their most relevant, albeit less tangible, role: that of creating the network effects, which are the ultimate source of value in network industries. Platformization usually starts as platforms intermediate traditional services substituting the service provider in the direct relationship with the users. Platforms use technology to reduce transaction costs and empower users to navigate a complex and fragmented supply side. Platforms are increasingly displacing traditional players as system coordinators. Platforms do not merely act as neutral matchmakers between supply and demand, but they actively nudge demand to a specific service provided by a specific supplier. They have become the new editors of a newspaper potentially formed by all the news pieces in the world. They can become the organizers of the mobility system in a city. They might even organize the electricity system in a region. They have the potential to become the system coordinators as they select, as they prioritize, and as they have the power to shape the ecosystem around them.

The book then analyzes in detail how platforms create networks on top of each network industry: first communications, then transport, and finally energy. The platform strategy is taken to the extreme when the algorithms interconnect not just random assets and services, but assets that were already structured into a network – sometimes more than a century ago, as it is the case with telephony, railways, and electricity. Platforms are disrupting the network industries, both by substituting and by platformizing the traditional infrastructure traditional players.

The parallelisms between the old and the new network industries will be underlined. The stories of Mr. Vail and AT&T, Edison and General Electric, and the role of J.P. Morgan and the "morganization" of the US network industries will be presented. The objective is to extract lessons that will help us understand the strategies of the leaders of the new network industries today. Standardization, investment ahead of demand in order to build network effects and the pursuit of monopoly power can be identified in the new network industries, just as they worked a century ago in the traditional network industries.

Then, for each of the network industries, the most relevant digital platforms will be analyzed, always from the perspective of network theory: Hotmail, Skype, WhatsApp, YouTube, Amadeus, Uber, BlaBlaCar, Tesla, and so on. We will identify the complementarities exploited by the platform, the growth strategy to make network effects real, the ways to monetize the value created, and, finally, the disruption of the traditional players. Such an analysis reveals the pervasive relevance of network effects, the similarities in how platforms active in different industries have grown to relevance, and even the repeated presence of a short list of individuals and venture capital firms implementing the same business strategy. Such similarities justify a homogenous approach to the analysis of digital platforms.

We devote particular attention to analyzing the vertical relationship between platforms and the traditional players in network industries. We will see how such vertical relationships are often determined by regulation. As traditional players are heavily regulated, platforms can often piggy-back on existing regulation to force traditional players to engage with the platforms and have their services intermediated by them. However, as platforms are becoming larger and more powerful, thanks to their larger and more powerful network effects, there are growing calls for a review of the existing regulation and the definition of a level playing field for the competition between the traditional and the new players in the network industries. We will describe how this debate is evolving in each of the network industries, both in the United States and in the European Union.

Finally, the book concludes with the study of the regulation of the new network industries, and in particular the regulation of their market power. Our main argument is that platform regulation must build, and already is building, on the experience of the regulation of the traditional network industries, particularly the pro-competition regulatory model implemented over the past 30 years in the US and more vigorously in the European Union. This should not come as a surprise, as platform regulation faces the same fundamental challenge as the traditional network industries has over the past few decades: how to reconcile the benefits derived from the full exploitation of network effects, which lead to market concentration if not monopolization, with the benefits of competition.

The apparent conundrum between network effects and competition is not new. Back when the network industries appeared during the second half of the nineteenth century, particularly when telephony and electricity networks emerged, it took around 30 years to understand the dynamics created by network effects and to design the regulatory tools to harness them. The decision back then was to monopolize the industries, to exclude competition, and either to subject the monopoly to regulatory control (in the US) or simply nationalize it (in the EU and most of the world).

More insightful has been the approach after the 1980s and the liberalization and deregulation of the network industries. We now have 40 years of experience in the exploitation of network effects in competitive markets, thanks to pro-competition regulatory frameworks. The industries have been fragmented, both horizontally, as newcomers compete with the previous monopolist, and vertically, as infrastructures have been sometimes separated, unbundled, from the provision of services over them. This has been the case in aviation, road, and maritime transport, and has been imposed in the EU on the electricity and railway industries. Regulatory tools have been put into place to reduce barriers to entry, eliminate lock-in effects, and ultimately share network effects through interoperability and network access obligations, thus somewhat avoiding the pressure towards concentration. Regulation has been adopted for each network industry in order to address the specific market failures in a proportionate way, each time developing specific institutions and procedures.

Digital platforms raise similar challenges, as network effects always lead to market concentration: digital oligopolies or, in some cases, digital monopolies thanks to a winner-takes-all dynamic. This is a pervasive phenomenon in the platform economy. As Eric Schmidt, Google's ex-CEO, put it: "All big successes in the Internet century will embody large platforms that get better and stronger as they grow."[2] Peter Thiel, the don of the so-called PayPal mafia, has exposed the situation even more bluntly: "Monopoly is therefore not a pathology or an exception. Monopoly is the condition of every successful business."[3]

Authorities around the world, led by the European Union, are considering the adoption of a new regulatory framework to govern digital platforms. There is a role for antitrust authorities, and EU authorities in particularly are increasingly active on this front. But there is also a growing consensus around the idea that further regulation will be necessary. Antitrust fines will not modify the market structure. They will not increase competition. They will not modify the industry structure once an oligopoly or a monopoly has emerged. The experience of the pro-competition regulation in the traditional network industries shows the way. We propose considering the role of platforms as intermediaries, and imposing upon platforms more of the types of obligations which were traditionally imposed upon intermediaries. We believe that platforms act like what we call superintermediaries, which would recommend reinforcing such traditional types of obligations.

During the early days of the internet it was advanced that technology would create a utopia of disintermediated, non-hierarchical, peer-to-peer interactions between humans in commercial, political, and all other human relations. This was encapsulated in the "Declaration of the Independence of Cyberspace" issued in 1996: "We will create a civilization of the Mind in Cyberspace. May it be more humane and fair than the world your governments have made before."[4]

Twenty years later, it has become obvious that reality evolved in a very different way. Effectively, the most fundamental effect of digitalization has been the creation of new forms of coordination of human activities. Algorithms enable new and more sophisticated forms for human beings and their organizations to interact with each other and with the world around them. In other words, algorithms support the creation of increasingly complex technological systems engaging a growing number of persons and organizations. Digital platforms are at the center of this transformation, and it is undeniable that they are creating major benefits for users and for society as a whole, as we will describe in this book. This is what lies behind the terminology and the theoretical framework developed by economists for the analysis of digital platforms. This is what lies behind network effects and multi-sided markets.

The power of traditional intermediaries and gatekeepers has certainly diminished, if not vanished. We will describe how the music record industry and newspapers lost their role as gatekeepers or system coordinators, while new gatekeepers – the digital platforms – have emerged, concentrating a previously unknown amount of power. Facebook has more advertising revenue than all newspapers in the world combined. It is expected that in 2021 Google will have more advertising revenue than all the TV networks in the world combined. The same evolution could be under way in transport and energy.

This evolution might have come as a surprise for Californian techno-optimists, but it was fully in line with the previous conclusions of scholars of large technological systems, namely the network industries. Sociologists and historians had already identified that "Inventors, organizers, and managers of technological systems mostly prefer hierarchy, so the systems over time tend towards a hierarchical structure."[5] Founders, investors and managers of digital platforms prefer a centralized hierarchical system that enables them to capture at least part of the value created by the different types of network effects. Building platforms requires large investments and certainly entails a large risk, but both have to be rewarded. Highly centralized platforms give an extraordinary power to their managers as system coordinators. A handful of platforms have a key role in the communications industries around the world, and the transport and energy industries might evolve in the same direction as platforms are in the position to grow larger network effects than traditional players. Being industries of general interest for the well-being of citizens and society, necessary for trade and even a preconditions for democratic participation, the regulation of the new network industries cannot be further delayed.

Notes

1 Bill Gurley, Partner in Benchmark Capital, in Lashinsky, A. (2017). *Wild Ride. Inside Uber's Quest for World Domination*. Penguin, New York, p. 102.
2 Schmidt, E. & Rosenberg, J. (2014). *How Google Works*. John Murray (Publishers), London.
3 Thiel, P. (2014). *Zero to One. Notes on Startups, of How to Build the Future*. Virgin Books, New York, p. 34.
4 Barlow, J. P. (1996). *Declaration of the Independence of Cyberspace*, retrieved from www.eff.org/es/cyberspace-independence
5 Hughes, T. P. (1987). The Evolution of Large Technological Systems. *The Social Construction of Technological Systems. New Directions in the Sociology and History of Technology*. MIT Press, Boston, 2nd ed. 2012, p. 49.

Part I

The Digitalization Dilemma in the Network Industries

Digitalization is disrupting the traditional network industries: communications, transport, and energy. As infrastructures are being digitalized, costs for the construction and operation of infrastructure are being reduced. Smart networks empower infrastructure owners to better manage their assets. However, digitalization also opens the door to more profound transformations.

Once an industry is digitalized, it can be disrupted by new players – the digital platforms – as they build new complementarities on top of pre-existing assets by making use of algorithms to manage the available data, whether they are generated by the company itself or not. Digital platforms build new network effects by coordinating the traditional industries in new and creative ways. We call this process "platformization."

The network industries are being platformized as well. The infrastructures and services that are owned and operated by traditional players are increasingly intermediated by the new players in the data layer (the digital platforms) to create new and larger network effects on top of the network effects that had already been built by the traditional players in traditional network industries.

However, the ways in which network effects are built differ between the two cases. Traditional infrastructure players build network effects by pooling assets and services under one hierarchical structure that owns the infrastructures, provides the services, and coordinates them to fully exploit the complementarities among all these assets. The larger the scale, the larger the network effects, and the bigger the competitiveness of the operator. Digital platforms, by contrast, do not own the assets and do not provide the services, but merely identify the complementarities in the assets and services operated by third parties, subsequently aggregate them, intermediate them with the users, and coordinate the new ecosystem. Such aggregation, intermediation, and coordination is made in the data layer, using algorithms to exploit big data. Investment is necessary to create the network effects, but not as much as to build and integrate all infrastructure assets into one firm.

From an economic point of view, digital platforms exploit a new form of industrial organization that has been subsumed under the term "multi-sided markets." As Silicon Valley platforms are disrupting traditional network industries, it is useful to have a better understanding of what a platform in a multi-sided market is. In this first part of our book, we will analyze some early examples of off-line platforms such as newspapers, payment cards, and videogame consoles, as well as how digitalization multiplied the relevance of multi-sided markets. We will analyze the rise of the digital platforms, particularly that of the leaders: Amazon, Google, and Facebook.

From there, we will follow with an analysis of the first cases of disruption in the music and newspaper industries. Platforms can disrupt traditional industries in different ways. We differentiate between disruption by substitution and disruption by "platformization." There are cases of traditional players being substituted by platforms in the network industries (such as email substituting letter mail, platforms substituting traditional media, and Uber substituting taxis), but substitution is exceptional, as infrastructures are usually indispensable and platforms have no ambition to substitute existing infrastructure. On the contrary, they need the existing infrastructure, so disruption in the network industries mostly takes the form of what we call platformization. Digital platforms create their network effects on top of the pre-existing infrastructure assets and services that continue to be provided by the traditional players. These players are aggregated, intermediated, and coordinated in a new way in order to exploit new complementarities, thus creating new and larger network effects.

The risk of being "platformed" presents traditional players with the "digitalization dilemma." Should they collaborate with platforms and offer their assets to be coordinated by the platforms in order to exploit the largest network effects possible and create new levels of efficiency, or should they refuse to work with the platforms in order to retain control of their assets and ensure that they can be properly funded and that their traditional public service objectives are met? In terms of regulation, this situation raises the question of how to deal with the new vertical relationship between traditional players and platforms.

Chapter 1

Digitalizing the Network Industries

Digitalization is transforming telecommunications, transport, and energy – the so-called network industries – but also all other infrastructures, such as water, buildings, and even green infrastructures, as it is transforming all other human activities. Sensors are being installed in infrastructures and data generated by such sensors are transmitted to the infrastructure managers, who can then use machine learning to manage infrastructures more efficiently.

Digitalization empowers infrastructure managers to reduce design, construction, and maintenance costs. Smart infrastructures are also more efficient, as they empower infrastructure managers to have a better control over supply and demand, controlling the load factor to make more efficient use of the capacity of the network.

At the same time, digitalization poses new challenges to infrastructure managers. Technology adds new costs and must be properly managed to deliver results. Customers are also empowered by the use of technology. Consequently, technology increases uncertainty, and it also generates new threats, such as in terms of security and privacy.

Most fundamentally, however, digitalization poses a threat to the position of the current infrastructure managers in the overall value chain. Digitalizing infrastructure paves the way for new data players, the digital platforms, and transforms the structure of the market. Network industries can now evolve into multi-sided markets, to be coordinated by the new data intermediaries. Digital platforms can use technology to become coordinators of infrastructure assets, in all network industries. As a result, traditional players could end up providing commoditized services as simple intermediaries to a digital platform, which will extract most of the value from the industry.

1.1 Network Industries

It is common to define telecommunications, transport, energy, and some other activities such as water and waste management, as network industries.[1] They all share common traits: relying on physical assets, usually referred to as infrastructures.[2] What is specific to network industries compared to other types of infrastructures such as buildings is that infrastructures in the network industries connect different points in space. Thus, physical assets in the network industries can be classified into two types: links and nodes. Links are the connections between two points, such as fiber-optic cables, rail tracks, air routes, electricity cables, or water pipes. Nodes are the points where links intersect, such as telecom switches, train stations, airports, transformers, or water pumping and cleaning stations. Assets are structured as a network, hence the term network industries.

Network industries require massive investment for the construction and maintenance of these infrastructures and such investments are usually sunk costs, as they cannot be recovered in case of market exit. They are also characterized by substantial returns to scale in production: the larger the scale of production, the lower the costs per unit. Although these are features can also be found in many other industries, network industries typically constitute an extreme example of sunk costs, also called asset specificity.

The physical assets in the network industries are used to convey different items. Telecom networks transport data encrypted in electronic signals. Transport networks convey passengers and goods. Energy networks convey electricity, gas, heat, hydrogen, and so on. Historically, the management of these networks and the provision of services on top of them were mostly bundled into a single organization, as it is typically the case in telecoms, rail, and electricity; in other network industries they were part of different organizations, as in air, road, and maritime transport. More recently, regulation, especially in the European Union, has come to impose a vertical separation (also called unbundling) between network infrastructures on the one hand and services on the other, such as in the case of electricity, gas, and railways. This has led to a distinction between an underlying infrastructure layer and a service layer on top of it.

Other than the management of the infrastructure and the provision of services, there is a third function in the network industries that tends to be overlooked. This is the role of coordinating the different elements of the network to make them complementary, a function that is often called "systems coordination." Historically, the network industries were born from investments in isolated assets, such as a telephone line connecting a few houses, a rail track connecting two towns, an electricity system serving a factory or a tram company. Only over time were these proliferating fragmented assets coordinated into a network of complementary assets; in many parts of the world this was done through nationalization as part of establishing a nation-state's essential infrastructures. This coordination role includes standardizing the infrastructures to make them interoperable and coordinating the services in the form of schedules, rate structures, and so on. This coordination role is sometimes formalized, such as in aviation, where, for security and national sovereignty reasons, the air traffic control function is assigned to a specific organization, the so-called air navigation service provider. In most cases, however, coordination occurs within the organization.

As the infrastructures and the services layers are being coordinated to make them complementary, the coordinated system exhibits increasing returns to scale in consumption, which are commonly called network effects or network externalities: the larger the scale of consumption, the larger the benefits per unit. In other terms, the value of the service is determined by the number of users.[3] The relevance of network effects is the defining feature of the network industries and, as such, it is intrinsically dependent upon the successful coordination of the different elements that constitute the networked system. Thus, network effects are central to network industries.

It has been almost a century since economists identified the role of the firm as system coordinator. In 1937, in "The Nature of the Firm,"[4] one of the most quoted economic papers in history, Coase explained that, "outside the firm, price movements direct production, which is coordinated through a series of transactions on the market. Within a firm, these market transactions are eliminated and in place of the complicated market structure with exchange transactions is substituted the entrepreneur-coordinator, who directs production." The reason for this substitution is the high transaction costs in the market, which are mostly linked to poor information and uncertainty. The higher the transaction costs in the market, the larger the production that would be integrated and coordinated within the firm and the bigger the firm would become. Transaction costs were identified as the very reason for firms to exist.

Network industries might be the ultimate example of firms as system coordinators. In these industries, coordination was achieved by integrating the complex and often costly assets into a single entity under the hierarchical control of the entrepreneur-manager, so as to create scale but also network effects by making the optimal complementary use of the different elements of the infrastructure system. Furthermore, concentration ensured the full exploitation of economies of scale (in the supply side) and of the network effects (in the demand side). This is how the different network industries became (public) monopolies in most countries during the first decades of the twentieth century.

Regulation was adopted to foster scale and concentration. The very same institution of the corporation was made possible by the authorization to create legal persons and the adoption of sophisticated legislation to govern them. Capital accumulation was facilitated by new legislation on banking and stock-exchanges. Regulation was later adopted in order to govern the negative effects of industrial firms: antitrust rules to protect competition, sector-specific regulation to control monopolistic network industries, labor laws to govern the relationships between corporations and workers, consumer protection legislation, public service regulation, and so on.

In recent decades, two fundamental evolutions have taken place: globalization and liberalization in the network industries. In general terms, this model of industrial organization has been taken to the extreme by globalization, which has created the largest of firms and the largest of markets, rewarding firms with even bigger economies of scale. The pursuit of scale has produced behemoths in one industry after another: larger financial institutions, larger media groups, and larger industrial production plants.

Liberalization and deregulation, by contrast, have fragmented the network industries. Competition was introduced into the network industries to increase efficiency and foster innovation. The network industries were fragmented both horizontally (several competitors) and in some cases vertically (unbundling; that is, separation of infrastructure management and service provision). The introduction of such competition led to substantial regulation in order to be able to continue the coordination of the different elements of the networked system, which had previously been integrated (coordinated) within one firm.

1.2 Digitalization

Digitalization can be defined as the conversion of information (text, pictures, sound, and basically any type of information) into a series of the digits 0 and 1, which can be automatically processed by a computer. Transistors in integrated circuits are the building blocks for the transmitting, processing, and storing of such digits, and therefore of digitalization. The transistor was invented in 1947 by John Bardeen, Walter Brattain, and William Shockley, three researchers working at Bell Labs, the research branch of the American telephony monopoly. They were awarded the Nobel Prize in physics for an invention that would transform the world. Shockley moved to his family's hometown, Palo Alto, to launch his transistor manufacturing company and attracted some of the world's best talent for his venture. Shockley's venture did not succeed, but as his first employees (the so-called "traitorous eight") started new ventures, the history of Silicon Valley cluster kicked off. Increasing computer power has dramatically increased the capability to process and store huge amounts of data at lower and lower cost.

Digitalization allows the creation of "digital twins" that simplify the coordination of complex systems such as network industries. A data layer is emerging on top of physical reality, which virtually recreates it *in silico*. Algorithms are then able to identify the opportunities to improve the organization of the system, thus increasing efficiency. The underlying reality can subsequently be transformed and improved. This is also the case of infrastructures and network industries,[5] such as aviation,[6] railways,[7] and electricity.[8]

Sensors, cameras, meters, and other devices can be installed in the physical assets that capture and transmit data to the infrastructure manager. Such data can recreate the status of the infrastructure (location, damage, collapse, and so on) in the data layer, as well as the usage of the infrastructure for the provision of services (capacity, traffic flows, and payments).

Just as important as the processing and storage of digits by computers is the transmission of such digits between computers. On October 29, 1969, at 10:30 pm, the computer network of the University of California in Los Angeles and the computer network of the Stanford Research Institute were interconnected, creating ARPANET, the network of computer networks of ARPA, the research arm of the US Department of Defense. This was the seed for what would later become the internet,[9] an abbreviation of "interconnected networks." In its essence, the internet is a protocol (Transmission Control Protocol/Internet Protocol – TCP/IP), developed by Vinton Cerf and Bob Kahn, that enables the interaction of independent computer networks. The very nature of the internet is to facilitate the interaction of previously fragmented and isolated computer networks.

Digitalizing infrastructure depends on the availability of connectivity for the transmission of data, as high-speed internet access is a key enabler for other digital technologies.[10] The development of 5G wireless networks, not only in densely populated areas, but also in the remote areas crossed by infrastructure, is a fundamental condition for the digitalization of infrastructures.

Algorithms are enabling the full exploitation of Big Data.[11] Sophisticated algorithms are necessary to give order to the massive amounts of data generated by sensors and other data-collecting devices, thus making them relevant. Furthermore, algorithms are incorporating machine learning tools, or "artificial intelligence." They are no longer a set of fixed commands rigidly linking a fact to a consequence. On the contrary, algorithms peruse through the available data in order to learn from previous experiences and subsequently transform data into information and link it to consequences. Algorithms improve upon themselves with each interaction; they are becoming more and more predictive.[12] Automation is the ultimate goal of many digitalization projects.

1.3 Smart Networks

Digitalization can reduce the cost of infrastructures. It can reduce the cost of the design, construction, and maintenance of the physical infrastructure and it can also reduce the cost of operating the infrastructure, particularly by managing the load factor and optimizing the available capacity in the network.

Automated computer design can reduce design costs and can further optimize design and construction by better coordinating all the participants in a network. Building information modeling (BIM), defined as "a set of interacting policies, processes and technologies generating a methodology to manage the essential building design and project data in digital format throughout the building's life-cycle,"[13] can reduce construction costs. In the case of roads, cost reductions have been estimated to be 15–20 percent compared to a traditionally designed system.[14]

Digitalization can also reduce the cost of maintaining infrastructures. The Internet of Things (IoT) allows the installation of sensors and other data-collecting devices in all elements of any infrastructure. This enables infrastructure managers to monitor the status of these elements and tailor maintenance to the real condition of the infrastructure, thus enabling "conditions-based maintenance."[15] Furthermore, intelligent algorithms can make use of existing data to predict the need of maintenance, enabling so-called "predictive maintenance."[16] As an example, for railways rolling stock, it has been estimated that condition-based maintenance can reduce costs by 10–15 percent, while predictive maintenance can further reduce costs by another 10 percent.[17]

Digitalization can also reduce the cost of charging for the use of the infrastructure. The charges for the use of infrastructure are usually related to the volume of use of the infrastructure (the number and duration of telephone calls, the kilometers traveled, the kilowatts transported, etc.). So-called "smart meters" offer a much more accurate picture of the usage of the infrastructure, thus generating information that can optimize loads and therefore reduce costs. In Taiwan, for example, it has been identified that electronic tolling in road transportation can reduce congestion (by 60.1 percent) and CO_2 emissions (by 12.4 percent).[18]

Digitalization is not only transforming the construction and maintenance of infrastructure, but is also transforming the provision of services, particularly the key feature, which is the control of the load factor. Digital technologies empower infrastructure managers to have more control of demand, in the form of traffic flows, but also supply, in the form of dynamic management of network capacity. Efficiency can be increased, as spare capacity is reduced while avoiding congestion.

Controlling demand in order to avoid congestion is a key element in infrastructure management. Network industries often display negative network effects, particularly in the form of congestion. One mechanism for reducing congestion without diminishing the use of the infrastructure is to distribute traffic more evenly across time and space. Digitalization helps adapt demand to capacity. As infrastructure managers have new tools to predict traffic flows (such as predictive algorithms), they can incentivize the use of the infrastructure in off-peak periods against the use in peak periods. They can also build more sophisticated pricing schemes, based on real-time metering and billing. Infrastructure managers can respond to fluctuations in demand through dynamic pricing.

Digital technologies not only make it possible to adapt demand to capacity, but also to adapt capacity to demand. Software-defined networking (SDN) can dynamically adapt capacity to demand by virtualizing infrastructure,[19] thus providing "capacity as a service," rather than as a fixed asset. SDN decouples the physical infrastructure layer from the control layer and uses software to dynamically adapt the capacity in the infrastructure layer to the existing demand. If a customer demands more capacity, it is provided in real time. If a customer demands less capacity at a given time, the vacant capacity is used to serve other customers. This is particularly useful for managing bandwidth in large data centers. The concept of SDN is being exported to other infrastructure industries such as electricity, where SDN is being proposed as a solution to dynamically manage distribution networks. For example, at times of low load, the voltage and operating frequency of the network can be reduced, lowering the cost of operation of the network. In transportation, a smarter management of capacity can take the form of smaller vehicles to provide mass-transit services when demand is low. More and larger vehicles can be dynamically assigned when a peak in demand is identified in real time through the sensors and the predictive algorithms.

As a result, more control over the load factor empowers infrastructure managers to adapt the existing capacity to cope with growing demand without congestion. This ability can also be used in case of unpredictable events, such as black-outs and accidents. In this way, costs can be reduced substantially, while the resilience of the network can be increased.

Thus, network industries managers are investing in the digitalization of their networks. By doing so, they expect to substantially reduce costs and to increase efficiency; in short, networks are becoming smarter.

1.4 Challenges in Digitalizing Networks

However, digitalization not only brings benefits for infrastructure managers; it also creates new risks. To begin with, digitalization is a cost in itself and a burden if the wrong choices are made. Furthermore, digitalization is empowering not only infrastructure managers, but also infrastructure users, who can modify their usage patterns and find alternatives to reduce their expenditures. They can sometimes also compare. Finally, digitalization brings new security risks in the form of cybersecurity threats.

Investment in digital technologies will not automatically deliver results. Incompatible solutions might be deployed in such a way that efficiency is not created. Deployment of digital technologies may bring benefits, but these benefits will be limited as their lifecycle tends to be much shorter than that of traditional infrastructures. Constant investments and upgrades might become necessary. Furthermore, increases in productivity depend on managerial and technical skills, as well as on incentives, with strong complementarities between the two.[20]

Secondly, digital technologies not only empower infrastructure managers; they also empower users, giving them more information about alternatives, as well as the ability to switch to such alternatives. Infrastructure managers have traditionally operated with little or no competition, either from direct competitors or from other infrastructures. Digitalization is now offering such new alternatives.

Digitalization is not only enabling users to choose among different infrastructure-based services, but also enabling them to optimally combine infrastructures. This is particularly the case in transportation, as different infrastructure networks provide partial services and often lack a door-to-door service (for both passengers and goods). Digitalization and algorithms now allow users to identify the available options as well as the fastest/cheapest/greenest combinations of transport modes. Therefore, users can migrate to new transport modes, as they have better information.

Not all infrastructures have the same potential to reduce costs or to reduce them at the same pace or in the same amount as a result of digitalization. Road transportation costs might be substantially reduced as automation (autonomous vehicles) is implemented for both freight and passengers. As a result, road transportation might gain a competitive advantage over railways, which do not appear to benefit from similar cost reductions caused by digitalization and automation. Such secondary effects are beyond the control of the infrastructure managers and they add uncertainty in the planning of investments.

Finally, cybersecurity is a new risk associated with the digitalization of infrastructure. Indeed, digital technologies create new vulnerabilities. Privacy must be guaranteed to users. Services can collapse in case of bugs in the software or attacks by hackers. Infrastructure managers must invest in technologies in order to solve the uncertain vulnerabilities created by digitalization.

Overall, digitalization is introducing more uncertainty in the management of infrastructure networks. Even if it is providing new instruments to predict and manage traffic flows, it is also empowering users to make a more effective use of the existing infrastructures and to introduce more competition among infrastructure providers.

1.5 Disruption

Digitalization poses a much more fundamental challenge to the traditional network industries than what we have described so far, as it challenges the role of the established firms as coordinators of the complementary elements that make up each industry.[21] New digital players, most of which are based in Silicon Valley, are disrupting one industry after another. It has often been noted that these digital players disintermediate the disrupted industries. Network industries are no exception.

Let us remember that, in addition to managing infrastructures and services, network industry firms fulfill a third, most important, function, serving as the coordinators of the complex infrastructure system only thanks to which infrastructure services can be provided. As we have shown above, digitalization actually provides coordinators with new tools to better manage and coordinate all these infrastructure assets more efficiently.

Digitalization is more transformative than that, however, as it also reduces transaction costs. The internet reduces coordination costs. Algorithms reduce search, information, and bargaining costs. In Coase's terms, digitalization reduces transaction costs, which means that coordination in the market can become more efficient than coordination by the entrepreneur inside the firm. This is only part of the story: digitalization also enables a new type of coordination beyond the traditional coordination through the market,[22] namely coordination by algorithms.

Digitalization is also creating a new type of industrial players, now at the data layer. These players use digital technologies, particularly algorithms, to mediate between the different parties. Mediation can take place between businesses (B2B), between businesses and consumers (B2C), between consumers (C2C), or even between businesses or customers and third parties, such as governments, advertisers or others (B/C2X). In this sense, these new industrial players are intermediaries. The traditional direct link between supply and demand is being substituted by a new, digitally intermediated relationship.

This new form of industrial organization is spreading. It typically involves two or more distinct types of users interacting through a third party, called the platform. Network effects are key here. The more parties there are involved in the interactions, the higher the value for each of the parties. Platforms are central to this new model of industrial organization. First, they reduce transaction costs. They make use of the internet to reduce communication costs to almost zero, use algorithms to match supply and demand, and provide information about the counterparties. Second, platforms invest in growing the ecosystem around them to make the network effects possible. Third, they coordinate the "market place" – or, rather, the transaction space – by setting the standards, the rules and the distribution (or not) of the benefits created by the network effects to the different parties involved.

Scale can now be built by coordinating the assets that are managed and owned by third parties. An alternative is emerging to the large-scale firms that have coordinated industries for more than a century: platforms can put together more assets than individual firms and can coordinate them more efficiently.

As infrastructures are digitalized, and as a digital twin or mirror is created at the data layer, new players have the ability to take a leadership role in the management of such data and extend their leadership role to the management of the underlying systems, transforming the use of the infrastructure and substituting traditional infrastructure managers as coordinators of the systems. Traditional players will keep operating the infrastructure and manage the provision of the traditional services over the infrastructure. However, the new digital players take over the interface with the customers as they have the ambition to lead the third role of current network industry firms as system coordinators and to intermediate the network industries.

Economists have used the term *multi-sided markets* to describe this form of industrial organization. Even though there were precedents before the internet went mainstream, multi-sided markets have proliferated as digitalization has reduced transaction costs. With the popularization of the internet and big data, these new intermediaries, often based in Silicon Valley, have disrupted one industry after another.

Notes

1 Finger, M. (2019). *Network Industries. A Research Overview*. Routledge. New York.
2 Frischmann, B. (2012). *Infrastructure. The Social Value of Shared Resources*. Oxford University Press. Oxford.
3 Economiedes, N. (1996). The Economics of Networks. *International Journal of Industrial Organization*, 14(2), 673–699.
4 Coase, R. (1937). The nature of the firm. *Economica*, 4(16), 386–405.
5 Costa, P., Montero, J. J., & Roson, R. (2018). *The Impact of Disruptive Technologies on Infrastructure Networks*, 11th Meeting of the Network of Economic Regulators, retrieved from www.dirittoepoliticadeitrasporti.it/wp-content/uploads/2018/12/OECD-The-impact-of-disruptive-technologies-on-infrastructure-networks-2-Nov-2018.pdf
6 Abeyratne, R. (2020). *Aviation in the Digital Age*. Springer. Cham.
7 Montero, J. (2020). The Digitalization Dilemma in the Railways Industry. In M. Finger & J. Montero (eds.), *Handbook on Railway Regulation. Concepts and Practice*. Edward Elgar, Cheltenham, UK and Northampton, MA.
8 Brown, M., Woodhouse, S., & Sioshansi, F. (2019). Digitalization of Energy. In *Consumer, Prosumer, Prosumager. How Service Innovations Will Disrupt the Utility Business Model*, Elsevier, London.
9 Ryan, J. (2010). *A History of the Internet and the Digital Future*. Reaktion Books, London, p. 30; and Blum, A. (2012). *Tubes. A Journey to the Center of the Internet*. HarperCollins, New York, p. 36.

10 OECD (2019). *Digital dividend: Policies to harness the productivity potential of digital technologies.* Economic Policy Paper No. 26, OECD Publishing, Paris. Retrieved from https://doi.org/10.1787/273176bc-en

11 Iansiti, M. & Lakhani, K. R. (2020). *Competing in the Age of AI. Strategy and Leadership when Algorithms and Networks Run the World*, Harvard University Press, Cambridge, MA.

12 Agrawal, A., Gans, J., & Goldfarb, A. (2018). *Prediction machines. The Simple Economics of Artificial Intelligence.* Harvard Business Review Press, Boston.

13 Succar, B. (2009). Building information modelling framework: A research and delivery foundation for industry stakeholders. *Automation in Construction, 18(3)*, 357–375.

14 Blanco, F.G.B., & Chen, H. (2014). The implementation of building information modelling in the United Kingdom by the transport industry. *Procedia–Social and Behavioral Sciences*, 138(2014), 510–520.

15 Jardine, A. K., Lin, D., & Banjevic, D. (2006). A review on machinery diagnostics and prognostics implementing condition-based maintenance. *Mechanical Systems and Signal Processing*, 20(7), 1483–1510.

16 Daneshkhah, A., Stocks, N.G., & Jeffrey, P. (2017). Probabilistic sensitivity analysis of optimised preventive maintenance strategies for deteriorating infrastructure assets. Reliability. *Engineering & System Safety*, 163, 33–45.

17 McKinsey & Co. (2016). *Huge Value Pool Shifts Ahead–How Rolling Stock Manufacturers Can Lay Track for Profitable Growth*, Report September 2016.

18 Tseng, P. H., Lin, D. Y., & Chien, S. (2014). Investigating the impact of highway electronic toll collection to the external cost: A case study in Taiwan. *Technological Forecasting and Social Change*, 86, 265–272.

19 Knieps, G. (2017). Internet of Things, future networks, and the economics of virtual networks. *Competition and Regulation in Network Industries*, 18(3–4), 240–255.

20 Andrews, D., Nicoletti, G., & Timiliotis, C. (2018). *Digital technology diffusion: A matter of capabilities, incentives or both?* OECD Economics Department Working Papers, No. 1476, OECD Publishing, Paris, retrieved from http://dx.doi.org/10.1787/7c542c16-en

21 Montero, J. & Finger, M. (2017). "Platformed! Network industries and the new digital paradigm," *Competition and Regulation in Network Industries*, 18(3–4), 217–239.

22 Munger, M. C. (2015). Coase and the "Sharing Economy." In *Forever Contemporary: The Economics of Ronald Coase*. Institute of Economic Affairs, London, 187–208.

Chapter 2

Platforms in Multi-Sided Markets

Silicon Valley did not invent platforms in multi-sided markets as a model of industrial organization. Newspapers, payment cards, and videogame consoles were already using this business model; they are all platforms coordinating ecosystems around themselves to create and monetize network effects. The stories of these platforms illustrate how direct and indirect network effects work in multi-sided markets.

At the turn of the twenty-first century, economists – but mostly investors – started to focus their interest on multi-sided markets as the internet significantly empowered their creation. Since then, abundant literature has defined and classified the fundamental characteristics of platforms in multi-sided markets. Regulation has often proven to be central in such markets.

2.1 Newspapers

On 7 December 1995, just as the internet was going mass-market, Jim Pavelich and Dave Price launched a start-up in Palo Alto, the heart of Silicon Valley. It was not a portal or a search engine. It was the *Palo Alto Daily News*, a newspaper offered at no cost to readers, and only in paper format. It was not the first free newspaper, but would become one of the leading ones in the United States, with a daily circulation of 67,000 copies. Free newspapers shared the same business model as the most popular digital platforms – Google, Facebook, Uber, and so on – but, over time, digital platforms certainly proved to be more successful.

The readers of the *Palo Alto Daily News* would not pay for the newspaper, but advertisers would pay to have their ads printed in a newspaper that had a large circulation due to the fact that it was free for the readers. This is an illustrative example of how multi-sided markets work. An intermediary (the newspaper) facilitates the interaction between two different groups of users (readers and advertisers). The larger the number of users on each side, the higher the benefits for the other side. On one side, advertisers would be more interested in the newspaper as it reaches a wider audience. On the other side, readers would be more interested in a newspaper if content is enriched with the revenue generated by a larger pool of advertisers. Newspapers were already an example of a platform in a multi-sided market, but free newspapers have taken this business model to the next level, as they rely only on revenue from one of the sides they serve (the advertisers), making it free for the readers (zero-pricing).

Free newspapers were first received with suspicion, and then with anger by well-established newspapers. They initiated legal actions against them for all kinds of reasons: unfair competition, illegal use of the word "newspaper," and even littering.[1] Legal action would become a standard strategy that traditional companies used against platforms that disrupted them.

Ten years later, in 2005, the *Palo Alto Daily News* was acquired for $25 million by *Mercury News*, the leading San Francisco Bay area newspaper. As explained the new publisher, the purpose of the acquisition was "to provide additional ways for our advertisers to reach their customers."[2]

The story of the *Palo Alto Daily News* does not have a happy ending, however. In 2015, the new publisher of the newspaper had to cut its print edition to one day a week. Free newspapers, like traditional newspapers, were disrupted by the internet and the more powerful multi-sided markets that the internet was making possible.

2.2 Payment Cards

In 1949, Frank McNamara was having lunch in a New York restaurant when he realized he had left his wallet at home. While waiting for his wife to bring cash to pay the bill, the idea of a payment card came to his mind. The following year, 1950, he incorporated Diner's Club with €1.5 million of start-up capital, 14 restaurants, and a group of friends as purchasers. One year later, 42,000 card-holders and 330 merchants, mostly restaurants and hotels, accepted the card, paying a 7 percent commission to Diner's Club.[3] The American Hotel Association reacted in 1956 by creating its own card, and the National Restaurant Association signed on. This was an early attempt to avoid the 7 percent commission charged by the intermediary, Diner's Club.

The payment card industry is a classic example of a multi-sided market. An intermediary (the payment card company) facilitates the interaction between merchants and purchasers. The key to the success of payment card companies is the number of users, both merchants and purchasers. The wider the use by each group, the higher the benefit for the other group. On one hand, merchants will be increasingly interested in the card as it is more widely used by purchasers. On the other hand, purchasers will be increasingly interested in a card if it is accepted by a substantial number of merchants.

Diner's Club was the first payment card, but this did not guarantee a monopoly to the first mover, not even a leadership position, as other companies entered the market. A company created 16 years later, VISA, eventually became the market leader. Strong competition exists between payment card companies, and both merchants and purchasers usually accept more than one card, a practice called "multi-homing."

The key benefit for both sides making use of payment cards is the reduction of transaction costs. Payment cards make payments easy and eliminate the risk of cash being stolen or lost. The payment card company has a key role as it defines the distribution of the benefits derived from the new interaction among all the players: the merchants, the purchasers and the platform itself, including the so-called interchange fees to be paid for the use of the card in each transaction. Despite competition, the distribution of the benefits created by payment cards has raised numerous conflicts. After decades of such conflicts, regulation was adopted both in the US and in the EU to set a price cap on interchange fees. This is an early illustrative example of how platforms, like intermediaries, are often subject to regulation to exclude abuses.

2.3 Videogame Consoles

On September 23, 1889, Fusajiro Yamauchi created Nintendo, a company that produced and distributed playing cards. In the middle of the twentieth century, the company experimented with new ventures such as a taxi company, a love hotel chain, and then toys. However, the most successful venture was the production of videogame consoles, which started in 1972.

Videogame consoles are another classic example of a multi-sided market. An intermediary (the producer of the consoles) facilitates the interaction between game producers and players. The key to the success of a console is the number of users, both game producers and players. The wider the use by each group, the higher the benefit for the other group. On one hand, the game producers will be increasingly interested in a console as it is more widely used by players, in such a way that the fixed cost of producing a game can be split among a larger number of buyers. On the other hand, players will benefit from a larger pool of games.

Consoles present the typical "chicken and egg" problem.[4] Game producers are reluctant to develop games for a new console, given the uncertainty about whether it will attract a large number of players. However, players will not acquire a console if a sufficient number of popular games is not available. In the crudest form: "Developers go where the money is, and the money is where people are."[5]

The classic console wars[6] – Atari vs. Nintendo, Nintendo vs. PlayStation, PlayStation vs. Xbox, etc. – cannot be understood merely as competition between increasingly sophisticated consoles. In order to understand competition between console producers. It is necessary to understand that competition takes place between entire ecosystems. Players choose their console based not only on the features it displays, but also on the quality and the quantity of the available games. For this reason, each console always tries to attract the best game producers and the largest number of gamers; this leads to more game producers being attracted to work with the console producer, thus creating a virtuous cycle.

It is not common for players to use more than one console, due to the high price of each. As "multi-homing" is not common, consoles that benefit from a large pool of game developers and players have always tended to dominate the market. This tendency has been termed "winner-takes-all." However, the console wars demonstrate that, in dynamic markets, such market power can evaporate over time as more sophisticated consoles come to the market, which attract new games and new players and erode the dominant position of the previous console producers.

2.4 Multi-sided Markets and Network Effects

Multi-sided markets are not a new phenomenon. Newspapers financed by advertising have existed since the nineteenth century.[7] Payment cards were introduced in 1950 and the explosion of videogames took place during the 1980s. However, it was only when the internet became widespread, in the mid-1990s, that multi-sided markets proliferated, grew multibillion memberships, and attracted the interest of investors and economists.

Early this century, Nobel Laureate Jean Tirole and his peer Jean-Charles Rochet published their seminal work *Platform Competition in Two-Sided Markets*.[8] In that paper, seven "mini cases" were analyzed, including media, credit cards, and video games. The new paradigm built by Rochet and Tirole has spawned a substantial amount of industrial organization literature.[9]

Over the years, a consensus has developed that multi-sided markets typically involve two or more distinct types of users, interacting through a third party, the platform.[10] Key elements are the relevance of positive indirect network effects, which have to be strong enough to affect business conduct,[11] as well as the leading role that the platform plays in the distribution of such benefits across the ecosystem of interacting parties through pricing and other design decisions. However, let us briefly explain, step by step, how this theory has been built.

It is not by chance that Tirole had originally worked on, among other things, the regulation of network industries, particularly telecommunications. In fact, in *Platform Competition in Two-Sided Markets*, Rochet and Tirole explicitly stated that their theory builds on network economics.[12] We have already explained how, in network industries such as telecommunications, transport, and electricity, the larger the number of users of a specific infrastructure or system, the higher the benefit for all other users. This is the so-called "network effect." It is considered an externality as it is a benefit that does not derive from the mere single decision of the party that contracts a service.

The benefits derived from network effects can easily be underestimated. They were early identified by the most successful leaders in network industries, such as Theodore Veil, the president of AT&T at the turn of the twentieth century (see his story in Chapter 6) and David Sarnoff, leader of the radio and television network NBC. Sarnoff observed that the value of his network seemed to increase proportionately to the number of users, but he still underestimated the value of the network effects. Robert Metcalfe, working on the development of Ethernet in the 1980s, later concluded that the value of a communications network would be not be the number of users, but the square of the number of users. A network of ten telephones would not have a value of ten, but a value of 100 – the number of potential connections between users. This has been known as Metcalfe's Law. Furthermore, it was proposed in the early 2000s that as different sub-groups with more than two users could be formed, the value of a network of ten users could be as high as 501, the total number of potential connections including the variable numbers of users (Reed's Law).[13]

The key novelty in the analysis of multi-sided markets is that network effects are not limited to systems with a single category of users, as is the case with telephone networks. These are called direct network effects. The interaction of different types of users can also generate network effects. This is the case with advertisers and newspaper readers, merchants and purchasers using payment cards, and videogame developers and players. These are the so-called "indirect network effects."

Indirect network effects build up in a different way than the more traditional direct network effects. The key difference is that a balance must be reached and then sustained that takes into consideration the interests of all the parties involved in the multi-sided markets. It is not sufficient to sell an innovative good or service at an attractive price, as is the case in traditional industries. It is not sufficient to attract a single category of users to generate the direct network effects. An attractive offer must be provided to all the parties in the multi-sided-market, including to the intermediary itself.

On one hand, negative same-side effects can emerge. If one side of the multi-sided market grows too large, positive network effects may become jeopardized, as is the case with nightclubs, for example. Nightclubs can be understood as multi-sided markets that facilitate the interaction of opposite-sex individuals. A growing number of men might produce negative same-side effects. This is why it is an established practice to allow women free entry to clubs (another example of zero-pricing), while men are charged a fee to enter.

On the other hand, negative cross-side effects can also be identified. In some multi-sided markets, a growing number of users can make matching more complex, more expensive, or slower, or it can just degrade the service provided by the intermediary. This is the case with advertising in the media industry. Advertisers can be attracted by large audiences, but the introduction of too much advertising, such as in radio or TV shows, can discourage audiences from participating in the market; that is, from following the shows.

For a multi-sided market to succeed, positive network effects must be created inside each group (positive same-side effects) and across the participating groups (positive cross-side effects). Furthermore, some benefits must be captured by the platform that created the market and that curates it to keep the right balance. Thus, the active role of the platform is key in multi-sided markets.

2.5 Platforms

Platform is the term that Rochet and Tirole gave to the structures that make the interactions between different groups in multi-sided markets possible. Newspapers, payment cards, and videogame consoles were considered to be platforms.

The word "platform" is defined by the Oxford English Dictionary as "The surface or area on which anything stands; esp. a raised level surface."[14] The term was used in computer science to refer to the hardware on which a software application could be run. Software had to be specifically developed for each hardware platform. This is very clear in the case of videogame consoles and in the case of computer operating systems, two of the mini-cases analyzed by Rochet and Tirole. It is understandable that they proposed the word "platform" to refer to the central role of the structure that makes interaction between different groups possible, even when hardware does not play such a role.

The term "platform" has been adopted in the industrial organization literature, with a flow of publications including the word platform in their title, such as: *Platform Leadership*,[15] *Platform Economics*,[16] *The Age of the Platform*,[17] *Platform Ecosystems*,[18] *Platform Scale*,[19] *Platform Revolution*,[20] *Platform Capitalism*,[21] and *The Business of Platforms*.[22] Such an enthusiastic adoption might be explained by a second meaning of the word in English. As early as the sixteenth century, a more abstract use of the term had emerged: "A plan, design; something intended or taken as a pattern, a model."[23] Platforms create patterns for the interaction of different groups.[24]

Multi-sided markets are created by entrepreneurs who invest in the structure that makes the interaction of different groups possible. This was the case with Jim Pavelich and Dave Price and the *Palo Alto Daily News*, with Frank McNamara and Diner's Club, and with the heirs of Fusajiro Yamauchi and Nintendo. Platforms are an economic activity in themselves. Platforms provide services to the interacting groups that make use of the platform. Two major business models can be differentiated: transaction and non-transaction platforms.[25]

On one hand, in non-transaction platforms, the different groups do not conclude transactions among themselves, but transactions take place only between the groups and the platform. This is the case with newspapers and, in general, with all the platforms that rely on advertising. Advertisers contract with the platform to include ads in the newspaper. Readers contract with the platform to acquire the newspaper. No contract is concluded between the advertiser and the reader.

In transaction platforms, on the other hand, the different sides of the market interact directly with one another other, concluding contracts for the provision of the service. This is the case with videogames: videogame developers conclude contracts with the platform; players buy the console; and then a direct interaction takes place between game developers and players, as players acquire the games from the developers. A very relevant model of transaction platform is the matchmaker. The matchmaker does not develop a specific service (a newspaper, a payment card, a console), but merely facilitates the interaction of third parties interested in an interaction. There are lots of examples of traditional matchmakers: stock exchanges, real estate agencies, dating agencies, etc. These are all platforms.

It has been underlined that some platforms provide the common technological building blocks for third parties to develop new complementary products and services. These have been called "innovation platforms."[26] This is the case of videogame consoles and operating systems for computers, smartphones etc.

A common characteristic of all platforms is that they reduce transaction costs by making the interaction between third parties easier, safer, and cheaper. Transaction costs can be defined as the costs involved in a market exchange. Examples of transaction costs reduced by platforms in multi-sided markets are search costs, information costs, monitoring costs, communications costs, and bargaining costs. As a conclusion, by pooling together different groups of users, under specific terms and conditions defined by the platform, transaction costs are reduced and a benefit in the form of an indirect network effect is generated.

All of these business models require a substantial investment in order to create the structure that sustains the interaction (the newspaper, the payment card, the console). However, this is only part of the necessary investment and is rarely the key to the success of a platform in a multi-sided market. Furthermore, it is necessary to invest to attract the critical mass of users to trigger or ignite the network effects. All kinds of creative techniques are used to attract users to the platform, but it is mostly a matter of investment in the form of advertising and providing reductions in the price of the provision of the service, even setting the price below cost. The larger the critical mass that is pursued, the larger the investment. A small investment might be necessary to establish a local free newspaper in Palo Alto, but a much larger investment would be necessary to establish a videogame console on a global scale.

Finally, it is not sufficient to fund the creation of the platform and to invest in the attraction of large pools of users. For a platform to succeed, it is not sufficient to be able to reduce transaction costs and generate indirect network effects. As important as the benefit is the balanced distribution of such a benefit across the parties in the multi-sided market. It is the role of the platform to propose and to maintain such a balance. An appropriate balance will allow the multi-sided market to flourish and to be sustainable over time.

Benefits derived from the network effects must be carefully distributed among all the sides of the multi-sided market, making sure that they all have the incentive to join the platform and keep working with it. It is the role of the platform to understand the needs of the different sides and to ensure that they have the proper incentives to continue to use it and work with it.

If a newspaper tries to set very high prices for its readers, it will fail to build a large pool of readers and will not be able to attract enough advertisers to sustain the newspaper. If the newspaper tries to sell advertising at a very low price, but fills the newspaper with ads, readers will desert it. Innovation in the definition of the balance was exemplified by free newspapers like the *Palo Alto Daily News*. Zero-price services can actually be explained as being just one side of a multi-sided market, whereas the other side bears the full cost.

If a payment card sets very high fees for the purchaser in the form of an annual payment, it will attract few users and merchants will have little incentive to accept the card. High fees for the merchants will also discourage the use of the card, even though a large pool of purchasers might be attracted. In the case of payment cards, innovation in the definition of the balance was exemplified by rewards for purchasers.

If the platform tries to monopolize the benefits derived from the network effects, in the form of large fees from the beginning of the activity, the multi-sided market will not attract users and the platform will never ignite and grow. This is why it is common for platforms to start with discounted prices and even with the free provision of the service: examples include complementary newspapers, subsidized games, and payment cards with no fees for users.

However, for a multi-sided market to be stable, it is necessary to ensure the necessary reward for the investment of the platform. A multi-sided market will sometimes flourish, as the platform provides a valuable service to all sides, but the platform itself is not able to secure a profit (or sometimes even revenue) from its activity. Consequently, the multi-sided market never comes to exist, or it is terminated as the platform runs out of funds.

In conclusion, platforms have a central role in multi-sided markets. They are the creators and then curators of this multi-sided market. They identify the possibility to create the market. They invest to create the structures that make the market possible. They invest to attract large pools of users on all sides of the market (which is a fundamental barrier to market entry, as identified by antitrust authorities).[27] Finally, they define the dynamic balance in the distribution of the benefits generated by the multi-sided market.

Public intervention in multi-sided markets has been common. Intermediaries are often mistrusted, as conflicts of interest are common. Regulation of media and payment cards is an example. Regulation is common in stock markets, real estate agents, and many other traditional matchmakers and is designed to create transparency and protect the rights of users.

From a regulatory perspective, platforms present specific challenges. Network effects require scale, so only a limited number of platforms are able to reach such a scale in a given multi-sided market. Oligopoly is the most common structure in multi-sided markets, and the oligopoly sometimes becomes a monopoly. Limits to the concentration of newspapers and other media have been common. Even if pluralism has traditionally been the main reason behind restrictions, the oligopoly structure of media, as multi-sided markets, has triggered such a concern. Market power of platforms in payment card markets has triggered the intervention of public authorities in many jurisdictions, notably by setting limits to the commissions.

At the same time, incumbents have often asked for regulation when their markets are being transformed into new multi-sided markets. This was the case that traditional newspapers made against free newspapers and merchants made against payment cards. The above are a few examples of how regulation has traditionally played an important role in multi-sided markets.

Notes

1 Picard, R. G. (2001). Strategic responses to free distribution daily newspapers. *The International Journal on Media Management*, 3(3), 167–172.
2 Helft, M. (2010, Feb 27). *In a Country of Monopoly Newspapers, Palo Alto is Awash in Competition, New York Times*, retrieved from www.nytimes.com/2010/02/28/us/28sfnewspaper.html
3 Evans, D. S. & Schmalensee, R. (2005). *Paying with Plastic. The Digital Revolution in Buying and Borrowing*. MIT Press, Cambridge, MA, 2nd ed, p. 54.
4 Caillaud, B. & Jullien, B. (2003). Chicken and egg: Competition among intermediation service providers. *RAND Journal of Economics*, 309, 328.
5 Wortham, J. & Wingfield, N. (2012, April 5). *Microsoft is Writing Checks to Fill Out Its App Store, New York Times*.In M.A.Cusumano, A.Gawer, &D. B. Yoffie (2019). *The Business of Platforms. Strategy in the Age of Digital Competition, Innovation, and Power*. Harper Business, New York, 134.
6 Harris, B. J. (2014). *Console Wars. Sega, Nintendo and the Battle that Defined a Generation*. HarperCollins, New York.
7 Wu, T. (2017). *The Attention Merchants*. Atlantic Books, London.
8 Rochet, J.C. & Tirole, J. (2003). Platform competition in two-sided markets, *Journal of the European Economic Association*, 1(4), 990–1029.
9 These titles have been particularly influential: Evans, D. S. & Schmalensee, R. (2016). *Matchmakers. The New Economics of Multi-sided Platforms*. Harvard Business Review Press, Boston; Libert, B., Beck, M., & Wind, J. (2016). *The Network Imperative. How to Survive and Grow in the Age of Digital Business Models*. Harvard Business Review Press, Boston, MA; and Parker, G. G., Van Alstyne, M. W., & Choudary, S. P. (2016). *Platform Revolution*. W. W. Norton & Company, New York.
10 OECD (2009). *Two-Sided Markets*, Policy Roundtables, retrieved from www.oecd.org/daf/competition/44445730.pdf
11 Evans D. & Schmalensee, R. (2019). *Antitrust Analysis of Platform Markets. Why the Supreme Court Got it Right in American Expre*ss. Competition Policy International, Boston, p. 59.
12 Rochet, J.C. & Tirole, J. (2003). Platform competition in two-sided markets. *Journal of the European Economic Association*, 1(4), 991.
13 For a critical analysis of these estimates, see Briscoe, B., Odlyzko, A., & Tilly, B. (2006). Metcalfe's law is wrong-communications networks increase in value as they add members-but by how much? *IEEE Spectrum*, 43(7), 34–39. Some antitrust researchers have also warned "not to overestimate the novelties of two-sided commerce"; Hovenkamp, E. (2018). Platform Antitrust. *Journal of Competition Law*, 2019, 49.
14 *The Oxford English Dictionary*, 2nd edn, 1989.
15 Gawer, A. & Cusumano, M.A. (2002). *Platform Leadership*. Harvard Business School Publishing, Boston, MA.
16 Evans, D. S. (2011). *Platform Economics: Essays on Multi-sided Businesses*. CPI.
17 Simon, P. (2011). *The Age of Platform*. Motion Publishing, Henderson, NE.
18 Tiwana, A. (2014). *Platform Ecosystems*. Elsevier, Waltham, MA.
19 Choudary, S. P. (2015). *Platform Scale*. Platform Thinking Labs Pte, Ltd.
20 Parker, G. G., Van Alstyne, M. W., & Choudary, S. P. (2016). *Platform Revolution*. W. W. Norton & Company, New York.
21 Srnicek, N. (2017). *Platform Capitalism*. Polity Press, Cambridge.
22 Cusumano, M.A., Gawer, A., & Yoffie, D.B. (2019). *The Business of Platforms. Strategy in the Age of Digital Competition, Innovation, and Power*. Harper Business, New York.

23 *The Oxford English Dictionary*, 2nd edn, 1989.
24 For an analysis of the different uses of the term "platform," see Baldwin, C. & Woodard, C. J. (2009). The Architecture of Platforms: A Unified View. In A. Gawer (ed.), *Platforms, Markets and Innovation*. Edward Elgar Publishing, Cheltenham, UK and Northampton, MA.
25 On the differences between transaction and non-transaction markets, see van Damme, E. E. C., Filistrucchi, L., Geradin, D., Keunen, S., Klein, T. J., Michielsen, T. O., & Wileur, J. (2010). *Mergers in Two-Sided Markets–A Report to the NMA*. Netherlands Competition Authority, pp. 1–183.
26 Cusumano, M.A., Gawer, A., & Yoffie, D.B. (2019). *The Business of Platforms. Strategy in the Age of Digital Competition, Innovation, and Power*. Harper Business, New York.
27 Commission Decision of 27.6.2017 relating to proceedings under Article 102 of the Treaty on the Functioning of the European Union and Article 54 of the Agreement on the European Economic Area (AT.39740 - *Google Search (Shopping)*), OJ 18.12.2017.

Chapter 3

Digital Platforms

As we have shown above, platforms in a multi-sided market are not a new phenomenon, as direct and indirect network effects existed before the internet and companies had already exploited them for their benefit. Now that the basic elements of this model of industrial organization have been identified, we are in a position to better understand how digital technologies have been used to create new multi-sided markets with even more massive network effects.

Network effects have multiplied in scale as more people become connected to the internet. The inventors of the transistor and the individuals who envisaged and built the internet while working for public institutions such as ARPA, NSF, or CERN – have empowered the most sophisticated tool for human interaction. However, they did not "monetize" their efforts, as the first generation of the internet, connecting large mainframe computers, was a fully distributed network, open to the research community with no hierarchies, and it grew at a slow pace during its first 20 years.

As early as 1964, the visionary Marshall McLuhan predicted the revolutionary nature of electronic communications in his famous book, *Understanding Media: The Extensions of Man*, where he introduced the somehow mystifying but popular assertion: *"the medium is the message. This is merely to say that the personal and social consequences of any medium, that is, of any extension of ourselves, result from the new scale that is introduced into our affairs by each extension of ourselves or by any new technology."*[1]

The internet only became massive when a new medium, the personal computer (PC) was popularized in the late 1980s and millions of computerized users were ready to be interconnected. A small cluster of individuals in Silicon Valley understood that businesses could be built by growing specific purpose communities inside the new massive internet, and that those creating the communities would monetize the network effects. This led to the second generation of the internet. Platforms proliferated, expanded both into traditional activities like selling goods (eBay, Amazon, Alibaba) and newer ventures such as internet searches (Google), as well as digital social networks (Facebook). Digital platforms built "communities" of previously unimaginable scale.

The size of the network effects grew as smartphones were used to connect to the internet (the third generation of the internet), empowering a new generation of platforms (Airbnb, Uber). Finally, as not only humans but also objects are being connected (Internet of Things, 5G and so on), multi-sided markets in the form of smart-homes and even smart-cities are emerging (the fourth generation of the internet). The scale of the network effects created by the interaction of billions of humans and devices is difficult to comprehend.

The stories of the leading digital platforms demonstrate the relevance of network effects. Communications technologies have multiplied direct and indirect network effects. Data-fed algorithms are creating a new type of network effects, which we call *algorithmic network effects*. In this chapter, we will demonstrate how the founders of and investors in the leading platforms were aware of the relevance of network effects and, unsurprisingly, built their companies around them.

3.1 eBay, Amazon, and Marketplace Platforms

As the internet went massive in the mid-1990s, e-commerce companies proliferated. By the end of the century, two leaders had emerged: eBay and Amazon. They had very different business models, and one would become dominant.

On September 4, 1995, the auction site eBay was launched by Pierre Omidyar, a computer engineer who had previously worked in several start-up companies in Silicon Valley.[2] According to the narrative built by eBay, the first item to be auctioned was a broken laser pointer. The bidder was a collector of broken laser pointers, probably the only one in the world. This is the beauty of digital platforms: they facilitate the interaction of individuals who previously had limited chances to interact.

As the number of available goods grew, the platform became more attractive for bidders. As the number of bidders increased, it made it more interesting for sellers, as even the most improbable item might be of interest for at least one person in a large crowd. In this way, indirect network effects benefited both sellers and bidders. Furthermore, any item could be traded, no matter how low the value, thanks to the low transactions costs.

eBay did not develop any infrastructure for the management of goods. eBay does not manage inventory, storage, and distribution; it only provides a marketplace for individuals to make their own transactions. The seller takes care of sending the item to the bidder and eBay takes a commission from the bidder. Thus, eBay, and digital marketplaces in general, are transactional multi-sided markets.

Instead, eBay invested in the development of soft infrastructure to manage transactions. In 2005, eBay acquired Skype, a communications platform (see Chapter 8) so that sellers and bidders could communicate easily, even across borders. In 2002 eBay acquired PayPal, a payment platform already popular among eBay users. PayPal was a digital payments system that facilitated payments between individuals without the intervention of a bank or a payment card company. It was inspired by the peer-to-peer model that was popular for the exchange of music files (see the next chapter). Skype and PayPal were themselves platforms that were creating massive network effects by pooling millions of individuals.

PayPal CEO Peter Thiel correctly identified the potential of network effects, but he also identified the "chicken-and-egg" challenge. A payment system requires a critical mass of participants. In order to solve this challenge, his solution was to focus on a small, closely connected community that could adopt the new service almost as a block. eBay heavy users were such communities.[3] PayPal grew on top of a pre-existing network (eBay) and was built over still another network, the internet. It is indeed easier to build a network on top of another one.

The group of individuals who created PayPal, based on their early and deep understanding of network effects in the internet, ended up having a long-lasting influence on the evolution of digital platforms. The "PayPal mafia," as it came to be called, would be key in developing platforms such as Facebook, YouTube, LinkedIn, and Yelp, which involve individuals and companies that will repeatedly appear along this book.

The initial marketplace for the peer-to-peer exchange of collector items evolved over time to include any saleable good, fixed price sales, and an international reach. As of 2019, eBay had more than 180 million active buyers and revenue close to $11 billion.

The main competitor of eBay was Amazon, originally incorporated in 1994 by Jeff Bezos, an electronic retailer. It started selling books and music CDs. The fact that Amazon did not have physical stores was turned into a competitive advantage as the lack of physical constraints allowed Amazon to manage the largest inventory on Earth.[4] That was Amazon's winning proposition that enabled it to grow. Eventually Amazon expanded to sell basically any relevant saleable item.

Amazon's strategy was the opposite of eBay's. The objective was not to invest in soft infrastructure such as communications and payment. On the contrary, Amazon invested heavily in the development of the physical infrastructure to improve the buyer's experience by reducing delivery times. Amazon has developed the world's densest network of automated warehouses, where goods are stored and packed, as well as the most efficient parcel delivery network.

Over time, Amazon proved more successful than eBay – its revenue in 2019 was $280 billion, ten times more than that of eBay. Investing in infrastructure proved a winning bid. Such a large size, along with Amazon's aggressive pricing strategy, have attracted criticism for potential antitrust breaches.[5]

In any case, Amazon's success relies on adopting a multi-sided market strategy, which is not so different from eBay's strategy. In November 2000 Amazon launched Amazon Marketplace. The retailer would also support third parties to sell products on Amazon's site and make use of Amazon's physical infrastructure. Amazon became a digital platform in a multi-sided market, intermediating between customers and other retailers and thus exploiting indirect network effects. The more retailers there were, the more attractive the platform was for customers. The more customers there were, the more attractive the marketplace became for retailers.

Amazon Marketplace created tensions inside the company, as the department selling books had to compete with the department managing the marketplace for the sale of used books by third parties – all on the same webpage.[6] However, scale was necessary to make Amazon's logistic network of warehouses and parcel delivery the most efficient logistic network in the e-commerce market.

Amazon's vertical integration has always raised suspicion among third parties using the marketplace, as accusations of discrimination and predatory pricing strategies have been common. These are common issues when digital platforms vertically integrate and start competing with their customers. Significant challenges exist to ensure the right incentives for third parties to participate in the platform, but it is not impossible, as Amazon Marketplace has shown. In fact, more than half of the items sold on Amazon are sold by third-party sellers, not by Amazon itself. However, suspicion continues, and the European Commission opened an antitrust investigation against Amazon in 2019 for self-preferencing: Amazon would be exploiting access to third-party data as the manager of the marketplace, in order to provide a competitive advantage to its own activities as retailer. Self-preferencing is becoming one of the more common criticisms leveled against digital platforms.

In the international market, the Chinese marketplace Alibaba is a powerful competitor,[7] with 654 million annual active buyers and annual revenues in 2019 of $56 billion, even if most of its users and revenue are concentrated in China.

3.2 Google and Searches

The idea of Google was formed in January 1998, when two PhD researchers at Stanford University, Larry Page and Sergey Brin, presented a paper describing an alternative model for managing web searches. The internet had grown so much that a search tool had become necessary. InfoSeek, Yahoo, and Excite already had their own search engines. Instead of searching keywords in web pages, as the existing search engines did, the Stanford students built an algorithm that would take into consideration hyperlinks to measure the relevance of a web page. The search engine was called Google, after the word *googol*, which represents the number 10^{100}, or one followed by 100 zeros, coined in 1938 by the nine-year-old nephew of the English mathematician Edward Kasner.[8] As the domain name with the word googol was taken, they settled for Google.

Google Inc. was incorporated on September 7, 1998, with capital of $1 million provided by four investors: Jeff Bezos, founder of Amazon; Ram Shriram, a member of the executive team in Netscape (the first popular browser) and then vice-president at Amazon; and Andreas von Bechtolsheim and David Cheriton (professor of distributed systems and networks at Stanford and mentors of Page and Brin), who were both founders of SUN Microsystems. Sequoia, the well-known venture capital firm, joined during the next round to finance the start-up.

The search engine grew a large customer base of searchers. As of the beginning of 1999 it was managing half a million searches a day; by the start of 2000 this number had reached seven million searches a day, and then 100 million daily searches by the end of 2000, reaching a share of 40 percent in the global search market.[9]

Google would become the most popular search engine in the world, with a market share beyond 90 percent globally, except for China (where Baidu has a 70 percent market share) and Russia (where Yandex has a 36 percent market share). In the Western world, the only alternative search engine is Bing by Microsoft. Companies such as Yahoo now merely resell Google's search results.[10] As will be described, network effects in search seem to be so relevant as to tip the market to a near monopoly.

"Given the internet's enormous breadth and constant evolution, establishing and maintaining a commercially viable general search engine is an expensive process. Google's search index contains hundreds of billions of webpages and is well over 100,000,000 gigabytes in size. Developing a general search index of this scale, as well as viable search algorithms, would require an upfront investment of billions of dollars. The costs for maintaining a scaled general search business can reach hundreds of millions of dollars a year."[11]

Google does not have a subscription fee or charge per search. Again we find a zero-pricing service in a multi-sided market. The company's founders were particularly averse to the advertising model followed by competitors like Yahoo and Excite, but in 2001 AdWords was launched. Google decided that it would not display advertising on its web page, as other portals did, and would not accept payments from companies to manipulate search results. On the contrary, revenue would be generated by a mixed service, in which companies would pay to have their ads introduced as results to searches, but separated from so-called "organic" results, and ads would be displayed as a result of a sophisticated evaluation that considered the relevance of the result and the fee to be paid by the advertiser, determined by an automated auction, and only when users clicked on their ads. The model of ad-financed services was extended to email in 2004, when Gmail was launched. Users received a free email account with a large capacity. Google would scan the content of the emails and would introduce personalized ads in the display of the emails.

Google has created an advertising category within itself – so-called search advertising – to be differentiated from more traditional display advertising up to the point of being considered a different product market for antitrust purposes. In search advertising, the ads are presented as paid results for a search, together with but separated from organic results. These ads are very effective, as they are directly connected to the consumer's declared interest in a product. Consequently, search advertising produces more revenue than digital display advertising. For instance, in the UK in 2019, search advertising raised £7.3 billion in revenue, 90 percent of which was through Google. Google's revenue was larger than that generated by all companies in the digital display market (£5.5 billion) as well as all TV advertising revenue in the UK (£4.9 billion).

Google is also active in the digital display market. Google displays ads in its email service (Gmail), its map service (Google Maps), and so on. Google's acquisition of YouTube in 2006 was a natural extension of this business model (see Chapter 9). Google's share in display advertising is also relevant; for the UK, it has been calculated that Google has a market share of close to 10 percent (Facebook has more than 50 percent market share).

Furthermore, Google provides intermediation services for publishers in the digital advertising market. AdSense was launched in 2002. Google's business model became even more sophisticated as the platform was extended to consider content producers and benefit them in the form of the payment of a percentage of the advertising revenue generated by their sites. AdSense allowed small ads to be displayed in web pages, blogs, and other sites produced by third parties. Google would sell these ads to advertisers following the AdWords model and share revenue with publishers, as much as two-thirds of the fee paid by the advertiser.

Google creates indirect network effects like the ones we already know from newspapers, namely by connecting advertisers and eyeballs. In the words of the European Commission: "The indirect network effects stem from the link between the attractiveness of the online search advertising side of a general search engine platform and the revenue of that platform. The higher the number of advertisers using an online search advertising service, the higher the revenue of the general search engine platform; revenue which can be reinvested in the maintenance and improvement of the general search service so as to attract more users."[12]

However, the scale of the indirect network effects created by Google has no precedent in the history of media. The largest TV audience in the US, the Super Bowl, reaches around 100 million people once a year. Google manages more than 5 billion searches per day, throughout the year. It is not surprising, then, that the advertising revenue of Google in 2019 was $134 billion, while the full-year revenue of the network broadcasting the Super Bowl (CBS) was around $15 billion. Worldwide combined TV advertising revenue in 2019 was approximately $166 billion. If Google keeps growing as much as it has in previous years, by 2021 it wouldattract more advertising revenue than all TV stations in the world combined.

AdWords and AdSense have grown to such a strong position that many advertisers and publishers rely on them for their business. Such dependence creates conflicts, as Google is not always clear about the rules; rules are often amended without previous notice and accusations of discrimination have often been leveled, including discrimination in favor of Google's own services (self-preferencing). Regulators are putting increasing attention on platforms.

In June 2017, The European Commission imposed a €2.4 billion fine on Google for self-preferencing:[13] in its general search results' page, Google positioned and displayed its own comparison shopping service more favorably than the competing shopping services. This was one of the first cases on self-preferencing by a platform intermediating its own downstream services. Complaints against Google for self-preferencing were later filed by specialized travel search engines, against Amazon for self-preferencing its own products in Amazon Marketplace, and against Apple for self-preferencing its own apps in the app store. Self-preferencing is one of the main accusations against dominant platforms; the remedies are non-discrimination obligations, or even vertical unbundling.

In March 2019, the European Commission imposed a €1.49 billion fine on Google for abusing its dominant position in the commercialization of AdSense for searches. Google first imposed exclusivity on publishers and prohibited them from placing search adverts from competitors on their own websites. Exclusivity was eventually modified to allow the display, but only after written approval from Google regarding how rival ads would be displayed, reserving the most visible and most clicked parts of the websites for Google. Imposing exclusivity is becoming another common accusation against dominant platforms, with multi-homing defined as the remedy.

In December 2019, the French antitrust authority imposed a €150 million fine on Google for abuse of dominant position. A publishing company, Gibmedia, which charges users for weather and telephone information, had its account suspended without previous notice for violating Google's rule about not displaying ads on sites that charge for services that are usually provided for free. The authority concluded that Google's rules on advertisers are established and applied under non-objective, non-transparent, and discriminatory conditions. The opacity and lack of objectivity of these rules make it very difficult for advertisers to apply them, while Google has full discretion to modify its interpretation of the rules in a way that is difficult to predict, and to decide whether the sites comply with them or not. This allows Google to apply them in a discriminatory or inconsistent manner and leads to damages for both advertisers and for search engine users. The lack of fairness and transparency in the management of the multi-sided market is another common accusation against digital platforms.

3.3 Facebook and Social Networks

"I think that network effects shouldn't be underestimated with what we do." These were the words of Mark Zuckerberg, Facebook founder, in an interview in 2007.[14] Facebook is one of the most successful network-effect-builders in history.

Zuckerberg created Facebook in 2003 as a sophomore at Harvard. Zuckerberg, a computer science major, enjoyed producing small software programs based around the notion of social relationships, such as "Course Match" a software program for selecting courses based on which students had already enrolled in them. Facebook started as one of these small experiments. The idea was to produce an online version of the popular paper face books that universities produced with the pictures of freshmen students as they arrived at campus. The first version of the software, "Facemash," invited users to compare student pictures to vote on who was sexier. The software was an instant hit at Harvard dorms: in a matter of hours 22,000 pairs of pictures were voted. However, the experiment was short-lived, as the university sanctioned Zuckerberg for a violation of the college's code of conduct for breach of security, copyright, and privacy, given that the pictures had been hacked from the university's system. Facebook had committed its first privacy breach even before it was incorporated!

On January 11, 2004, Zuckerberg registered the web address "thefacebook.com," and a few weeks later, on February 4, the website went live. Facebook invited users to upload their own content (without any more hacking), like pictures and basic information about themselves. It also facilitated communication between registered users, empowering them to invite their friends to join what was already named as the "social network." Access was limited to individuals with a Harvard email address. In four days, 650 Harvard students had registered. After three weeks, more than 6,000 students – three-quarters of Harvard undergraduates – were members.[15] By the end of the month similar services were launched at Columbia, Stanford, and Yale, with similar success.

The biggest source of new users was a program that imported people's contacts from Microsoft's Hotmail (see an analysis of Hotmail in Chapter 7). Microsoft objected: "They were [...] trying to build their social network on the backs of others."[16] Facebook ended 2004 with 1 million registered users. It was not only numbers that ramped up very quickly; even more relevantly, users would spend hours browsing through the site and improving their own profiles.

Network effects were key. The more registered users there were, the more value the network would have for each user. The growth strategy was perfectly designed.[17] Instead of opening Facebook to any user, they opened universities one by one, to ensure that a significant proportion of each community would enroll; this made it attractive for users as they would be able to connect with a substantial number of their friends. Had Facebook opened the network to any user, new members would have found out that none of their friends were connected and would probably never return to the platform, making the network useless. By focusing on elite and closed environments like Ivy League universities, one by one, Facebook cracked the "chicken and egg" challenge of users expecting other users to join the network. Thiel at PayPal had used the same strategy.

Discussions about network effects were common in the early days of Facebook, just as McLuhan's *Understanding Media* was favored reading. After all, the founders of Facebook were not just college dropouts, they were Harvard dropouts. Facebook exhausted the US university market, then moved to universities in other countries, and then to high schools (September 2005); it was only in September 2006 that Facebook opened itself up to the general public. At that point in time, it had already around ten million active users. It was a perfect lesson in how to grow a community.

Facebook attracted the attention of the early champions of network effects in the internet. The company appointed Sean Parker as president in its first months. Parker had started another digital platform in Napster, the famous file-sharing service (discussed in the next chapter). In September 2004, Parker invited his Silicon Valley contacts to invest the first $500,000 in seed capital, in exchange for a 10 percent of the corporation. The leading investor was Peter Thiel, who had founded and then sold off PayPal and had also invested in other platforms such as LinkedIn and Spotify (see Chapter 8). The other investors were Reif Hoffman, co-founder of PayPal and subsequently of LinkedIn, and Mark Pincus, who was also an early investor in Napster, Twitter, and Snapchat. These men all had a deep understanding of the power of network effects, which is precisely why they invested in Facebook.

Over the years, Facebook has built the largest community in the world, with 2.7 billion active monthly users as of 2020. Facebook acquired Instagram, a popular alternative social network, as well as WhatsApp, the largest instant messaging app in the world (see Chapter 8). "[M]ake no mistake, growth tactics are how we got here." "The best products don't win. The ones everyone uses win." These statements in an internal memo under the title *The Ugly* encapsulate Facebook's strategy.[18] Facebook creates massive direct network effects, as the largest pool of users exchange their content, comment, give likes, and so on.

Having the largest customer base provides a large advantage to Facebook. Once customers have opened an account, created the network of connections, spent hours curating their profile, and years and years uploading their pictures and videos, they are locked in. "[T]here are network effects around social products and a finite number of different social mechanics to invent. Once someone wins at a specific mechanic, it's difficult for others to supplant them without doing something different". These words from Zuckerberg are repeatedly quoted in the antitrust suit filed by the US Federal Trade Commission against Facebook in December 2020.

In 2007 Facebook launched Facebook Platform. It encouraged developers to build apps and tools (games, page-design tools, e-marleting apps and so on) on top of Facebook. Developers would be allowed to use Facebook's Application Programming Interfaces (APIs) to extract content from Facebook, including the Find Friends critical API. Facebook successfully built an ecosystem of apps, which grew on top of Facebook's social graph. Interoperability was encouraged when it strengthened the platform (more content and apps), but it was restricted when it would allow the development of competing products or multi-homing with other social networks (individuals in small social networks cannot be reached from Facebook, content in Facebook cannot be accessed from other networks, and exporting it to other networks is certainly not facilitated). The US FTC found these restrictions anticompetitive and they were included in the antitrust suit filed in December 2020.

From the start, advertising was Facebook's business model. Facebook included so-called flyers or small ads for a low cost ($10–$40), mostly around university activities, such as school material and parties. Facebook reached $400,000 in revenue in its first year of operations. This is how Facebook builds indirect network effects, as the platform connects 2.7 billion viewers with millions of advertisers, including small companies and individuals that were previously excluded from the advertising market. Over time, Facebook has built a non-transactional platform, just as traditional media have, but at a much larger scale of operations and therefore with much larger network effects. The 2.7 billion active users multiply the audience of any newspaper, radio station, or TV network in the world. This is one reason why Facebook makes more revenue that all the newspapers in the world, combined.

Nonetheless, data is as relevant for Facebook's success as direct and indirect network effects. Data feed algorithms optimize the matching of users and advertisers, thus empowering the network effects. Algorithms target the ads to be displayed to users. Not only does Facebook allow the largest audience to be reached, but it also has such a deep knowledge of its audience that ads can be personalized. It is well known that that ads can double in price when they are personalized.[19] The more data you have about your audience, the better you can personalize the ads, creating so-called algorithmic network effects (see below).

Creating algorithmic network effects poses obvious risks in terms of privacy protection. It is not by chance that the largest fine due to privacy breaches, $5 billion, was imposed on Facebook by the US Federal Trade Commission in July 2019, due to the Cambridge Analytica scandal around voter profiling with data extracted from Facebook.[20]

The value created by Facebook in terms of direct, indirect, and algorithmic network effects multiply the benefits of traditional media. In 2019, Facebook had revenue of more than $70 billion. It has reached a market share beyond 50 percent in digital display advertising, for instance in the UK, where the second-largest player's Google's YouTube market share is below 10 percent.[21] Even more extraordinary is Facebook's net income, which has reached 40 percent of its revenue. The margins enjoyed by Facebook are exorbitant, proving that the company's network effects are creating a lot of value, but also that the company has succeeded in capturing a substantial amount of it, not distributing it across its ecosystem.

3.4 Airbnb and the Sharing Economy

Smartphones caused the rise of the third generation of the internet. When the iPhone was launched in June 2007, 1.1 billion users were connected to the internet, mostly through their PCs. Ten years later, and thanks to the smartphone, 3.8 billion people – 51 percent of the world's population – were connected. Again, the medium was the message. The smartphone provided a new experience and thus a quantum leap in the connection to the internet. Being cheaper, personal, and portable, smartphones connect individuals anywhere and anytime.

The mobile internet is very different from the PC internet. Apple developed a closed operating system that would not allow any third-party software to be downloaded to the phone without Apple's permission (see Chapter 8). Every app available in the app store has been previously approved by Apple. Google developed a model that was similar, but not so rigid. This contrasts with the PC ecosystem, as PC users are free to download any software they please, with no vetting by the company that runs the operating system: Microsoft. This results in more freedom, but also in more bugs, privacy breaches, and so on.

Apps were developed by third parties to fully exploit the network effects of the new medium. WhatsApp facilitates communications between a growing pool of users and a more personal and intense engagement (Chapter 8). Individuals can easily share their thoughts in real time, but also their houses (Airbnb), their work (Taskrabbit), and their transportation (as explained for Uber in Chapter 13 and BlaBlaCar in Chapter 14). Network effects empowered by platforms with billions of users transform more and more sectors.

In October 2007, Brian Chesky and Joe Gebbia, two unemployed designers, built a web page to sublet three air mattresses in the living room of their apartment during a big design conference in San Francisco. The ambition was to create an app so that guests could book accommodation in a private residence as easily as they could in a hotel. Airbnb built an app where homeowners (hosts) could list the space they wanted to offer for rent (a room, a full house, a castle), with all the relevant information, including pictures, price per night, etc. Guests could identify the accommodation they were interested in, communicate with the host through the site, book the accommodation and then pay for it, always through the app. Airbnb takes a commission of 6–12 percent.

The app grew slowly during its first months, mostly through so-called "guerrilla marketing." It was only in April 2009, when the venture capital firm Sequoia invested $585,000 in the firm, that the platform started to grow, reaching weekly bookings of $100,000 in August 2009.[22]

Creating trust was key to the growth of Airbnb. Guests and hosts both provide evaluations of each other, which are available for the rest of the community to check before contracting. As the number of transactions grows, a significant number of evaluations are available, creating a digital reputation for each player. All payments take place through the platform. Once the guests make the payment, it is escrowed by Airbnb, which only pays the host 24 hours after the guests' arrivals and only in case no complaint has been raised by the guest. Airbnb offers secondary insurance to cover liabilities for hosts and guests.

Airbnb has grown impressively, as a result of the power of indirect network effects. As more homes are available for rent, more guests join the platform, which leads to more house-owners being interested in joining the platform. As stated in the IPO filing, Airbnb is in the business of building networks: "Hosts and guests attract each other to Airbnb, creating a global network across more than 220 countries and regions."[23] The multi-sided market has grown to cover the globe. Travelers using the service as guests in San Francisco would then join the platform as hosts in their home city in Paris, attracting guests in other European cities. The wider the geographic availability of homes, the more attractive the platform became for guests, further reinforcing the virtuous cycle.

By 2018, Airbnb was listing more than 5 million accommodation options in more than 81,000 cities in 191 countries, had revenue of $2.7 billion, and profits of $93 million.[24] The $585,000 investment made by Sequoia in 2009 multiplied to a valuation of $11.76 billion in 2020 when Airbnb went public. The whole of Airbnb had a value of $100 billion.

In 2017 Airbnb expanded its product range from accommodation to other travel services, coined as "experiences," such as walking tours, cooking lessons and a whole set of services to be locally provided to travelers. Once a digital platform ignites, it can easily expand to neighboring markets, building on the existing network effects.

Airbnb is a digital platform in a multi-sided market. It does not build hotels, own real estate, manage the accommodation, or even contract the provision of the accommodation service with guests. Airbnb is merely an intermediary, a matchmaker that facilitates the contracting between hosts and guests: the two sides in the multi-sided market. The Court of Justice of the European Union had to confirm this, as the French association of professional accommodation providers challenged it, claiming that Airbnb digital services "forms an integral part of an overall service, whose main component is the provision of an accommodation service." The Court confirmed that "the compiling of offers using a harmonised format, coupled with tools for searching for, locating and comparing those offers, constitutes a service which cannot be regarded as merely ancillary to an overall service coming under a different legal classification, namely provision of an accommodation service."[25]

Airbnb is a leading example of the so-called "sharing economy." Thanks to digital platforms and the reduction of transaction costs, private individuals can share their assets (housing, vehicles, ability to provide small services, etc.) with other private individuals, in competition with traditional corporations. Similar platforms are Uber or Taskrabbit.

3.5 Algorithmic Network Effects

Digital platforms build on direct and indirect network effects, as previous platforms, but they also rely on a new type of network effects: algorithmic network effects. Big data is necessary for the provision of some popular digital service, the best example of which is search. Google relies on the massive amount of previous searches to respond to a specific search.[26] In the words of the European Commission: "because a general search service uses search data to refine the relevance of its general search results pages, it needs to receive a certain volume of queries in order to compete viably. The greater the number of queries a general search service receives, the quicker it is able to detect a change in user behaviour patterns and update and improve its relevance."[27]

Algorithmic network effects mean that a service improves as it obtains more data from users: the more the service is used, the more data is captured; and the more data is captured and processed, the better the service becomes.[28] For this reason, data has been considered "the new oil."[29] Amassing data, if properly managed, can provide a powerful competitive advantage. Since Google has the largest pool of search history on the planet, it has a competitive advantage. Bing's technology might be as good as Google's, and it is only "a click away," but being way behind Google's pool of data, it cannot replicate the quality of Google's searches. This is again a network effect, and is so powerful that tips the market in favor of the largest player.

Algorithmic network effects play a fundamental role for digital platforms as they are key to overcoming the most relevant risk of any popular network: congestion. As the number of users of a service or an infrastructure increases, it can become so crowded as to hinder or even prevent its use. Attention is usually focused on positive network effects, but negative network effects can be just as relevant and can counterbalance the benefits derived from pooling large pools of users. Congestion is the most relevant negative network effect. In case of congestion, a new user does not add value, but actually detracts value, as it further prevents the use of the service for others. We are all familiar with road congestion, as well as the usual policies to fight it: increase capacity, raise prices for the use of the service to reduce demand, and so on.

Technology has traditionally helped to reduce the problem of congestion in network industries. Mechanical automatic switching was introduced in the 1920s, solving the problem of congestion in early telephone networks. Digital switching, which made use of computers, further empowered telephone networks to switch millions of telephone calls.

Digital platforms have always faced the risk of congestion. It is not only that many start-ups could not always cope with exploding traffic by installing more and more servers. In a more fundamental way, the larger the pool of users, the more difficult it is to ensure a coherent and fruitful interaction between them. In a marketplace like eBay, having more sellers is key to success, but having millions of sellers and even more items for sale makes it complicated to ensure buyers can find the items they are looking for (and even more difficult to trigger the availability of things they might desire, but that they are not actively searching). In social networks such as Facebook, as the amount of users reaches the billions and each user is uploading more and more information, it is important to select which information is displayed to the viewers browsing Facebook. The same applies to videos proposed by YouTube, to apartments shown by Airbnb, and to drivers matched by Uber.

Sorting presents diseconomies of scale. Sorting becomes more difficult and expensive as the number of items to be sorted increases. "The unit costs of sorting, instead of falling, rises."[30] Automation has always been a key to the performance of digital platforms. Platforms use algorithms to automatize the matching decisions that allow interaction between the different sides in the platform. The Oxford Dictionary defines an algorithm as "a process or set of rules to be followed in calculations [...] especially by a computer." Algorithms are increasingly using machine learning technology. They are not a set of fixed rigid commands predefining the links between the parties in the platform; on the contrary, algorithms peruse through the stored data in order to predict the most useful link between the different sides in the platform.

Algorithms improve upon themselves with each interaction.[31] Google, Facebook, and Airbnb identify users' reactions, for instance in the form of clicks in one of the search results, clicks in one of the videos proposed, or ratings by riders after using Uber. These reactions are incorporated into the algorithm in order to inform future services. In this way, algorithms are in a position to predict which information will be of most interest for future users. They predict the best result of a search, such as which YouTube video will be of most interest for the viewer or which apartments an Airbnb guest will prefer. They predict which driver will arrive soonest to pick up an Uber rider and provide the best service. More data makes predictions better.

Algorithmic network effects are a precondition for and reinforce direct and indirect network effects. Platforms will only succeed if they are able to efficiently intermediate between users. Large pools of users will only be attracted to the platform if the intermediation service is provided smoothly, adding value to the users. Only the most sophisticated algorithms are in the position to overcome the challenges of ensuring a productive interaction between billions of users. At the same time, the more users that are attracted to the platform, the better the algorithms can work and the better the service will get. The larger the pool of data and the better the algorithm that makes sense of it, the better the intermediation service provided by the digital platform can get.

Regulators have identified how algorithmic network effects create market power. The European Commission concluded that Google's dominance in the search markets relied on barriers to entry and expansion, such as the large investment required to create the search engine, but "also needs to receive a certain volume of queries in order to improve the relevance of its results for uncommon ('tail') queries. [...] The greater the volume of data a general search service possesses for rare tail queries, the more users will perceive it as providing more relevant results for all types of queries."[32]

The US Federal Trade Commission has identified the role algorithmic network effects in Facebook's business model: "Advertisers pay billions—nearly $70 billion in 2019—to display their ads to specific 'audiences'[...] created by Facebook using proprietary algorithms that analyze the vast quantity of user data the company collects regarding its users. This allows advertisers to target different campaigns and messages to different groups of users."[33]

European antitrust authorities have analyzed the role of data in potential anticompetitive practices. In 2016, French and German authorities published a report entitled *Competition Law and Data*.[34] In 2019, the Italian antitrust authority published a joint report with the national telecom and data protection authorities.[35] The Bundeskartellamt adopted an innovative decision in February 2019[36] declaring that Facebook had abused its dominant position in the market of social networks for private users; that is, the zero-pricing service provided by Facebook to users. The authority concluded that Facebook imposed exploitative business terms in relation to data, particularly imposing the combination of data extracted from other corporate services (like WhatsApp) with data extracted from the Facebook service, in order to display targeted ads in Facebook. No fine was imposed, but the German antitrust authority imposed a data separation remedy. Data from different sources can now only be combined with the explicit consent of the user.

Notes

1   McLuhan, M. (1964). *Understanding Media. The Extensions of Man*. MIT Press edition, 1994, p. 7.
2   Gitlin, M. (2011). *eBay. The Company and its Founders*, ABDO Publishing Company, Edina MI.
3   Thiel, P. (2014). *Zero to One. Notes on Startups, of How to Build the Future*. Virgin Books, New York, p. 53.
4   Stone, B. (2013). *The Everything Store. Jeff Bezos and the Age of Amazon*. Corgi Books, London.
5   Kahn, L. (2017). Amazon's Antitrust Paradox, *The Yale Law Journal*, 126(3), 564–907.
6   Stone, B. (2013). *The Everything Store. Jeff Bezos and the Age of Amazon*. Corgi Books, London, pp. 368–376.
7   Clark, D. (2016). *Alibaba, The House that Jack Ma Built*. HarperCollins, New York.
8   Asimov, I. (1982). *On Numbers*. Random House, New York.
9   Auletta, K. (2009). *Google. The End of the World as We Know It*. Penguin Books, New York, pp. 56 and 62.
10 Competition & Markets Authority (2020). *Online platforms and digital advertising*, Report, retrieved from https://assets.publishing.service.gov.uk/media/5efc57ed3a6f4023d242ed56/Final_report_1_July_2020_.pdf
11 US Department of Justice versus Google, antitrust complaint filed on 20 October, 2020.
12 Commission Decision of 27.6.2017 relating to proceedings under Article 102 of the Treaty on the Functioning of the European Union and Article 54 of the Agreement on the European Economic Area (AT.39740 –*Google Search (Shopping)*), OJ 18.12.2017, para. 296.
13 Ibid.
14 Ammirati, S. (2016). *The Science of Growth. How Facebook Beat Friendster and How Nine Other Startups Left the Rest in the Dust*. St. Martin's Press, New York, p. 180.
15 Kirkpatrick, D. (2010). *The Facebook Effect*. Virgin Books, New York, p. 34.
16 Levy, S. (2020). *Facebook. The Inside Story*. Penguin, New York, p. 215.
17 Ammirati, S. (2016). *The Science of Growth. How Facebook Beat Friendster and How Nine Other Startups Left the Rest in the Dust*. St. Martin's Press, New York.
18 Levy, S. (2020). *Facebook. The Inside Story*. Penguin, New York, p. 441.
19 Competition and Markets Authority (2019). *Online Platforms and Digital Advertising, Market study interim report*, Report, p. 15.
20 Levy, S. (2020). *Facebook. The Inside Story*. Penguin, New York, p. 417; and Wylia, C. (2019). *Mindf*ck. Inside Cambridge Analytica's Plot to Break the World*. Profile Books, London.
21 Competition & Markets Authority (2020). *Online platforms and digital advertising*, Report, p. 10, retrieved from https://assets.publishing.service.gov.uk/media/5efc57ed3a6f4023d242ed56/Final_report_1_July_2020_.pdf

22 Gallagher, L. (2017). *The Airbnb Story*. Houghton Mifflin Harcourt, Boston, p. 35.
23 Airbnb Form S-1 for the IPO, November 16, 2020.
24 Zaleski, O. (2018, Feb. 6). *Inside Airbnb's Battle to Stay Private, Bloomberg*, retrieved from www.bloomberg.com/news/articles/2018-02-06/inside-airbnb-s-battle-to-stay-private
25 Judgement of the Court of Justice of the European Union of 19 December 2019 in case C-390/18, *Airbnb Ireland*, para. 54 ECLI:EU:C:2019:1112.
26 Argenton, C. & Prüfer, J. (2012). Search Engine Competition with Network Externalities. *Journal of Competition Law & Economics*, 8(1), 73–105.
27 Commission Decision of 27.6.2017 relating to proceedings under Article 102 of the Treaty on the Functioning of the European Union and Article 54 of the Agreement on the European Economic Area (AT.39740–*Google Search (Shopping)*), OJ 18.12.2017, paras. 287.
28 Iansiti, M. & Lakhani, K. R. (2020). *Competing in the Age of AI. Strategy and Leadership When Algorithms and Networks Run the World*, Harvard Business Review Press, Boston, MA.
29 Agrawal, A., Gans, J., & Goldfarb, A. (2018). *Prediction Machines. The Simple Economics of Artificial Intelligence*. Harvard Business Review Press, Boston p. 43.
30 Christian, B. & Griffiths, T. (2016). *Algorithms to Live By. The Computer Science of Human Decisions*. William Collins, London, p. 62.
31 Domingos, P. (2015). *The Master Algorithm. How the Quest for the Ultimate Learning Machine Will Remake Our World*. Penguin Books, New York.
32 Commission Decision of 27.6.2017 relating to proceedings under Article 102 of the Treaty on the Functioning of the European Union and Article 54 of the Agreement on the European Economic Area (AT.39740 - *Google Search (Shopping)*), OJ 18.12.2017, para. 288.
33 FTC, Complaint for injuctive and other equitable relief, December 2020, p. 34.
34 Autorité de la Concurrence and Bundeskartellamt (2016). *Competition Law and Data*. Report, March 10, 2016, retrieved from www.bundeskartellamt.de/SharedDocs/Publikation/DE/Berichte/Big%20Data%20Papier.pdf;jsessionid=DB4C45D515F8F7CF12F8AD39198B5F6F.1_cid381?__blob=publicationFile&v=2
35 AGCM, AGCOM & Garante (2019), *Indagine cognoscitiva sui big data*. Report, retrieved from www.garanteprivacy.it/documents/10160/0/Indagine+conoscitiva+sui+Big+Data.pdf/58490808-c024-bf04-7e4e-e953b3d38a9a?version=1.0
36 Bundeskatellamt, B6–22/16, Facebook, Exploitative business terms pursuant to Section 19(1) GWB for inadequate data processing, 15.2.2019, retrieved from www.bundeskartellamt.de/SharedDocs/Entscheidung/EN/Fallberichte/Missbrauchsaufsicht/2019/B6-22-16.html?nn=3600108

Chapter 4

Disruption by Digital Platforms
Substitution and Platformization

The explosive growth of digital platforms has often come at the expense of the disruption of traditional industries. However, not all industries have suffered the same type and the same degree of disruption. Building on management literature, we will analyze two of the earliest examples of disruption by digital platforms: Napster disrupting the recorded music industry and Google News disrupting newspapers. In doing so, we are able to identify two different types of disruption by digital platforms.

On one hand, digital platforms can substitute traditional players, as they grow larger network effects that make them more efficient. Substitutes (that is, digital alternatives) replace and compete with the services offered by the traditional players. Consequently, traditional players are marginalized to fringe segments or even ejected from the market.

On the other hand, digital platforms can "platformize" the products or services provided by traditional players by intermediating them. Subsequently, the platform moves to aggregate the goods and services provided by the different traditional players. In this way, platforms take over the direct relationship with the customer, and the goods or services produced by the traditional player become a mere commodity. At some point, the platform reaches a position thanks to which it can dictate the conditions for the provision of the services it intermediates. As a last step, digital platforms might vertically integrate: they no longer just provide the intermediation service, but also start to compete with the traditional players in the provision of the original product. Traditional players are "platformed" by becoming a mere side in a multi-sided market. Network effects are built on the data layer on top of traditional players.

4.1 What Is Disruption?

In the business management literature, disruption became popular through the 1997 book *The Innovator's Dilemma*, by Clayton Christensen.[1] Christensen postulated that some innovations would initially provide second-best solutions that were only of interest to niche users. Leading companies, listening to their mainstream customers, would not adopt the second-best solution, to be later excluded from the main market as the performance of the disruptive innovation would improve at a rapid rate. Christensen opposed the paradox of "disruptive innovations" to "sustained innovations"; that is, innovations, no matter how radical, which increased product performance and can therefore be identified and implemented by market leaders.[2]

Digital platforms are a good example of disruptive innovation, as they represent a new model of industrial organization empowered by technology. Digital platforms can be considered disruptive innovators on both the demand and the supply side.

On the demand side, performance tends to be low during the establishment of the platform, as the benefits derived from the network effects are small. As a result, by listening to their customers, established market leaders tend to ignore this new business model. This was the case with Airbnb. Accommodation at private residences provided by non-professional individuals was perceived by some as a crazy idea that might attract some fringe customers, but was certainly not an option for regular hotel guests. Consequently, it was not taken into consideration by well-established hotel chains. As network effects became relevant, the efficiency of the new model became clear and mainstream customers migrated to the Airbnb platform, thus disrupting the traditional accommodation industry.

On the supply side, it is difficult for incumbents to replicate digital platforms, as they create an entirely new model of industrial organization. In the case of Airbnb, entirely new and previously underutilized assets (private residences) were brought to market. Economies of scope, derived from the use of asset both as private residence and as an accommodation service, considerably brought down the cost of the service and constituted a competitive advantage that was not replicable by hotel managers. Most dramatic is the case of platforms that finance services with advertising, providing zero-pricing services to customers, whereas they previously had to be self-financing. An example of this is Google Maps; it is almost impossible for incumbents to compete with such disruptive services.

In the following pages, we will analyze two cases of disruption by digital platforms: the music recording industry, which was disrupted by digital platforms that provided alternative access to music; and newspapers, which were disrupted as digital platforms increasingly attracted advertising revenue that had previously been devoted to traditional media.

4.2 Napster and the Music Industry

In June 1999, Napster, the music file-sharing site, was launched by Northeastern University undergraduate Shawn Fanning, his uncle John Fanning, and his online friend Sean Parker, who would later become president of Facebook and an early investor in Spotify. Napster allowed the peer-to-peer exchange of music files in MP3 format. It would disrupt the music recording industry.

MP3 was a set of standards developed in the early 1990s around algorithms for the digitalization of audio recordings in a compressed way, by reducing the accuracy of certain parts of a continuous sound that are considered to be beyond human ability. The standard was developed by engineers working for telephone companies, but it proved useful for the storage of music files. In parallel, MP3 was adopted by companies such as Microsoft for Windows Media Player (1997). In 1998, portable MP3 players were launched on the market.

As the internet went mainstream in the 1990s, the flexible and distributed nature of the internet empowered users to exchange not only email messages, but also other digital content such as music files. Music files could be exchanged in many different ways. For instance, audio files could be attached to an email as early as 1992. As internet connections were still narrowband, files had to be compressed as much as possible. MP3 was the obvious solution.

However, Napster took file-sharing to the next level by creating a platform that would allow "music-givers" and "music-takers" to interact. Before Napster, peers had no obstacle to exchange music files and they did so, but on a small scale, as transaction costs were high. It was necessary to get into contact with another peer, exchange information about the music files each of them had, identify whether any such files were of interest for the other peer, agree on the conditions for the exchange, and then submit the files.

Napster created a new layer for the communication and exchange of information between peers. By developing specific software, Napster automatized the process for peers to display their available files, communicate the files they were interested in, and match givers and takers. This new layer was centralized on Napster's servers. Peers would then execute the file exchange through the internet, without Napster's intervention. In this sense, Napster was a so-called "structured peer-to-peer network." The entire process was automatized, making it so seamless that users would not really perceive it and would often assume they were downloading the music files from the platform.

As the number of users grew, network effects ignited. With a larger pool of users, it was easier to find any needle in an ever-larger haystack. Community building was incentivized, as the platform allowed groups with similar interests to be formed and discuss their musical interests. Growth was explosive: by the beginning of 2000, Napster had 30 million users, and it reached 80 million users in mid-2001. This is particularly impressive, given that the internet had only 479 million users at this time.

Certainly, access to music at no cost facilitated growth. There was no limit to the number of files that could be downloaded, and users were not reluctant to share their files, as there was no cost for them to share copies of their existing files. At no cost, demand was almost unlimited.

The party started to be spoiled on December 6, 1999, just six months after Napster was launched. The Recording Industry Association of America (RIIA) filed actions against Napster for contributory and vicarious copyright infringement.[3]

While it was clear to the music industry that Napster was not directly involved in the conveyance of the music files, it was directly contributing to the infringement by the third parties sharing the files and was fully knowledgeable of the fact such an illegal exchange was taking place thanks to the use of its software (contributory infringement). Furthermore, Napster had the ability to supervise the infringement and terminate it (vicarious infringement). The RIIA had a better case than a traditional newspaper suing free newspapers for littering or misusing the term *newspaper*.

Napster argued that sharing files was a traditional fair use of the copyright material, whether it took place through traditional exchanges among friends or electronically. In any case, Napster would not be in breach, as it had substantial non-infringing uses (just as VCRs were not infringing when making illegal copies). Finally, under the Digital Millennium Copyright Act, Internet Service Providers' liability would be limited to the removal of specific users found sharing copyrighted files.

The District Court for the Northern District of California firstly ruled against Napster. Later, it granted a preliminary injunction against Napster, as it was decided that peers were directly infringing copyright; as theirs was not a fair use, but a way to "get for free something they would ordinarily have to buy," Napster was liable as it found no substantial non-infringing uses and was contributing directly to the infringement, having full knowledge of it taking place.[4] However, a stay of the injunction and an expedited appeal was granted. It was only when the Court of Appeals for the Ninth Circuit confirmed the ruling that Napster suspended the service. The date was July 1, 2001 and Napster had just turned two years old.

Even if it was short-lived, Napster transformed the music industry. Napster suspended its services, but other platforms continued with similar services for most of the first decade of the twenty-first century: Gnutella, BitTorrent, Kazaa (whose founders would later create Skype), and Scour (where Uber's future CEO had a leadership position). They even included video services, extending disruption to Hollywood. All of these services were eventually closed following legal action.

Digitalization was an evident disruptive moment for the music industry. The first digital services in the early 1990s were niche services for geeks, as they required some basic computer knowledge. MP3 quality was not as good as alternative physical recording methods. Digital music was also a disruptive innovation on the demand and supply sides. Even if Napster demonstrated that digital music would go mainstream, the music industry turned out to be incapable of adapting its business model to this disruptive innovation.

In fact, it was particularly slow to adapt to digitalization. It was not the case that it failed to identify the risks at an early stage. The RIIA filed actions against Napster just six months after it was launched. As one of the frontrunners in the digitalization process, the music industry was very active in terms of filing suits against platforms breaching their copyrights, but it was slow to build an alternative business model adapted to the digital age. In fact, the music industry was not able to build a platform and it refused to join any platform developed by third parties. In the end, it was not the music industry that created the digital model for the industry, but technology champions: Apple's iTunes opened the way, followed by music streaming platforms, such as Spotify (see Chapter 8).

In the meantime, global sales of recorded music dropped from $38.67 billion in 1999 (when Napster was launched) to $15.7 billion in 2016 – a 59 percent reduction in revenue. However, in 2016, for the first time in years, the decline was halted and the music industry grew again, but this time thanks to digital distribution channels, which amounted to half of the global revenue of the industry, mostly streaming, against 34 percent of physical format sales.

4.3 Google and the "Link Wars"

In September 2002, Google launched the beta version of a new service, Google News, developed by the Indian engineer Krishna Bharat after the September 11, 2001 attacks. Google News is an efficient way to pool together news published by different sources and was the result of Google's policy of allowing engineers to devote 20 percent of their time to whatever project they thought would be Google's best interests.[5] The new service was a free news aggregator. An algorithm, with no human intervention, selects the news that are of relevance for a specific viewer, and then both the headline and a snippet (that is, a short introduction of around 200 characters) are proposed to the viewer. Viewers can then link to the news in the digital version of the newspaper.

The service was subject to various legal actions, particularly in Europe: the so-called "link wars." The conflict between Google and newspapers is a good example of the dilemmas faced by traditional companies when confronted with a digital platform.

As the internet went mainstream during the mid-1990s, newspapers and magazines decided to digitalize their content and to provide it free on the web. As a reference, *The New York Times'* web version started in January 1996. Registration was required, but readers had access to news at no cost. Digital editions had lower production costs than the paper editions, as there was no need to print and distribute the newspaper. At the same time, the strong brand recognition and reputation of traditional newspapers gave them a competitive advantage in relation to spreading digital-only news ventures. Newspapers were actually among the most popular websites in the early internet.

Newspapers did not suffer the innovator's dilemma. They reacted at the right point in time to the disruption created by digitalization. They did not ignore the internet, but embraced it from the early days, even though, from a publisher's perspective, the internet provided a second-best solution at that time. Remember that the internet had only 16 million users worldwide at the end of 1995, so the potential customer base was small, mechanisms for charging readers were not yet developed, and digital advertising was in its infancy. However, it was widely recognized that the internet was the future, as digital editions had the potential to substitute the printed versions.

Digital newspapers were not really a disruption from the supply side. News production was the bread and butter of the industry and migrating from printed to digital editions was perceived as a challenge, but not as a fundamental threat; particularly for a business that had been built as a non-transactional platform and financed to a large extent via advertising. The situation was quite different from the recorded music industry, which did not have a platform business model and mostly relied on selling physical recordings.

However, migration proved more challenging than expected. As readers migrated to the digital version, advertising revenue from the digital edition would not match the revenue loss from the printed edition (both sales and advertising). Advertising in the digital version did not grow as expected. The gap was getting larger and larger. What was happening?

The answer is that newspaper platforms were being substituted by larger platforms. As internet players such as Google and Facebook were reaching larger and larger audiences and displaying more and more targeted ads, the value of digital ads decreased. Newspapers could not compete with the much larger audiences of Google and Facebook, which were built mostly over searches and user-generated content. Such "attention traps" were cheaper to produce and were attracting more advertising dollars.

The situation became increasingly clear as Google started to ramp-up its advertising revenue. Remember that Google only introduced AdWords in 2001. While advertising revenue in that year only amounted to $70 million, it reached $410 million in 2002, $1.42 billion in 2003, and $3.14 billion in 2004. By 2004, newspapers had finally understood why digital advertising was not working for them: platforms were taking the ad dollars.

The non-transactional platform business model is not profitable for newspapers in the long term, as they cannot compete with larger and more sophisticated non-transactional platforms such as Google or Facebook. It turned out that the only viable long-term strategy is to rely mostly on direct payments from subscribers. However, only newspapers with the most attractive content, such as *The New York Times*, can effectively rely on this model.

There is indeed a paradox in the relationship between newspapers and the leading digital platforms. Google and Facebook have their own attention traps (mostly searches and user-generated content), but they also provide access to content generated by newspapers. This is the case of the Newsfeed in Facebook (introduced in 2006) and Google News.

The first injunction against Google News was received by Google in Belgium on September 5, 2006, from the Court of First Instance. Following a claim by Copiepress, the national French-language newspaper association, the Court ordered Google to withdraw all articles, photographs, and graphics produced by Belgian newspapers from all of its sites (Google News and "cache" Google or any other name) for copyright infringement. Google executed the order and ten days later it eliminated the content not only from Google News, but also from Google.be (the Google search web for Belgium). The injunction was confirmed by the Brussels Court of Appeals in 2007.

Belgian newspapers were not happy with the injunction. While they wanted Google to stop displaying headlines and snippets in Google News, the injunction was so broadly defined that it also affected regular Google searches. When Google stopped displaying links to Belgium newspapers in the search portal, traffic heading towards the newspapers dropped and ad revenue suffered.

When in 2011 the Brussels Court of Appeals finally adjudicated in favor of Copiepress and forced Google to pay a daily fine of €25,000 plus damages to the newspapers, Google decided again to exclude Belgian newspapers from the search engine. The exclusion lasted only a few days, but the reaction in the leading newspapers was immediate. *La Libre Belgique* described this action as an "*attitude brutale de Google*" (July 11, 2011) and *Le Soir*, in a front-page illustration, labeled it as censorship (July 16, 2011).

A final agreement was reached in Belgium in 2012. Headlines and snippets would be displayed both in the search engine and in Google News without Google making any payment to newspapers. However, Google agreed to work with the newspapers in research and actions to help newspapers to adapt to the new digital environment, as well as to insert some advertising in the newspaper for an undisclosed value.

Legal actions in France, Germany, and Spain were different as they took the form of bills imposing on Google and digital players the legal obligation to pay a fee to newspaper for the display of headlines and snippets in Google News. The French bill was not formally adopted. The German bill was adopted, but it granted freedom to German newspapers to opt out of the scheme, as most did. Legislation was particularly restrictive in Spain, as the fee was declared compulsory and newspapers could not opt-out from it. On December 16, 2014, Google News decided to suspend the services in Spain. As a result, it has been estimated that traffic to Spanish newspapers decreased by 11 percent, with the strongest effect on sports and regional newspapers, but no significant effect on larger national newspapers.[6]

European publishers insisted on their strategy of making platforms pay for the use of snippets. They convinced the EU authorities to adopt a new directive on copyright in 2019,[7] requiring Member States to adopt legislation at a national level to ensure "that authors of works incorporated in a press publication receive an appropriate share of the revenues that press publishers receive for the use of their press publications by information society service providers." The link war is not over.

The situation of newspapers has only worsened over the years. Advertising revenue has migrated from legacy media to digital platforms. For example, the *New York Times* lost 75 percent of its advertising revenue between 2006 and 2017.[8] Furthermore, an increasing number of readers have access to newspapers through the platforms themselves. Readers see headlines in Facebook's Newsfeed and then click to have access to the newspaper's site. Newspapers are losing their role as intermediaries that bundle different pieces of news to build their newspaper.[9] Increasingly, readers are only having access to unbundled news pieces created by newspapers, in competition with news pieces created by new digital media and even user-generated content.

Newspapers are trapped in a paradoxical situation as they increasingly rely on their competition – the digital platforms – to produce their advertising revenue. The UK anti-trust authority has calculated that only 43 percent of readers had direct access to the webs/apps of the large newspapers in the country, while most of the traffic to their websites was indirectly routed from Google (25 percent) and Facebook (13 percent).[10] Furthermore, newspapers rely on digital intermediaries, mostly Google, to distribute their advertising space, with such intermediaries taking a 35 percent cut on the revenue (see Chapter 9).

Newspapers face a dilemma. They are fully aware that their participation in the digital platforms reinforces the platforms in the long term, as this is how they obtain quality content and even advertising revenue in their search sites and newsfeed features. Furthermore, their main asset and competitive advantage, the content, is commoditized for the benefit of the platforms, for which news are simply another way to attract eyeballs and advertising. As described in 2009 Ken Auletta's book, they have been "Googled."[11]

However, digital platforms also rely on newspapers as content producers. Digital platforms do not have the ambition of generating their own content in the form of news pieces. On the contrary, they merely intermediate between the content producers and the eyeballs. Digital platforms do not intend to substitute newspapers as content producers; they will continue to rely on third parties to produce content that attracts the attention of the public so that advertising can be displayed.

The question is whether the existing trend is sustainable. Platforms are capturing such a large part of the value created by content generated by newspapers that newspapers are being forced out of business (that is, the number of newspapers has been severely reduced all over the world). The alternative is to reduce the volume and quality of the content, or simply to abandon the advertising model and to charge readers for content, keeping it beyond the platforms' reach.

4.4 Disruption by Substitution

The most severe disruption that digitalization can cause is the substitution of traditional players by digital platforms. Goods and services that were previously provided by companies, some for centuries, are now being provided by digital platforms. As a result, traditional players are being expelled from the market or reduced to niche players. Such substitution can take different forms.

The most extreme case of substitution takes place when a new digital product substitutes the traditional physical products. It is not only the service provider that is substituted, but the product itself. Such a new product might prove a substitute for traditional offline product and be more efficient in terms of functionalities and pricing. In these circumstances, disruption for the traditional providers of offline services is unavoidable and such providers might just try to delay the transformation and look for new market opportunities.

An example of this kind of substitution is the sale of music records and CDs, which has been largely replaced by digital music distribution. Another example is email, one of the first and more transformative services created in the internet. Email is cheaper, faster, and more reliable than letter mail as distributed by the traditional postal service.

Substitution can also be the result of digital platforms empowering newcomers to compete with traditional players. The product is always the same, but platforms foster new players. This is the case of user-generated content and the sharing economy. As transactions costs are reduced by platforms, non-professional service providers and small companies, under the coordination of a digital platform, are in a position to compete with well-established corporations.

Platforms such as Facebook, but also the HuffPost, are displaying user-generated content, substituting newspapers. This is also the case of the so-called sharing economy and platforms such as Airbnb displacing hotels, Uber displacing taxis, and BlaBlaCar displacing trains and coaches. New service providers, empowered by a platform, can substitute traditional players, at least partially.

Furthermore, substitution can be triggered by platforms transforming a traditional market into a multi-sided market financed with advertisement. A new service provider develops the product in competition with a traditional player, but finances itself not by consumers (at least not totally), but by advertising. Traditional players usually do not have the chance to replicate such a multi-sided market and are forced out of the market.

This is the case with services such as Google Maps. In January 2012, the Commercial Court of Paris ruled in favor of Bottin Cartographes, a map manufacturer, and imposed €500,000 damages on Google France for predatory pricing, as it was providing its mapping services for free. The French *Autorité de la Concurrence* delivered its opinion[12] and helped the Paris Court of Appeals to overturn the judgment in January 2016. No predatory pricing existed; it was just that Google was able to obtain revenue from a different side (advertisers) in a multi-sided market.

Finally, substitution might the result of larger and more efficient multi-sided markets out-competing traditional platforms. Digital platforms can create and curate larger multi-sided markets with more powerful network effects than traditional offline platforms. This seems to be the case of newspapers in relationship to Google and Facebook. The same situation seems to exist with traditional matchmaker intermediaries such as taxi dispatchers, real estate agents, and dating agencies.

All these cases of substitution by digital platforms are examples of *creative destruction*, which Schumpeter defined as the "process of industrial mutation […] that continuously revolutionizes the economic structure from within, incessantly destroying the old one, incessantly creating a new one."[13] Better products empowered by digitalization substitute pre-existing goods and services along with the companies producing them. In all of these cases, a platform replaces the traditional service provider, which becomes redundant and expelled from the markets, or at least reduced to a fringe position in a market niche.

4.5 Disruption by "Platformization"

Disruption can also take a more subtle, but not less transformative form. Most often, traditional players become "platformed" as their markets are "platformized." This is the process that transforms a traditional market in which supply directly meets demand into a multi-sided market, in which traditional players are "intermediated" by a digital platform. The traditional player is not substituted by the platform; it is not expelled from the market, but its status in the market is modified. The platform takes over the leadership and extracts a largest portion of the value out from the market. This is a process that evolves over different phases. These phases might take different forms, as not all platforms evolve in the same way and do not display exactly the same features, but they are common in the most mature platforms.

In the first phase, the platform starts intermediating the goods or services provided by a traditional player. The platform has no intention to substitute the traditional player in the provision of the service. Such a provision would require a substantial investment that the platform does not intend to make.

The intermediation can be the result of an agreement between the platform and the traditional player, but also a situation that the traditional player tolerates. Traditional players often perceive the platform as a new distributor that might bring them new business. This is what happened with newspapers as they started to receive traffic from Google and other platforms: such traffic increased their advertising revenue. The situation is similar with many small and medium enterprises, such as local craft shops, which, thanks to the platform, obtain access to a much larger market.

Platforms can also play the role of interloper, however, running between supply by traditional players and demand by customers, intercepting the business of the traditional player. Just like seventeenth-century traders interfering in the monopoly of the East India Company, platforms interfered in the distribution of traditional goods and services. This was certainly what happened when Napster breached music copyrights, and when Google reproduced snippets. To the surprise of no one, the motto "move fast and break things"[14] was coined in Silicon Valley. The chicken-and-egg challenge requires the construction of a large pool of supply and demand, if necessary by ignoring the refusal to participate in the platform.

In the second phase, the platform aggregates the goods and services provided by a large number of competitors. Intermediating a large number of traditional players is necessary to build the large pool of supply that is required (together with a large pool of demand), so as to create relevant direct network effects. Furthermore, platforms can aggregate different products and services to create more powerful indirect network effects.

The aggregation of supply in highly fragmented markets can become a competitive advantage in itself. Amazon was popular originally because it displayed a larger catalogue of books than even the largest bookstores.[15] Napster was popular not only because it was free, but also because it provided easy access to basically any recorded song; something that was just unimaginable by any other means at the beginning of the twenty-first century.

Furthermore, algorithms might identify new complementarities between traditional goods and services, which can amount to the creation of a new service. This was the case of Google News. Algorithms not only aggregated news from different newspapers, but these newspapers were automatically selected by Google's algorithms in order to provide a personalized newspaper with only the news that is of interest for each reader. Such a proposition would be in a good position to compete with a traditional newspaper.

In the third phase, platforms displace traditional players in their relationship with the customer. The value proposition of the platforms becomes so powerful that customers prefer the direct contact with the intermediary, rather than consuming the good or service from the traditional players. This does not mean that platforms produce the good or the service. They will always only be intermediating the good and service of the traditional player.

Over time, platforms such as Google and Facebook became so efficient at making the best personalized selection of news that readers trusted platforms to feed them the daily news. The news are drafted by the traditional newspaper, and platforms merely route the reader to the specific newspaper, but readers rely on the selection made by the platform rather than on the aggregation of news made by the editor of the newspaper. This is already a reality in many markets: only a minority of readers directly access the digital edition of a newspaper to browse through the news compiled by the editor.

Such intermediation could work so smoothly that consumers might not even realize that the platform is only intermediating the service and not providing it. This was the case of Napster. Most users thought they were downloading the songs from the platform, when they were actually downloading them from a peer's computer. Similarly, with Uber, consumers often think the platform is providing the service, when it is actually being provided by a large pool of independent drivers.

In the fourth phase, the good or service produced by the traditional player becomes a mere commodity, as customers do not differentiate between one underlying provider and another. As a result, the traditional service providers are no longer in a position to impose the economic terms for the provision of their services. Eventually, platforms evolve into a position to extract the largest portion of the value from the market.

When consumers contract directly in the platform, which they find more efficient, they might not even be aware that the platform is not the provider of the service but only an intermediary; traditional players are then no longer in a position to differentiate their service. In fact, platforms often implement an active policy to standardize supply. Standardization facilitates transactions, as consumers expect a standard quality in the service in the platform. However, standardization also makes the service undistinguishable among the different provider; Amazon is a good case in point here.

When platforms intermediate, standardized undistinguishable services provided by different providers, they can impose pressure on traditional service providers to reduce prices (or extract larger commission). Traditional players are no longer in a position to sustain prices based on higher quality or on the customer's perception that they are the best provider. The algorithm will privilege the service providers with a lower price, forcing a price war between the service providers.

The ultimate outcome of this situation is that the platform sets the price for the customer. Platforms have an incentive to reduce prices as a way of growing a larger pool of consumers. Prices might be temporarily below cost, but the platform takes the difference as an investment to grow the platform. Prices might actually be reduced as a result of cost reductions due to network effects. However, if the platform grows, particularly if it tips into a dominant position, pricing pressure on the services provided by traditional player can be maintained, prices for customers can be increased, and the platform can be in the position to extract most of the value from the market.

In the fifth and final phase, the platform reaches the position at which it can dictate the conditions for the provision of the services it intermediates: from exclusivity, to arbitrary and discriminatory conditions, to taking full control as coordinator of the market.

Network effects foster concentration, as a large scale is necessary to exploit direct, indirect, and algorithmic network effects. As a result, oligopoly is the most common market structure in platformed markets. Network effects might even lead to a platform reaching a tipping point and building a dominant position. This is the case of Google in search, Facebook in social networks and Amazon in marketplaces. Once a platform assumes a dominant position, it can impose conditions on the underlying service providers.

For instance, some platforms try to impose exclusivity on service providers so that they will not be able to use other platforms (multi-homing), thus creating an obstacle to competition between platforms. In the most extreme cases, platforms even try to obstruct the direct provision of services by traditional players outside the platform. See the case of payments in the Apple store in Chapter 9.

The platform dictates the terms and conditions for the use of the platform. As platforms achieve market power, the conditions become more inflexible and interpretation of the terms and conditions become more arbitrary, sometimes discriminatory. Changes in the algorithm are implemented without notice or transparency, making traditional players lose business against competitors. Traditional players can be excluded from the platform from one day to the next without a clear explanation. Complaints against the largest platforms for this kind of practices are increasingly common.

The ultimate outcome of this evolution is when the platform determines the market. The algorithms no longer select the more competitive supplier to be matched with the consumer. Rather, the algorithm simply defines the kind of service that will be provided to the consumer. Algorithms are not neutral, just as platforms are often not neutral either. They drive the ecosystem around them and are in a position to promote one type of underlying service against the other. For instance, they can promote professional content against user-generated content if they think it will be more attractive for advertisers.

It has been identified that digital platforms sometimes vertically integrate: they no longer provide the intermediation service, but start competing with the traditional players in the provision of the product. Traditional players often complain that, under these circumstances, platforms self-preference their own service against those provided by third parties.

While the ambition of the digital platform might not be to substitute traditional players in a market, it might have the opportunity to provide some specific products in competition with the traditional service providers. We have already provided some examples: Amazon Marketplace intermediates goods sold by third-party retailers and producers, but at the same time Amazon sells a range of Amazon-produced goods. Apple has its own apps in competition with apps intermediated by Apple in the app store. Google has launched services such as Google Shopping and Google Travel that compete with price comparison and online travel agency services that are intermediated by the search service.

Digital platforms have incentives to discriminate in favor of their own services when providing the intermediation service. In fact, third parties have increasingly complained about self-preferencing and discrimination. These are practices that are considered illegal; that is, an abuse when implemented by platforms in a dominant position and if there is no objective justification. Amazon, Apple, and Google have open cases before antitrust authorities for such practices and Google was already fined for self-preferencing by the European Commission in 2017 in the Google Search (Shopping) case.

Self-preferencing is facilitated by the deep pool of data and sophisticated algorithms that platforms have regarding transactions in the markets they intermediate. As intermediaries, they have full information about all of the transactions taking place in their market – what has been termed as "God's view." They can process all the data they gather to identify trends in the market, popular products and effective marketing strategies, and can replicate them. They can also react to the pricing policies of the competitors thanks to the information they have about these pricing policies.

As a market is platformized and evolves into a multi-sided market coordinated by a digital platform, traditional players are not substituted; they are always relevant as the platform needs providers of the intermediated service. Nonetheless, traditional players lose their central position in the market, their ability to independently operate in the market, and their market power. They are platformed.

Notes

1　Christensen, C. M. (1997). *The Innovator's Dilemma. When New Technologies Cause Great Firms to Fail.* Harvard Business Review Press, Cambridge MA.
2　Gans, J. (2016). *The Disruption Dilemma.* The MIT Press, Cambridge MA.
3　For the discussion of the legal arguments on the case, see Berschadsky, A. (2000). RIIA v Napster: A Window onto the Future of Copyright Law in the Internet Age. *The John Marshall Journal of Information Technology & Privacy Law,* 18(3), 755.
4　*A&M Records, Inc. v. Napster, Inc.,* 114 F.Supp.2d 896 (2000).
5　Auletta, K. (2009). *Google. The End of the World as We Know It.* Penguin Books, New York, pp. 101–102.
6　Calzada, J. & Gil, R. (2016). *What Do News Aggregators Do? Evidence from Google News in Spain and Germany,* retrieved from https://papers.ssrn.com/sol3/papers.cfm?abstract_id=2837553
7　Directive (EU) 2019/790 of the European Parliament and of the Council of 17 April 2019 on copyright and related rights in the Digital Single Market and amending Directives 96/9/EC and 2001/29/EC.
8　Parcu, P. L. (2020). New digital threats to media pluralism in the information age. *Competition and Regulation in Network Industries,* 21(2), 91–109.
9　Carr, N. G. (2008). *The Big Switch: Rewiring the World, from Edison to Google.* W. W. Norton, New York.
10　Data for year 2019, as in Competition & Markets Authority (2020). *Online platforms and digital advertising.* Report, p. 17.
11　Auletta, K. (2009). *Google. The End of the World as We Know It.* Penguin Books, New York.
12　Autorité de la Concurrence, Avis 14-A-18 rendu à la cour d'appel de Paris concernant un litige opposant la société Bottin Cartographes SAS aux sociétés Google Inc. et Google France, 16.12.2014, retrieved from www.autoritedelaconcurrence.fr/fr/avis/rendu-la-cour-dappel-de-paris-concernant-un-litige-opposant-la-societe-bottin-cartographes-sas
13　Schumpeter, J. A. (1942). *Capitalism, Socialism and Democracy.* Routledge, London, pp. 82–83.
14　"Move fast and break things" was Facebook's motto until 2014, when it was displaced by "Move fast with stable infrastructure."
15　Auletta, K. (2009). *Google. The End of the World as We Know It.* Penguin Books, New York.

Chapter 5

Network Industries Disrupted by Platforms

Network industries are being disrupted by digital platforms trying to build multi-sided markets and new network effects on top of the traditional network industries. There are some cases of substitution of traditional players by platforms. Letter mail is being substituted by email. Media is being substituted by platforms such as YouTube and Facebook, taxis are being substituted by Uber, and traditional electricity generators are being substituted by prosumers. However, the impact of substitution is limited, as physical goods and especially infrastructure is always required; flat substitution is expensive and platforms have no intention of building infrastructure.

Therefore, disruption – at least in the case of the network industries – is mostly taking the form of platformization. Platforms are intermediating traditional players and creating new complementarities between traditional services, for the benefit of consumers. In the network industries, however, platformization is somewhat paradoxical, as platforms compete with traditional players in the creation of network effects, which is the business model in the network industries.

The digitalization dilemma is particularly acute in the network industries. Traditional players in network industries, being aware of the importance of network effects, are predominantly hostile to being intermediated by platforms. They have few incentives to have their services pooled with other services so that platforms can build new complementarities on top of them. Their reluctance is reinforced by the abuses committed by the most mature platforms in the content industries, which are already under the investigation of the antitrust authorities.

Thus, regulation is going to play a fundamental role in the relationship between networks industries and digital platforms. Regulation will determine the conditions for platforms having access to the infrastructures and the services of the traditional network industries. Furthermore, regulation will determine the conditions for traditional players to have access to platforms. It is important to acknowledge that the balance of power in this vertical relationship is in transition: power is passing from the traditional network industries to the digital platforms.

5.1 Substitution

Just like any other industry, traditional network industries can be disrupted by digital platforms. Disruption can take the form of substitution, but the special characteristics of the network industries makes substitution more difficult than in other industries.

It is only in some exceptional circumstances that digitalization has produced a substitute for the service provided by a network industry; the most illustrative example here is letter mail. One of the very first applications developed over the TCP/IP protocol was electronic mail (email). Subsequently, online platforms started to offer email services (Hotmail, Gmail, Yahoo Mail, etc.) and traditional telecommunications operators also started to provide email services. Email has substituted traditional letter mail services to a substantial extent. In the US, for example, total postal volume decreased by more than 27 percent between 2006 and 2016 (from 213 billion letters to 154 billion).[1] Similar substitution, and sometimes even more drastic effects, can be observed among postal operators worldwide.

Email is a prime example of what we will call the product substitution effect. Online platforms develop a new product that substitutes the traditional service provided by a traditional network industry. The new service uses an alternative infrastructure, not the infrastructure of the affected network industry.

However, this effect is of limited relevance in infrastructures, as it is only possible when digitalization completely substitutes the physical element in the provision of the service. This was the case of letters, but it cannot be the case of the provision of physical services such as the transport of goods, people, electricity, or bits. A physical element is indispensable. No purely digital product can substitute the traditional service.

However, digital platforms can substitute traditional players in the network industries, as they build the services on top of new service providers. In fact, it is not the product that is substituted, but the provider. There is always a physical element, which does not disappear, but the provider of the physical service is no longer the traditional service provider; a new one takes its place.

Digital platforms have the power to mobilize small service providers, even non-professional ones. The reduction in transaction costs empowers even the smallest provider to compete with traditional well-established players. This phenomenon can be identified in all network industries, even if under different names: user-generated content in communications, the sharing economy in transport, and prosumers in electricity.

In the media industry, user-generated content aggregated by platforms such as YouTube and HuffPost compete with traditional media. The product itself is not substituted, but it is now provided by new players intermediated by digital platforms. Platforms aggregating user-generated content are substituting traditional media.

In transport, taxis are being substituted by small, often non-professional, service providers intermediated by Uber and other platforms. Uber is not deploying its own infrastructures in the form of roads and streets or even acquiring its own vehicles. New service providers (individuals driving their private vehicles or Private Hire Vehicle service providers), aggregated by platforms, are substituting the traditional transport service providers (taxis).

This might also be the case in the electricity industry. Platforms cannot substitute the traditional service providers with a purely digital service. Consumers will always consume electricity. However, platforms can substitute the traditional electricity generators by mobilizing new electricity providers, creating a new network of distributed generators (prosumers who generate their own electricity in their properties).

Platforms are in a position to substitute traditional service providers by mobilizing new service providers or service providers that already exist in the market but are active in a fringe segment on a small scale. The "sharing economy" and the creation of distributed networks are good examples.

In any case, the traditional network industries display some features that make disruption by substitution rather exceptional. Traditional network industries rely heavily on infrastructure that is very expensive to deploy and then to maintain. Such infrastructure will only be substituted in exceptional circumstances. This is the case with messages that can be digitalized and sent as emails rather than letters. However, passengers, cargo, and electricity cannot be digitalized and transported over the internet, so they will always be physical and will require a physical infrastructure to be transported.

For the same reason, there are limited chances that alternative infrastructures will be deployed under the coordination of platforms, in competition with traditional service providers. This can be the case with media content and taxis, but not with roads, railways, airports, electricity grids, and so on.

5.2 Platformization

Infrastructures are being disrupted in a more subtle way, as the market structure for the provision of the infrastructure evolves into multi-sided markets intermediated by a digital platform. Traditional infrastructure managers are not substituted, but their services are being commoditized as they are aggregated and intermediated by digital platforms. Platforms have the ambition to monopolize the relationship with the customer and end up coordinating the provision of services by the underlying traditional players. Thus, infrastructures are platformed or at least in danger of being platformed.

This is the most relevant – but, at the same time, the most elusive – effect of digitalization in the traditional network industries. Traditional players will not be substituted; rather, they will still manage the underlying infrastructures and services. But their role in the market will be transformed, their power reduced, and their revenue diminished.

Platforms are transforming the traditional network industries into multi-sided markets. Traditional service providers become mere "sides" in a multi-sided market. Traditional service providers merely work for a platform's algorithm. The algorithm dynamically determines for a user what the best service is: the service provided by a traditional service provider, the service provided by a competitor, or even an alternative service that would not have traditionally been a direct competitor.

Transport provides the most mature example: Mobility-as-a-Service. The platform coordinates all the available transport modes (from taxis to bikes and scooters, but also public transportation offerings) and nudges passengers into the most efficient provider, which will be an incumbent such as a taxi, or a newcomer like a private driver, or perhaps a public bus, which can be an efficient alternative. If the algorithm identifies a nearby bus stop and a bus approaching in a matter of seconds, the bus will, overall, be faster and cheaper than a taxi. Alternatively, the algorithm may not offer this alternative because it could result in lower revenue for the platform.

Platforms control the direct relationship with the customer. Traditional service providers (taxis, bus fleets, subway, etc.) start working for the algorithm. Consequently, traditional service providers have more difficulty implementing dynamic management tools such as yield management and discounts, as well as customer relationship management more generally. They become a mere commodity coordinated by the platform.

At the same time, platforms become the coordinators of the system. It is now the role of the platform to match supply and demand, and also to dynamically balance supply and demand. Finally, it is the platform that must ensure the availability of supply.

The platforms create network effects. They appropriate the benefits from these network effects and, subsequently, determine how the so appropriated value is distributed across the ecosystem, including their own share. Still, the ecosystem in and from which online platforms live need to be sustainable. Sustainability requires sufficient funding for the deployment and maintenance of infrastructures; otherwise, the services provided by platforms will eventually collapse.

As platforms are acquiring a central position in more and more traditional network industries, they are impacted ever more profoundly, with consequences that are not yet fully understood and cannot yet be fully predicted.

Digital platforms are transforming the traditional network industries into multi-sided markets. The traditional direct link between service providers and users is being substituted by an intermediated relationship, in which a digital platform intermediates in the relationship and matches service providers and users. Service providers become just one side in a multi-sided market, the service is "commoditized," and the online platform takes the lead in the coordination of the system. There are already multiple examples of such a trend in the infrastructure network industries.

However, online platforms might pose risks for the funding of infrastructure. First, they can reduce revenue generated by traffic. Second, they can reduce revenue overall, as they allow users to hack the existing pricing structures, as in the case of OTTs in the telecommunications industries. Third, platforms are businesses that require their own revenues to operate. Platforms in the content industries have successfully financed their operations with advertisement. In the infrastructure network industries, however, platforms usually take a commission from the intermediated services. Such a commission will detract some of the value that was previously captured by the infrastructure manager.

Thus, a more structural effect of platforms on the funding of infrastructure can be identified. As infrastructure managers become mere sides in a multi-sided market, the platform gains influence in the provision of the service. The platform has the ability to nudge users from one infrastructure to another, from one transport mode to another, from one telecommunications infrastructure to another, from one electricity producer to another. The services mediated by the platforms become commodities, subject to new competitive pressures.

Infrastructure managers have traditionally taken a coordinating role in the management of the complex systems associated with infrastructures. They build and operate the physical infrastructure and often vertically integrate and provide the transport services over the infrastructure (this was the case in telecommunications, railways, electricity, etc.). However, they also play a more important role as the coordinators of the system, determining the capacity of the infrastructure, prices, managing congestion, etc., often under the supervision and control of regulatory authorities.

Online platforms are now transforming infrastructure industries into multi-sided markets, taking over this central role as coordinators of the systems. They have an increasing role in managing capacity, setting prices, arbitrating among substitutes, etc. – roles that were traditionally played by infrastructure managers under the control of regulators.

In fact, regulation is often fostering this trend. On one hand, deregulation has fragmented the infrastructure industries with new competitors and the unbundling of the vertically integrated monopolies.[2] On the other hand, and perhaps surprisingly, regulation sometimes ignores the market power of new actors in the data layer. Examples are regulatory obligations such as "net neutrality" in telecommunications, obligations to share data for transport service providers, and even obligations regarding ticketing, which could give platforms an advantage over traditional players.

It should be noticed that the new coordination role played by the online platforms is often managed through "black box" algorithms.[3] In the analogue world, the decisions of the infrastructure managers (capacity, prices, management of congestion, etc.) were public and subject to public discussion and the scrutiny of the regulators. In the digital world, platforms coordinate industries through algorithms that are not public, are constantly evolving, can be different in each jurisdiction and even for every user and, most importantly, are driven by the interest of the platform.

In short, there is clear evidence that platforms can diminish the value traditionally captured by infrastructure managers, either because such value is captured by the platforms themselves (commissions) or because it is eroded by the new competitive pressure created by the platforms. Even if platforms bring efficiency to the infrastructure industries, they might increase the difficulties for funding the creation and maintenance of infrastructure. Furthermore, platforms are gaining a leadership position as coordinators of the infrastructure networks, substituting both infrastructure managers and regulators as coordinators of the system.

5.3 The Digitalization Dilemma

Digitalization poses a dilemma to traditional network industry players, including infrastructure managers. On one hand, digitalization increases efficiency in the management of their infrastructures. As described, digitalization, algorithms, and automation reduce the cost in the design, construction, maintenance, and operation of infrastructures. On the other hand, as digitalization creates a data layer on top of the infrastructures, new players can use the generated data to transform the structure of the industry. Online platforms can create multi-sided markets in which infrastructure managers become just one side of the market, a commoditized provider of services under the coordination of the online platform. The platform reduces the value traditionally captured by the infrastructure manager, as new competitive pressure destroys value and the platform captures some of the value. This is particularly relevant in the network industries, as the availability of funds for the construction and maintenance of infrastructure must be ensured.

How should infrastructure owners, which are often governments, and managers react to digitalization? Let us recall some precedents. Newspapers embraced digitalization and voluntarily launched their own digital versions. They tried to exploit digital content through advertising, as they had done in their traditional printed versions. However, digital platforms (Google and Facebook) have outsmarted newspapers, growing much larger audiences and more sophisticated systems to target ads to larger audiences than newspapers ever could. Today, a substantial portion of the digital audience of newspapers comes from Google Search/Google News and from Facebook.

Newspapers are facing the digitalization dilemma. They are fully aware that their participation in the digital platforms reinforces the platforms in the long term, as they obtain quality content and even advertising revenue in their search sites and newsfeed features. Furthermore, their main asset and competitive advantage – the content – is commoditized for the benefit of the platforms, for which news is only another way to attract eyeballs and advertising. Ultimately, they risk becoming irrelevant and being expelled from the market altogether.

In the short term, however, newspapers' digital editions are heavily reliant on search results and newsfeeds in platforms. According to several sources, 40 percent of traffic directed towards the digital editions of general information newspapers derives from platforms, with Google clearly leading.[4] When newspapers tried to get their content excluded from digital platforms, as *The Times* did in 2010, or the Belgian newspapers experimented in 2007 and 2011, they realized how substantial the drop in traffic and digital advertising revenue was for them and were subsequently forced to ask Google to re-include them in their platform. They had been "Googled," as described in 2009 Ken Auletta's book of the same name.

Traditional companies are increasingly facing this dilemma as they must define their digital strategies. Google is not the only platform benefiting from third parties' digitalization strategies. Amazon raises the same dilemma for retailers and producers of goods when they have to decide whether to participate in Amazon's Marketplace. Retailers see an increase in sales as they join Amazon Marketplace, but often start to face competition from Amazon itself or other retailers selling below their prices. Participating in Amazon's Marketplace reinforces Amazon as a competitor.[5] Nevertheless, the alternative is to become irrelevant if limited to physical stores. Retailers can also be "Amazoned," which means "to watch helplessly as the digital upstart from Seattle vacuums up the customers and profits of your traditional brick-and-mortar business."[6]

As traditional industries are digitalizing, digital platforms are disrupting them. Content providers are "Googled," traditional retailers are "Amazoned," taxi drivers are "Ubered," and hotels are "Airbnbed." More generally, traditional companies are being "platformed."

This digitalization dilemma also applies to the network industries: when traditional players digitalize their operations, they are facilitating the transformation of their industry into a multi-sided market, with a third party, the online platform, eroding the value traditionally captured by the infrastructure manager and taking over the role of the coordinator of the market.

Traditional organizations might be tempted to delay digitalization altogether, to avoid being substituted or "platformed." However, this is a risk that most organizations are probably unable to bear, as competitors might get a competitive edge if a traditional organization does not benefit from the efficiencies derived from digitalization, and furthermore, if the traditional company remains isolated from the new market structure and ecosystems in the digital world. However, this strategy should not always be excluded. As an example, the founders of the *Palo Alto Daily News* launched another free newspaper as late as 2008. Most of the content can only be read on paper; it is not digitalized. While the audience might be smaller, it is there – people traveling in mass-transit, when there is no coverage, etc. – and the editor monopolizes the (small) advertising revenue it generates.

Traditional network industries players might be tempted to reduce the speed of digitalization, or even not to digitalize their infrastructures at all, in order to delay the rise of platforms in their industries. This might not be a wise strategy either and does not appear to be in the general interest. Efficiencies derived from digitalization are too significant to be ignored, as described in the previous sections. Even more importantly, this strategy might not work in the long run. As in a traditional prisoner's dilemma, competitors (where they exist) might embrace change and monopolize the benefits of a good relationship with the platform operator, making the position of the traditional player even weaker.

Even monopolistic infrastructure networks might not manage to avoid the rise of a platform by delaying digitalization. The infrastructure manager installing sensors can extract data about infrastructure, but third parties can also extract it in the most creative ways. For example, data on traffic can be extracted from passengers' smartphones, from sensors installed in vehicles produced by third parties yet using the infrastructure, from sensors installed in the cargo being transported, from meters used by the users of electricity networks, etc. Platforms can be built over data generated by third parties, not only data generated by the infrastructure managers themselves.

Traditional players can vertically integrate into the data layer and build platforms for their industries. This is a common strategy and there are many examples of infrastructure managers creating platforms, such as railway undertakings, shipping companies, telecom operators, electricity utilities, etc. They have the ambition to intermediate not only in the provision of their services, but in the provision of services by third parties, sometimes close competitors. Obviously, other service providers are suspicious and tend not to participate in platforms managed by their competitors, as they are afraid they could be discriminated against; that is, in favor of the operations of the competitor managing the platform. Successful platforms have been led by vertically integrated companies, such as Amazon Marketplace. However, it seems clear that not all members of an industry can become platforms. This is clearly not the way forward for all infrastructure managers.

Finally, traditional players might try to pool resources and build a common platform to aggregate and coordinate their services. In this way, they can be somewhat in control of their own platformization. This is what a group of European airlines did in 1987 when they created Amadeus, one of the first digital platforms (see Chapter 12). This is increasingly common in media, as traditional players pool resources to directly interface with the demand side in advertisement. Building this kind of platform requires a lot of good will among industry players (content creators in the book industry, music producers in the music industry, transport services suppliers in the transport industry, etc.). There are multiple examples of failed attempts to build this kind of common platform, and competition law poses further challenges.

Infrastructure managers, as well as regulators, are not bound to be mere spectators in the process of digitalization and the emergence of new market structures. Lessons can be learnt from other industries that have already been platformed. Infrastructures will always be necessary, as they will only rarely be substituted by digital services (as is the case in letter mail). Infrastructure managers must adapt to the evolution of the market structure and find their rightful place in the new ecosystems.

The challenge for all actors (infrastructure managers, candidates to become platforms, users, public authorities funding infrastructures, but also regulatory authorities) is to ensure the emergence of a balanced and sustainable competitive environment. However, the system will only be sustainable if the new value created is fairly distributed, and particularly if infrastructure managers are not deprived of the necessary funding for the maintenance and construction of their assets. This is a difficult balance to achieve and a challenge to address.

5.4 The Role of Regulation

Regulation plays various roles in the digitalization and platformization of the network industries.[7] First, regulation is fragmenting the market in the desire to introduce competition in the different network industries. Such fragmentation significantly facilitates the position of platforms as market integrators and coordinators.

Over the past 30 or 40 years of deregulation and liberalization, traditional network industries have been fragmented (unbundled) both horizontally and often vertically. Network industries have been horizontally fragmented, as newcomers have been allowed to enter the market. The monopolist incumbents now compete with a number of newcomers, but the number is limited. In some network industries such as telecoms, newcomers have been able to install their own competing networks. Regulation already imposes on all market participants the obligation to interconnect their networks to ensure full interoperability. Every user can make a phone call to any other user, even if they are served by different networks. The objective is to ensure the full exploitation of network effects in a fragmented market.

Network industries have also been vertically fragmented. In some cases, full separation of the infrastructure owners and the providers of services over the infrastructure has been mandated. This is the case of electricity in many countries. Entities managing the transmission networks are separated from the entities generating the electricity. Network effects are still being exploited in the monopolistic network of electricity transportation, but are not extended to the electricity generation market, where a larger number of players can compete. A similar approach has been followed in Europe in the railway industry. This is how the maritime and air transport industries already work: infrastructure management (ports and airports) are unbundled from the shipping companies and the airlines.

Even in industries where such a vertical unbundling does not exist, such as postal services and telecommunications, the incumbents usually manage the largest infrastructures benefiting from the larger network effects, and it is common to develop regulation to grant access to such infrastructure to the newcomers.

This has led to two immediate consequences in terms of regulation, both pertaining to the newly created interfaces between the monopolistic and the competitive elements (also called 'layers') of these network industries; that is, regulation of the access to the monopolistic/dominant infrastructure on one hand, and the financial flows between the two layers on the other. As for the monopolistic infrastructure, regulation was no longer seen as pertaining to market power, but is now seen as a question of regulating an efficient monopoly.

Further complications and needs for additional regulation arose from public policy objectives (universal services obligations, etc.) and ensuing market distortions. However, most of these regulations were developed at the national level and applied to the various national network industries. Only the European Union developed a comprehensive supranational framework for network industry regulation.

Deregulation, and the unavoidable fragmentation of the markets was carefully crafted to protect the direct network effects of the traditional network industry players and to share them with all the newcomers through access regulation and a full set of specific instruments, such as number portability regulation in telecoms, access to ticketing and distribution systems in public transport, etc.

The imposition of fragmentation onto the different network industries has paved the way for digital platforms to enter the market as market coordinators. The more fragmented a market, the more relevant the role of digital platforms as a coordinator of the system for the benefit of consumers. Only in a fragmented market can a digital platform leverage network effects by coordinating complementary assets and services.

Regulation plays a second role ruling the vertical relationships between platforms and the underlying traditional service providers. A third layer emerges on top of the infrastructure and service layers: the so-called data layer. Platforms operate in this data layer. As with the unbundled traditional network industries, digital platforms create a new interface in the form of the data layer, which comes on top of the infrastructure and the service layers. A growing set of rules will have to regulate the vertical relationships between the three layers. And, as with the traditional network industries, regulation of these new interfaces will again have to pertain to access.

In an initial period, the regulation of the vertical relationship between the traditional players operating infrastructure and platforms active in the data layer has been determined by the traditional market power of infrastructure managers and the policy of fostering digitalization and the rise of the digital platforms. Consequently, most attention has focused on the regulation of access by digital platforms to traditional infrastructures, as well as to their data. The main regulatory conflict has been around the definition of the rules of access by platforms to telecom networks, which are indispensable for the provision of their services. Platforms succeeded by imposing so-called "net neutrality" rules on telecom carriers, both in the US and in the EU.

Following the same trend, platforms active in network industries have increasingly called for the regulation of access to the underlying infrastructure and services they want to intermediate, often under the label of "data sharing." For instance, transport platforms are asking for mandated access to transport services. Consequently, transport service providers would be obliged to share their data, but also to allow platforms to commercialize their services. Such legislation is increasingly common in Europe.

However, as platforms have matured and raised to market power, particularly in search and content markets, focus has shifted to also ensure that the traditional players have fair access to the platforms. In the vertical relationship between traditional players and platforms, platforms are increasingly at the strong end. Complaints about deplatformization (no access to the platform), lack of fairness in access conditions, discrimination, and self-preferencing are proliferating. Antitrust investigations have been initiated for anticompetitive access behavior by Amazon, Google, Apple, and Facebook. The first fines have already been imposed by the European antitrust authorities on Google for platform access conditions.

In 2019, the European Commission adopted Regulation 2019/1150 promoting fairness and transparency for business users of online intermediation services.[8] This is a first attempt to systematically regulate access by professional service providers to platforms. The scope of the Regulation also includes network industries' services.

Finally, the transformation of the traditional network industries into multi-sided markets intermediated by digital platforms requires the review of the existing public service obligations. The communications, transport, and energy industries present an obvious general interest that has triggered a copious regulation in each industry to ensure certain public service objectives, such as universality and affordability of services, accessibility to services, free speech, media, pluralism, and so on. As the structure of these industries evolves, the existing regulatory framework will have to adapt to this new market structure.

Notes

1 United States Postal Service (2017). Postal Facts 2016, p. 4, retrieved from https://about.usps. com/who-we-are/postal-facts/postalfacts2016.pdf
2 Montero, J. & Finger, M. (2017). Platformed! Network industries and the new digital paradigm. *Competition and Regulation in Network Industries*, 18(3–4), 217–239.
3 Pascuale, F. (2015). *The Black Box Society. The Secret Algorithms that Control Money and Information.* Harvard University Press, Cambridge, MA.
4 Competition & Markets Authority (2020). *Online platforms and digital advertising.* Report, p. 17.
5 Stone, B. (2013). *The Everything Store. Jeff Bezos and the Age of Amazon.* Corgi Books, London, pp. 368–376.
6 Ibid., p. 17.
7 Bundesnetzagentur (2017). Digital transformation in the network sectors Recent developments and regulatory challenges, Report, retrieved from www.bundesnetzagentur.de/SharedDocs/ Downloads/DE/Sachgebiete/Telekommunikation/Unternehmen_Institutionen/Digitalisierung/ Grundsatzpapier/KurzfassungDigitalisierungEN.pdf?__blob=publicationFile&v=1
8 Regulation (EU) 2019/1150 of 20 June 2019 on promoting fairness and transparency for business users of online intermediation services, OJ L 186, 11.7.2019.

Part II

Platforms in the Communications Industries

To date, the communications industries are the network industries that have been most deeply affected by the rise of the digital platforms. Communications platforms are the most mature platforms; their pools of users have grown into the billions and display the largest network effects in human history.

Network effects are particularly relevant in the communications industries. This part of the book starts by describing how network effects drove the business strategy of the early telephone companies. Lessons from the early days of telephony, and from the regulatory response to concentration driven by corresponding network effects, are still of interest today. Clear parallels can be drawn between the rise of the telephone monopoly and the rise of the digital platforms. We will also show how the new network effects driven by connecting computers to the network were the reason for the divestiture of the telephone monopoly in the US and the liberalization of telecommunications all over the world. In short, the telephone monopoly was the first victim of digitalization.

Furthermore, postal services are among the best examples of disruption by digital platforms. Web-based email services have substituted traditional letter mail services provided by postal operators. The US Post Service identified the threat of email as early as 1976, but could still not avoid the substitution of mail.

Telecom services, both telephony and messaging services, have also been disrupted by platforms such as Skype and WhatsApp. However, disruption has taken a different form. Traditional telecom carriers have not been substituted. Their infrastructures and services remain relevant, as they are used by platforms to build their own network effects on top of the network effects built by traditional carriers. Platforms extract value from the telecoms market and threaten the position of traditional carriers as organizers of the market. Finally, platforms such as Apple's and Google's app stores have excluded carriers from the value chain around mobile apps.

Media services are also disrupted by digital platforms. On one hand, they have been substituted as new platforms such as YouTube and Facebook have attracted the attention of larger audiences, and therefore an increasing share of advertising revenue, thanks also to the new algorithmic network effects. On the other hand, traditional media are platformed, as Google and Facebook are increasingly intermediating them. Traditional media receive a significant share of the traffic to their digital sites from the platforms. Platforms such as Facebook aggregate content from all sources, as well as traditional media, and drive audiences to the content of their choice. Content producers are increasingly working for Facebook' and Google's algorithms. At the same time, Google already intermediates in the commercialization of advertisement space by traditional media, capturing an important share of advertising value.

These evolutions are posing new regulatory challenges. As traditional networks and platforms compete, there have been calls for a more level playing field. The different liability regimes are an issue. An additional issue is the divergence in the way regulation reacts to concentration. While fragmentation is an explicit goal in the regulation of traditional telecom carriers and media, platforms have been allowed to grow to global domination, as no specific regulation limits their growth, and antitrust enforcement has been timid, to say the least.

As platforms get to intermediate traditional services, the vertical relationship between platforms and the underlying communications service providers becomes more relevant, particularly the conflicts of interest in the management of platforms, as they vertically integrate and self-preference their services. The net neutrality regulation imposed on telecom carriers seems increasingly misdirected, as platforms grow larger network effects and are in the position to commoditize telecom and media services. On the contrary, there are growing voices calling to regulate platforms and take into consideration the impact of platforms on the general interest that has always driven public intervention in the telecoms and media industries.

Chapter 6

Network Effects in Communications

Communications industries are the classical illustration of the power of network effects. The more users a communications network has, the more value it offers to every user. The early days of telephony show how network effects were central in the definition of the strategy of the first players in the industry. Telephony was monopolized to fully exploit network effects, as all telephone users in a country would be able to interact in the same network. Regulation certainly played a role. Understanding the centrality of network effects and regulation in the strategy of founders, managers, and investors that grew a small startup to become the telephone monopoly in the US should help explain the current evolution of digital platforms.

History also shows how communications networks were the first industry to be disrupted by digitalization. Telephone networks were the first infrastructure to be digitalized. The transistor, the building block of digitalization, was actually invented in the Bell Lab in 1947 and deployed for the most efficient transmission and switching of telephone calls. However, digitalization would eventually trigger the end of the telecommunications monopoly. Transistors empowered computers, as well as the demand for communications services to connect computers. Networks that were in a position to connect not only humans with other humans, but also humans to computers and computers between themselves, would produce much larger network effects than the traditional telephone networks. Regulators concluded that the telephone monopolist was not in a position to meet the challenge of connecting the computerized world.

The internet – the distributed network of networks conceived and financed by the Pentagon – would prompt the largest network effects in the human history. The internet has transformed the communications industry, all other network industries, and beyond. Let us take a closer look at how it all started, a century ago.

6.1 Network Effects at the Origins of Telephony in the US

As the nineteenth century was coming to its end, merchants in Quincy, Illinois were booming. An early railroad bridge over the Mississippi had made the town a regional transportation hub, connecting the flow of people and goods from Chicago and the Eastern cities of the State (Springfield and Peoria), deep into the western frontier. Quincy's population had grown to 36,000, making it the second largest town in the state. European farmers were rapidly settling in the rural counties around the town.

Being a progressive enclave, Quincy embraced telephony. In 1884, the patent rights of the first telephone company, the Bell System, expired. At that time, despite its high price, the service had 500 subscribers in Quincy, mostly merchants who used the phone to communicate with other merchants in Springfield and Peoria, as well as with other local merchants.[1]

As the Bell patent rights expired, new independent companies deployed telephone infrastructures in the rural areas, also around Quincy, an area that the Bell System considered non-profitable. These companies built a cable under the Mississippi river and Quincy was connected to Newark, Missouri, 40 miles west, and other 30 towns and villages in the rural area surrounding the town. This evolution was not specific to Quincy. By 1902 there were over 9,000 "independent providers" across the US.[2] The result of fierce competition was an impressive expansion of the service (from 270,000 lines in 1894 to 6, 100,000 lines in 1907),[3] particularly in rural areas,[4] and a substantial reduction in prices.[5]

The operation of a small independent network was relatively simple. Low-quality infrastructure in the form of copper cables was often installed by farmers themselves. Switching of the calls was more of a challenge. In the early days of telephony, the connection of the calling and the called subscriber was done manually. The larger the volume of users, the more complex and expensive the task became. It was calculated that an increase of 100 subscribers in an exchange would result in an increased cost of $5 for each subscriber.[6] Small rural exchanges were easier and cheaper to operate than large sophisticated urban exchanges.

As a result of the competition between the Bell System and the independents, merchants in Quincy faced what is today perceived as a perplexing situation. They had to contract with the Bell System to communicate with merchants in cities back east, and they had to contract with the independent company to communicate with the rural local communities around the city and further west. Even inside the town, two contracts, two monthly fees, and two telephone sets were required to reach all local telephone users.

The reason was that the two existing networks were not interconnected. Again, this situation was not specific to Quincy. By 1907, 57 percent of towns with populations above 5,000 had what was called a "dual system"; that is, two or more competing telephone networks. Independent operators had more than 2 million customers and a market share above 35 percent nationwide.[7]

For a merchant, while having to contract with two (or even three) telephone providers to reach all his customers was a nuisance, it was not worse than contracting with two (or even three) local newspapers to advertise the merchant's services and reach all the potential audience. Furthermore, competition had reduced prices so severely that contracting with two carriers was often cheaper than contracting at the high rates charged by the monopolist before competition existed. Competition lowered prices and multiplied the penetration of the service, even in rural areas, so the service made it possible to reach a much greater pool of users. Dual service – or, to use today's terms, "multi-homing" – had its opponents, but it also had supporters among subscribers fearing an increase in rates if the monopoly would return to town.

The main opponent of the dual system was the Bell System, particularly after 1907 when the financier JP Morgan gained control of the group and forced a change of strategy to consolidate the market, reduce competition and put an end to price wars. Morgan had already implemented the same strategy with great financial success in railroads, electricity, and steel.[8] The term "morganization" was coined to describe such a strategy.

Morgan hired his most respected manager to run the company: Theodore Vail.[9] Vail had made his early career as manager of the railway division in the US Postal Service; he had experience with the telegraph and had already led the Bell System in its early phase, until 1885. Even back in 1885, Vail already understood that the Bell System should develop the largest network as soon as possible, to deter competitors. The main competitive advantage of a telephone network would be the ability to communicate with the largest number of subscribers. "A telephone without a connection at the other end of the line is not even a toy or a scientific instrument. It is one of the most useless things in the world. Its value depends on the connection with other telephones and increases with the number of connections."[10]

Vail was fully aware of what is today called Metcalfe's law: the effect of a telecommunications network is proportional to the square of the number of connected users of the system (n^2). This nineteenth-century entrepreneur understood network effects.

However, the board of the Bell System, even at the eve of competition as patent right expired in 1884, preferred dividends for shareholders over network expansion. The early Bell System limited the deployment of the network to dense urban areas, like Quincy, with a large number of merchants ready to pay high prices. Network deployment in rural areas was not the priority.

Vail resigned in 1885 as shareholders were not ready to make the investment to expand the network beyond urban centers. However, the evolution of the market in Quincy and all around the US proved Vail right. When patent rights expired, the aggressive expansion and low prices of the independents threatened the very existence of the Bell System. JP Morgan was to put an end to this situation.

Once reappointed as president of the Bell System in 1907, and with Morgan's financial support, Vail started an aggressive strategy to undermine the independent competitors. Prices were reduced in competitive markets. Acquisitions were made, both of competing long-distance services (the telegraph giant Western Union) as well as local independent companies, in what has been described as a Genghis Khan strategy – "join the network and share the wealth, or face annihilation"[11] – not so different from the growing strategy of other moguls at that time, such as Rockefeller.[12]

According to Morgan, a man always has two reasons for doing anything: a good reason and the real reason. The good reason for the new strategy of the Bell System is summarized in the slogan "One system, one policy, universal service," which was repeated again and again in the annual reports of the company and through one of the first and most successful public policy campaigns in corporate history. "It is believed that the telephone system should be universal, interdependent, and intercommunicating, affording opportunity for any subscriber of any exchange to communicate with any other subscriber of any other exchange. [...] It is not believed that this can be accomplished by separately controlled or distinct systems nor that there can be competition in the accepted sense of competition. It is believed that all this can be achieved to a reasonable satisfaction of the public with its acquiescence, under such control and regulation as will afford the public much better service at less cost than any competition or government-owned monopoly could permanently afford."[13]

Vail proposed to guarantee that any subscriber would be able to communicate with any other subscriber without the need to contract with different competing carriers. In this way, the system would be universal, not because it would necessarily reach any corner of the country, but because it would allow to reach all the existing subscribers. Full network effects would be exploited by a single entity. Such an objective would be ensured by a single system, with a single policy: one network built around the principles of standardization and centralization.

Universality could have been achievable through the interconnection of the different networks, which would allow a subscriber to communicate with any other subscriber regardless of which company provided the service to each subscriber. The networks could be connected and function, from the perspective of the subscribers, as a single network. This is the situation today. However, the Bell System refused to interconnect with the networks owned by independent companies. Interconnection was perceived as an infringement on the Bell System's ownership rights for its own infrastructure. It would diminish the value of the company as the competitive advantage of having a larger network would have to be shared with the smaller networks. At the same time, interconnection would limit the freedom of the networks to set the technical standards of the networks, pricing strategies, and so on.

6.2 Network Effects and Regulation

Vail also understood that monopolization would only be accepted politically if government would be in control of the rates charged by the monopolist. Subscribers did not like the dual system, but they were afraid a return to monopoly would increase telephone rates. In fact, "morganization" in railroads had created popular resistance to the concentration of economic power, the adoption of the first federal antitrust act, the Sherman Act, in 1890, and the divestiture in 1903 of the railroad conglomerate Northern Securities backed by JP Morgan. To avoid this fate, the Bell System actively promoted regulatory intervention as a counterbalance to the monopolization of the telephony service.

Linking regulation to monopolization was fully successful as a strategy, since the adoption of the Willis-Graham Act in 1921, which allowed telephone operators to acquire any competitor with only the approval of the Interstate Commerce Commission and the exclusion of the antitrust authorities. In 1934, the Federal Communications Commission (FCC) was created, with the mandate to enforce the entry-and-price regulation model originally envisaged by the Bell System 30 years earlier.

National monopolies in telephony were the prevalent market structure not only in the US, but also in Europe and the rest of the world for most of the twentieth century. A standardized system of telephone numbers, implemented on a global scale by the International Telecommunications Union (ITU), identified any subscriber in the world. The telephone would eventually also achieve universality in geographic terms, at least in Western countries, even if at a slower pace than expected.

What was the economic rationale behind monopolization? The experience of the independents proved that a monopoly was not necessary to ensure investment in infrastructure. On the contrary, competition between literally thousands of independent carriers had advanced the penetration of the telephone service far more than in Europe, where state-owned monopolies had been the norm from the origin of the industry in most countries.

From a network theory perspective, the monopoly facilitated the harvesting of all the network effects derived from a single infrastructure providing service to all existing users.

As relevant as the infrastructure monopoly was, the role of the monopolist was to ensure the universality of the service; that is, the possibility for any user to communicate with any other user. As a counterbalance, regulation would then ensure that the benefits would be fairly distributed in the form of low rates, and not monopolized by the network manager.

As Mueller brilliantly described in his analysis of the early days of the telephone,[14] economies of scale were not the driver behind the success of the Bell System. It was network effects and the aggressive pursuit of the largest network effects driven by Vail that led to monopoly. This was not specific to telephony, as a similar evolution can be identified in electricity, railways, aviation, and other network industries around the world.

Vail was probably one of the first individuals to grasp the power of network effects. His experience in the management of the three leading communications networks of his age (posts, telegraphs and telephones) was definitely helpful. Vail would have been a great leader for a digital platform in Silicon Valley in the twenty-first century.

6.3 Digitalization and the End of the Telephone Monopoly

The monopoly that Vail built at the beginning of the twentieth century was the first company to be disrupted by digitalization. The reason behind it, paradoxically, was the network effects that Vail understood so well at the earliest stage of the communication industry. The management of the Bell System after the Second World War did not see it coming.

Computers can be considered a spin-off of the telecommunications industry. Telephone networks have always relied on the transmission of electricity. Devices to control electric currents were key to the deployment of telecommunications networks. Vacuum tubes were developed as early as 1907 to improve long-distance telephony. It was not by chance that the first transistor was developed in the research arm of the Bell System, the Bell Labs, in 1947. The transistor substituted the previous mechanical vacuum tubes for the management of electricity.

Transistors empowered all devices using electricity, from hearing devices (remember, the reason for Alexander Graham Bell's research in electricity was to help his deaf wife) to portable radio sets (the first killer application).[15] Transistors were used in telecommunications networks and later in integrated circuits empowering the IBM computers, as they evolved from mechanic to electronic. Transistors, made of silicon, rapidly became a multi-million dollar industry, and Silicon Valley became the cluster of the leading electronic companies.[16]

Computers were the most powerful application of the new technology. The binary status of electricity (on/off), expresses data as series of the digits 0 and 1 and is the basis for the transmission, management and recording of such digits, being the origin of "digitalization." Digitalization and computing changed the game, first of all in the telecommunications industry itself. It is important to understand the different ways in which digitalization transformed telecommunication.

Digitalization improved the existing telecommunications networks. Transistors increased the reliability and reduced the cost of long-distance transmission of electricity signals. In parallel, computers were used for switching calls. Centralized networks, like the telephone network developed by the Bell System, create powerful network effects, but such network effects create the risk of congestion. This was the case in the early days of manual switching. The larger the number of active lines in an exchange, the more expensive it was to run it. Efficient switching had always been a challenge for telephone networks. Computers substantially improved the switching capabilities, dramatically reducing the cost of switching communications among the hundreds of millions of telephone users across the world. But this is only one of the effects of digitalization, and not the main one.

Even more importantly, digitalization increased the number of points to be connected by communications networks. Communications networks not only connected people to talk (telephony), but also computers to exchange data. As computers developed after the Second World War thanks to the transistor, a point was reached when the existing telecommunications networks could support the transmission of voice, but also the transmission of data between computers. As Metcalfe's law stated, the more points there were to be connected, the larger the network effects.

It was understood early on that computers would be interconnected. The first IBM electronic computers were large devices, weighing into the hundreds of tons. They were very limited in number, as the cost of production was very high. But as technology improved and costs were reduced, the number of computers increased rapidly. Organizations that made use of an increasing numbers of computers, such as universities or the military, rapidly started to connect the computers they owned. It was simple, as the underlying technology (electronics, transistors, and so on) was basically the same as the technology used in telecommunications networks.

The Bell System started to provide communications services to connect computers inside large organizations. Private lines connecting different locations owned by the same customer were initially used, for instance by broadcasters, which would use the Bell System's infrastructure to transport images from recording studios to the antennae from which they would be broadcast over the waves.

The profound implications of such a transformation did not escape the eyes of regulators. As early as 1966, the Federal Communications Commission (FCC) launched a prescient consultation under the title *In the Matter of Regulatory and Policy Problems Presented by the Interdependence of Computer and Communications Services and Facilities*. As described in the consultation, "Effective use of the computer is [...] becoming increasingly dependent upon communication common carrier facilities and services by which the computers and the user are given instantaneous access to each other."[17]

The FCC identified the risk of the Bell System monopolizing the provision of the new data services, thus obstructing innovation. The solution would be to limit the monopoly of the Bell System to the pure telecommunication service of transmitting data. By contrast, enhanced information services – data processing services that were layered on top of the basic telecommunications services, creating a new service for users – would be fully open to competition under a light regulatory regime. This would also occur a few decades later, with email, the World Wide Web, digital platforms, and even access to internet services.

6.4 The Internet: New and Larger Network Effects

Digitalization put an end to the monopoly of the Bell System. For the Bell System, disruption did not come from the use of computers to better run the telecommunications networks, nor from the new data processing services to be open for competition. Instead, disruption came from the new network structures that made the historic centralized network of the Bell System obsolete and from the larger network effects of a network that could connect both humans and computers.

In 1960, a psychologist turned computer scientist named J.C.R. Licklider envisioned the creation of a network of computers connected by telecommunications infrastructure in his seminal paper entitled "Man-Computer Symbiosis." In it, Licklider stated: "The picture readily enlarges itself into a network of such centers, connected to one another by wide-band communication lines and to individual users by leased-wire services. In such a system, the speed of the computers would be balanced, and the cost of the gigantic memories and the sophisticated programs would be divided by the number of users."[18]

Licklider's contribution was not only theoretical. In 1962 he was appointed head of the Information Processing Techniques Office at the United States Department of Defense Advanced Research Projects Agency, better known as ARPA.[19] ARPA wanted to save resources by pooling together all the computing capacity of the different projects financed by the office across universities and research centers, meaning that researchers at one university could remotely use the computer in another university.

One of the options for ARPA was to create a centralized network connecting computers, following the traditional structure of the telephone network. It was proposed to set up the switching point in Omaha, Nebraska, near the geographic center of the US,[20] where the Air Force was already centralizing communications.

Licklider's ARPA had a better idea: an innovative network model developed at the RAND Corporation by Paul Baran for the US Air Force and described in the paper entitled *On Distributed Communications*.[21] Instead of centralizing switching in a single node (centralized network), or even in a reduced number of nodes (decentralized network), Baran proposed establishing redundant links between the different points to be connected, in such a way that communications would be handled from point to point without being switched in a centralized node. In other words, every point in the network would have switching capabilities; every point would be a small node. The US Air Force had identified that the centralized network of the Bell System would collapse in case of a nuclear attack, and Baran was commissioned to identify an alternative to solve this problem. The distributed network was his solution.

The new ARPA network would still have another fundamental feature that Baran had envisaged: packet switching. Inspired by road transportation (another distributed network), a message would be divided into small blocks (packets), which would be independently transported across the distributed network, to be reconstructed at the destination using computing techniques.

Baran spent five years unsuccessfully trying to convince the Bell System to implement a packet-switched distributed network. His proposal was not only dismissed, but he was put though a training course in which 94 speakers would explain how a telephone network works, so he could understand why his idea was not feasible. He was told how a centralized network worked; how messages could not be broken down into packets, but could only be transported by opening a dedicated circuit making full use of the capacity of the circuit for the specific communication. Nevertheless, Baran was not convinced.[22]

The network financed by ARPA, called ARPANET, would prove engineers in the Bell System wrong. On October 29, 1969, the first connection of ARPANET, between UCLA and Stanford, started to operate.[23] This was the beginning of the internet, which would grow from a small network of computers connecting US universities to a network connecting basically everyone, everything, and everywhere.

The engineers of the Bell System were blinded by the nature of their network. A fully standardized and centralized network developed for voice transmission could not be easily transformed into a distributed network connecting computers owned by different institutions. There was no common operating system and no common standard for the computers to communicate with each other.

Had the engineers been more sensitive to network effects, as Vail had been decades earlier, they would have understood that a network connecting more points (not only humans to humans but also humans to computers and computers to each other) would be more valuable than a mere telephone network. Vail understood the power of converging the telephone and telegraph networks. His successors did not understand the power of converging voice and data networks.

Still, the challenges were enormous. Each university had developed its own communications system to connect their computers, so the universities were not interested in changing their systems to adopt a new common standard. The solution was to keep each independent network working under its own rules and to create a protocol on top of it so that different networks could communicate with each other. The solution was to create a network of networks.

A specific, simple protocol was developed so that the different networks could interact. The internet protocol (TCP/IP) created a new layer on top of the different networks that allows the interaction of previously isolated computer networks. It would not be necessary to fully standardize the pre-existing networks in order for them to interoperate. Each network would use its own standards, but they would all be able to interoperate thanks to a new standard on top of the legacy local standards. The prerequisite would be for the networks to use the TCP/IP protocol.

The internet would develop as a distributed network of networks. No central authority would run a non-existent central node. Eventually, everyone – universities, private companies, and even telephone carriers – was invited to join the network of networks. The larger the number of participating networks, the larger the value for them and for other networks. The internet would become the ultimate example of the power of network effects.

But before the internet outgrew the telephone network, digitalization would kill the telephone monopoly. Transistors empowered alternative communications systems such as microwave networks. Computers increased demand for point-to-point private services. Such services were not served by the regular centralized network that was the backbone of the Bell System, but by specific infrastructure, dedicated lines, provided by the Bell System at a high price. Alternatives started to flourish.

6.5 Deregulation and Fragmentation of the Market

On August 13, 1969, the FCC allowed Microwave Communications Inc. (MCI) to set up a series of microwave stations from Chicago to St. Louis, Missouri in order to provide to third parties telecommunications services along Route 66.[24] It was only a small experiment on the provision of a fringe specialized service. However, the company grew more ambitious and set its objective to compete with the Bell System in the provision of data services, and later on also voice services.[25] A decade of legal battles (its president called the company "a law firm with an antenna on top") gave MCI the right to compete with the Bell System and would end up with the deregulation of the industry.

In 1975, MCI initiated the provision of a new service, Execunet, which substantially competed with the common long-distance activities of the Bell System. The FCC ordered MCI to eliminate such a non-licensed service; however, the Court of Appeals reversed this order,[26] thus allowing MCI and the other specialized carriers to compete in all the long-distance telecommunications market.

In the meanwhile, the aggressive response of the Bell System to the new competition had led the Department of Justice to file an antitrust lawsuit in 1974 against AT&T, related to the vertical integration of the telephony service provider, and against Western Electric, the producer of telecommunications equipment.

The case was not settled until 1982, with a Consent Decree divesting the Bell System into seven regional monopolies for the provision of local telecommunication services, the "Baby Bells," unbundled from AT&T, which remained the provider of long-distance services. Long-distance telephone services would now be provided in competition with other carriers, such as MCI.[27]

Deregulation was finally introduced in the US with the adoption of the 1996 Telecommunications Act.[28] The Baby Bells consolidated around two groups, which would later acquire the leading long-distance providers AT&T and MCI, to form the companies that remain the leaders in the US telecommunications market: AT&T and Verizon.

The process in the European Union was led by the European Commission. Liberalization started with terminal equipment and value-added services, to be extended to all telecommunications services in 1998. No divestiture of the former national monopolies took place, but market entry was allowed for any interested carrier.

Liberalization put an end to decades of exclusive rights. Newcomers entered the market to compete in the provision of the old fixed telephony service, as well as in new services such as mobile telephony, broadband access to the internet, and data transmission services.

With a view to fostering competition, sector-specific regulation about network access was introduced in most of the world. Firstly, interconnection was required from all operators to ensure the interoperability of the telephone networks. In this way, the direct network effects were shared among all market players. Secondly, incumbents were forced to share their infrastructure with newcomers. Special access for competitors under regulated conditions was imposed on the incumbents on their old fixed-telephony networks, but also in the new broadband networks (unbundling of the local loop, access to fiber, etc.) and sometimes even in the mobile telephony networks (Mobile Virtual Network Operators, or MVNOs). In this way, a newcomer with a limited investment in infrastructures (or with no investment at all) would be able to enter the market and provide telecommunications services in competition with the incumbent.

Telecommunications had traditionally been a single product, a single network and a single company in each country. National monopolies harvested the network effects in telephony by benefiting from an exclusive right to deploy telephone infrastructure and provide the telephone service, according to Vail's view.

This model was unsustainable as digitalization multiplied not only the number connection points (computers), but also the number of transmission technologies and services. The traditional twisted copper pair had to compete with cable, optic-fiber, and wireless cellular networks. Traditional fixed telephony services had to compete with mobile cellular telephony and new data services. Traditional national monopolists had to compete with a full array of companies developing alternative infrastructures and providing new services.

Telecommunications became a fragmented industry, a puzzle of competing networks interacting under different rules. Regulated interconnection would apply to telephony services, both fixed and mobile. But new data networks and services would work under the distributed model, designed for networks and services in the internet arena, with no regulations on interconnection.

The evolution of the communications industries, postal services, media, and telecommunications was determined by the network effects created by the internet. The ultimate network effects empowered by the internet, as they were made effective by digital platforms, ended up disrupting all of the communications industries.

Notes

1 Mueller, M. (1997). *Universal Service. Competition, Interconnection and Monopoly in the Making of the American Telephone System.* MIT/AEI Press, Boston, p. 59.
2 Out of the 9,000 providers, 3,000 were commercial providers, 1,000 were mutualistic companies, and 5,000 were providers of services to a few dozens of grangers, Fisher, C. (1987). The revolution in rural telephony. *Journal of Social History*, 21, 6.
3 FCC (1939). *Investigation of the Telephone Industry in the United States.* US Government Printing Office, Washington DC, p. 151.
4 In 1912, 38 percent of the telephone lines were installed in rural areas, and in 1920, the telephone penetration rate was higher in rural areas (39 percent) than in urban areas (35 percent), *vide* Fisher, C. (1987). The Revolution in Rural Telephony. *Journal of Social History*, 21, 8.
5 See Gabel, R. (1969). The Early Competitive Era in Telephone Communication, 1893–1920. *Law and Contemporary Problems*, Spring, 343–346.
6 Mueller, M. (1997). *Universal Service. Competition, Interconnection and Monopoly in the Making of the American Telephone System.* MIT/AEI Press, Boston, p. 35.
7 Ibid., p. 57.
8 Strouse, J. (2000). *Morgan. American Financier.* Harper Perennial, New York.
9 Paine, A. B. (1921). *In One Man's Life.* Forgotten Books, 2012 ed.
10 AT&T Annual Report for 1908, p. 21.
11 Wu, T. (2010). *The Master Switch. The Rise and Fall of Information Empires.* Alfred A. Knopf, New York, p. 53.
12 Chernow, R. (1998). *Titan. The Life of John D. Rockefeller SR.* Vintage Books, New York.
13 AT&T Annual Report for 1910, p. 23.
14 Mueller, M. (1997). *Universal Service. Competition, Interconnection and Monopoly in the Making of the American Telephone System.* MIT/AEI Press, Boston.
15 Kaplan, D. A. (2000). *The Silicon Boys and their Valley of Dreams.* HarperCollins, New York, p. 41.
16 Berlin, L. (2017). *Trouble Makers. How a Generation of Silicon Valley Upstarts Invented the Future.* Simon & Schuster, London; and O'Mara, M. (2019). *The Code.* Penguin Press, New York.
17 FCC (1966). Notice of Inquiry in the Matter of Regulatory and Policy Problems Presented by the Interdependence of Computer and Communications Services and Facilities. Docket F.C.C. No. 16979, November 9, 1966.
18 Licklider, J. C. R. (1960). *Man-Computer Symbiosis.* IRE Transactions on Human Factors in Electronics, Volume: HFE-1, Issue: 1.
19 Weinberger, S. (2017). *The Imagineers of War. The Untold Story of DARPA, the Pentagon Agency That Changed the World.* Vintage Books, New York.

20 Hafner, K. & Lyon, M. (1996). *Where the Wizards Stay Up Late. The Origins of the Internet.* Simon & Schuster Paperbacks, New York, p. 71.

21 Baran, P. (1964). *On Distributed Communications: I. Introduction to Distributed Communications Networks.* RAND Corporation, Santa Monica, CA, retrieved from www.rand. org/content/dam/rand/pubs/research_memoranda/2006/RM3420.pdf

22 Hafner, K. & Lyon, M. (1996). *Where the Wizards Stay Up Late. The Origins of the Internet.* Simon & Schuster Paperbacks, New York, p. 62.

23 Ryan, J. (2010). A *History of the Internet and the Digital Future.* Reaktion Books, London, p. 30.

24 FCC, Decision, FCC Docket Nos 16509–19 "Applications of MCI for Construction Permits," FCC 2d 953.

25 Cantelon. P. L. (1993). *The History of MCI: 1968–1988. The Early Years.* MCI, Washington DC.

26 "The ultimate test of industry structure in the communications common carrier field must be the public interest, not the private financial interest of those who have until now enjoyed the fruits of *de facto* monopoly," MCI Telecommunications Corp. v. FCC, 561 F. 2d 356 (D.C. Cir 1977), p. 380.

27 Temin, P. (1987). *The Fall of the Bell System.* Cambridge University Press, New York.

28 See Krattenmaker, T. (1996). The Telecommunications Act of 1996. *Federal Communications Law Journal, 29*, p. 1.

Chapter 7

Email: Postal Networks are Substituted

If the telephone monopoly was the first victim of digitalization, postal services were the second, and they have been hardest hit so far. Email was first launched as an application in ARPANET in 1972, just three years after the network was established. Unexpectedly, email became the most popular application among engineers using the early internet, which was created to connect computers, not people. It was only in 1996, when Hotmail launched web-based email, that email went viral. Nevertheless, the US Postal Service had identified the threat of email as early as 1976, when the first study on the impact of email on postal services was commissioned. Such an early response did not stop disruption, however.

Email disrupted postal services by substituting the traditional letter service. Platforms use a different infrastructure – the internet and the underlying telecommunications infrastructures – to convey digital messages. They do not use the infrastructure of the traditional postal operators at all. Postal services are the prime example of substitution as one of the forms of disruption by digital platforms.

7.1 The (Originally) Glorious Days of Postal Services

It is fair to say that postal services were the first network industry, as they originated in the Middle Ages around courier services, which, at that time, were organized by city-states, merchant guilds, and religious orders. Typically, emperors or kings delegated the management of such courier services that covered many European cities to private operators against a concession fee. The originally Italian and then German Thurn and Taxi family managed to set up an almost postal monopoly between the fifteenth century and the end of the eighteenth century, controlling postal services in Spain, Germany, Austria, Italy, Hungary, and the Netherlands and employing at its peak 20,000 couriers and thousands of horses. Taking advantage of its geographical position at the heart of Europe, the Swiss aristocrat Beat Fischer (1641–1698) created the so-called "Fischer Post," which operated almost until the middle of the nineteenth century, yet had a hard time competing with the Thurn and Taxis that had even then put to profit the strong network effects that are operational in the postal sector today. Still, private courier services were highly profitable. As in telecommunications, the more senders are connected to the (postal) network, the better it is for the receivers, and vice versa.

During the nineteenth century, all postal services of the world were nationalized, a phenomenon that parallels the emergence of the nation-states. An additional wave of national postal operators emerged with decolonization. While the same network effects were also operational once postal services became nationalized, these network effects were put to profit for national public services purposes, such as universal coverage of a given country. At that time – that is, before the emergence of telecommunications – postal services became a strategic asset for nation states, just as railways and electricity did later. They ensured the communications infrastructure of a country, which is so vital for both military and political control.

At the turn of the twentieth century, first the telegraph[1] and then the telephone emerged. In all nation states – with the sole exception of the United States, where telegraph and telephone services were operated by privates (see above) – telegraph and telephone were integrated into the much older and already established postal services, thus the acronym PTT, with the *P* before the two *T*s. Postal services remained generally profitable until after World War II, when telephony, and especially highly lucrative international telephony grew rapidly and the *T* (for telephony, as the telegraph had since disappeared) started to cross-subsidize the *P*.

Starting in the 1980s, but mostly during the 1990s, the telecommunications sector was gradually liberalized, which logically led to the separation (unbundling) of the PTTs. As a result, postal operators came under heavy financial pressure (having been cross-subsidized by their telecommunications sisters). And, in the countries where this was possible (approximately 50 percent of the historical postal operators had financial services), postal financial services were subsequently split off from the postal services, and most were privatized.

Being under heavy financial pressure – first because of the lack of cross-subsidization and because their governments (as owners) did not want to support them financially, and secondly because of emerging competition as a result of globalization – postal operators worldwide started to restructure and to downsize whenever possible. This often resulted in a split between parcels and courier services on the one hand (rather global in nature, with strong international competition namely from privates such as FedEx and UPS) and letter mail on the other (rather national in scope). Needless to say, letter mail, because of its strong network effects (namely, the mail deliverer passes by each household regardless of volumes, whereas parcels and courier services are point-to-point deliveries) remained mostly monopolistic and lucrative, despite market opening, which did not see the emergence of competitors. In other words, after the split with telecommunications and following the emergence of global competition in the parcels and courier services, letter mail remained typically the only really profitable activity of historical postal operators. And this is when email comes into play.

7.2 The Creation of Network Mail

ARPANET, the original seed of what would later become the internet, was designed to share scarce computing time across different universities. It was not designed for communication among people. However, as ARPANET evolved, small applications were developed, like file transfer protocols and electronic mail, to facilitate the management of the network itself.

Ray Tomlinson, working for ARPANET, developed the software for electronic mail in ARPANET. In 1971 he wrote some code to exchange messages between computers owned by the same institution in a structured way. As in any communications system, the first step was to define the address for the exchange of communications. In a computer system, the address would include the name of the individual and the name of the computer system used by the individual. Tomlinson had the idea of introducing the '@' symbol to separate the name of the individual from the name of the computer system. This was the perfect sign; being very rare, it would not be included either in the name of the individual or the computer system. This is how the most popular digital symbol came to be created. Creating this protocol was not part of Tomlinson's job description, however. He was not instructed by his reporting line. He just thought it would be nice to facilitate the exchange of messages. One year later, the software developed by Tomlinson would be formally included in the ARPANET protocols that would be used not only inside one network, but in the ARPANET network of networks.

Network mail (as email was originally called) proved to be the most popular application in ARPANET. As early as 1973, three-quarters of the traffic in ARPANET was network mail.[2] This is a good example of how networks tend to grow organically beyond the expectations of their creators.

The use of email grew in parallel to ARPANET and the internet. Emailing was popular first among ARPANET users; it later expanded to the wider community of universities and research centers and was eventually commercialized for the use of any interested user by internet service providers (ISP), commercial ventures that provided the service for a profit. ISPs would provide an email address to each customer.

By the end of 1995, the internet was used by only 16 million people worldwide (0.4 percent of the world population), but it was mature enough to go mainstream, just like its most popular service: email.

7.3 Hotmail Makes Email the Killer App on the Internet

On July 4, 1996 two Stanford graduates, Sabeer Bhatia and Jack Smith, launched Hotmail, the first free web-based email service. While they did not invent electronic mail, they popularized the service. Web mail allowed individuals to have an address and to send and receive messages even without having a contract with an internet service provider.

The idea came to the Hotmail founders as they were working for Apple. They were exchanging ideas about the creation of their own firm and they were afraid their employer could have access to their corporate email accounts and would notice that they were using working time to develop alternative ventures. They could not use their personal email, as it could only be used from their internet connections at home. The solution to their problem, but also of interest for any user, was a web-based email account that could be used from any computer connected to the internet: from home, from work, or from an internet café for those without any other access to the internet.

A bright idea is never enough; investment is necessary to develop a product and a corporate structure is necessary to commercialize and monetize the product. The role of venture capital (VC) is key for the success of bright ideas. The first VC firm in Silicon Valley was established in 1959 by General William H. Draper Jr. and two partners.[3] The General's grandson, Tim, would later found Draper, Fisher, Jurvetson, the VC firm which invest in Hotmail. They funded the venture with $300,000, a 15 percent stake in an idea valued at $2 million.[4]

It is of interest to identify how platforms grow a customer base large enough to get powerful network effects. In the case of Hotmail, the trick was a line of text with a clickable URL that was introduced at the end of each email message sent with Hotmail: "Get your free e-mail at Hotmail." Each email became a powerful ad, proof in itself of the reliability of the service. The service ignited, adding 3,000 users a day. In six months, Hotmail reached 1 million users.[5]

In December 1997, just 17 months after it was launched, Hotmail was acquired by Microsoft for $400 million. It had nine million users, out of an estimated total of 70 million worldwide internet users at that time. Hotmail had built such a customer base faster than any other company in history. And this is indeed the result of the power of networks.

Web-based email became one of the most popular services in the internet. By 2020, the number of email users has been estimated to be 3.9 billion, 54 percent of the world population. Outlook.com (the successor to Hotmail) is used by 400 million users, but the market leader is Gmail, owned by Google, with more than 1.5 billion monthly active users.[6]

Web mail operators, such as Hotmail, are platforms in multi-sided markets. They make the interaction between different groups possible, for the benefit of all the groups participating in the platform. Web email is a multi-sided market in the sense that individuals are connected by different ISPs. They do not all contract the internet access service with the same company. The platform makes it possible to interact with the users connected to the internet by another ISP. In this sense, web mail users are not a homogenous group. Web mail platforms connect previously fragmented groups of users. Furthermore, web mail platforms introduce another group into the ecosystem, as the service is financed by advertising. The platform allows the interaction of different groups of email senders, recipients, and advertisers.

7.4 A Network on Top of Networks

The network mail software that Tomlinson developed and Hotmail share a fundamental trait: they empower communications without building dedicated infrastructure. Instead, they use pre-existing infrastructure and services, albeit in a different and creative way: they build new networks on top of previously existing networks.

The telephone network has traditionally been a centralized network managed by the same organizations that built the physical infrastructure supporting the service. Deregulation created a decentralized network of networks, as interconnection ensured that everyone could talk to everyone. The network of telephone networks is coordinated by regulators who define the standards and the conditions for interconnection. A standardized numbering system managed by an international institution (ITU) identifies all users in the world; standards for signaling and network interconnection are defined by regulators. Interconnection prices are regulated. Governance of this network of networks guarantees full network effects, but it is expensive and inflexible.

The internet is a fully distributed network of networks. Each independent network (a university, a public institution, a corporation, an ISP) runs independently and takes care of its own infrastructure. The Internet Protocol makes sure that different networks can communicate with one another. Network interconnection is decided on a commercial basis without governments intervening in the form of regulation. Packets are transmitted through the fastest route, across different networks. There is no centralized coordination. The network of networks is flexible and reliable.

The internet is more powerful than the old telephone networks, as it connects far more points. It connects not only humans who are willing to communicate with each other, but also connects computers with humans, and computers with computers – the number of computers is enormous (desktops, laptops, smartphones, sensors of all kinds in the Internet of Things, etc.). The internet, the network of all networks, includes the telephone network and many other networks, creating the ultimate network effects.

Applications like email exist on top of the internet. They do not create new infrastructure, but rather rely on the existing infrastructure of the internet, and on the protocols for the interconnection of the different networks. Email creates a new network on top of the internet. It can grow as large as the internet. Because it is built on the same principles of the internet, anyone using the protocol can connect with anyone else in the network. The internet is all about network effects, and so are the leading applications built on top of it.

The rapid development of Hotmail can only be explained by the fact it was built over pre-existing networks: the telecommunications infrastructure and the Internet Protocols interconnecting telecommunications networks over the internet.

7.5 Substitution of Letter Mail

The impact of email on traditional physical letter mail was spotted as early as 1976,[7] when a report was commissioned to the consulting firm Arthur D. Little by the White House Office of Telecommunications. The report estimated that 30 percent of all first-class mail in the US would be sent by email within a few years.[8]

The US Postal Service (USPS) cannot be accused of ignoring the risk to its services. It reacted swiftly, commissioning a $2.2 million contract to RCA to evaluate the feasibility of providing email services. Consequently, President Jimmy Carter (who started to use email during his presidential campaign in 1976) supported the proposal of using what would be called a hybrid mail service. Messages would be written by customers in their computers, sent electronically to the USPS, which would then print them and deliver them in paper form to the recipient.[9] Subsequently, many postal operators around the world copied the idea and actually perfected it by such means as transforming written documents into emails and delivering them to the customers.

It seems evident now that hybrid mail could not compete with email. It was slower, as delivery was not immediate as it is with email. And it was far more expensive than email (which was free). Consequently, hybrid mail did not deter email.

Over time, Arthur D. Little's predictions came true, even if not as fast as expected. The phenomenon became clearly visible for all postal operators as early as the year 2000,[10] the first year that the growth of letter mail volumes was no longer correlated with GDP growth. Ever since then, email has replaced and continues to replace a substantial part of letter mail. For example, in the US, the total postal volume decreased by more than 33 percent between 2004 and 2019 (from 213 billion letters to 142.57 billion).[11] Similar substitution, and sometimes even more drastic effects, can be observed among all the postal operators worldwide. Letter mail volumes typically decrease by 4–7 percent annually worldwide. The difference is explained by internet penetration, digital literacy (which is lower among ageing populations), and the more or less active promotion of digitalization by various national governments.

In terms of disruption, when the number of connected computers was low, email was a poor substitute for physical letter mail. Remember that the postal network was the first communications network to become universal in its reach. Email was initially popular among computer scientists, and then university communities, but that was not enough at the time to substitute postal services. Substitution became relevant as the internet went massive in the 1990s. Desktops and laptops with access to the internet became a standard in Western societies. Email was already a very good substitute for letters. Substitution was also facilitated by another factor – multi-homing – as individuals could send emails and physical postal letters. A young researcher in the computer science department at Stanford could use email to send written messages to his peers, and then use postal letters to communicate with her grandparents back in Quincy, Illinois.

As computers became increasingly universal and ubiquitous, in the form of smartphones, email achieved even larger network effects than postal letters. Emails could be received by almost everyone, in their personalized mailbox (not in a generic address shared with relatives/co-workers) and almost anywhere. The popularity of email has actually decreased as smartphones have become more popular, with other electronic communications services, such as instant messaging, replacing email use to some extent. As users have personalized terminals available all the time, they can substitute email with instant messaging services.

Recall the words of communications theorist Marshall McLuhan: "*The medium is the message*." Network mail was the message of the original internet, connecting large mainframe computers in research centers. Webmail was the message for the second generation of the internet, connecting desktops and laptops. Instant messaging services are the message for the third generation of the internet, built on smartphones.

Email is a prime example of what we call the substitution effect. Online platforms develop a new product that substitutes the traditional service provided by a traditional network industry, in this case the postal service.

The new service uses an alternative infrastructure, not the infrastructure of the affected network industry. In the case of email, it uses the telecommunications infrastructure, not the postal infrastructure. The new service fully bypasses the postal infrastructure and services.

Of course, email and electronic communications have limitations. They can convey electronic signals digitalizing written and oral messages, but they cannot digitalize physical parcels. Demand for the transport of letters has severely diminished due to the competition of electronic communications, but demand for the transportation of parcels has increased thanks to electronic commerce and the success of platforms such as Amazon.

Notes

1 Standage, T. (1998). *The Victorian Internet. The Remarkable Story of the Telegraph and the Nineteenth Century's On-line Pioneers*. Bloomsbury, New York.
2 Hafner, K. & Lyon, M. (1996). *Where the Wizards Stay Up Late. The Origins of the Internet*. Simon & Schuster Paperbacks, New York, p. 194.
3 Berlin, L. (2014). The First Venture Capital Firm in Silicon Valley: Draper, Gaither & Anderson. *Making of the American Century. Essays on the Political Culture of Twentieth Century America*, Oxford University Press, New York, pp. 155–170.

4 Bronson, P. (1999). *The Nudist on the Late Shift and Other True Tales of Silicon Valley*. Random House, New York.

5 Penenberg, A. L. (2009). *Viral Loop. The Power of Pass-it-on*. Hyperion Books, New York.

6 Number of email users worldwide, retrieved from https://financesonline.com/number-of-email-users/

7 Wood, A., et al. (1977). *USPS and the Communications Revolution: Impacts, Options, and Issues*. Final Report to the Commission on Postal Service, prepared by the Program of Policy Studies in Science and Technology, The George Washington University, Washington, D.C., Mar. 5, 1977.

8 Arthur D. Little (1978). *The Impact of Electronic Communications Systems on First Class Mail Volume in 1980–1990*, Cambridge, MA, April.

9 Hafner, K. & Lyon, M. (1996). *Where the Wizards Stay Up Late. The Origins of the Internet*. Simon & Schuster Paperbacks, New York, p. 212.

10 Finger, M., Bukovc, B., & Burhan, M. (ed.) (2014). *Postal Services in the Digital Age*. IOS Press Amsterdam.

11 US Postal Service's total mail volume from 2004 to 2019, retrieved from www.statista.com/statistics/320234/mail-volume-of-the-usps/

Chapter 8

Skype and WhatsApp
Telecom Carriers are Platformed

Telecommunications carriers have been disrupted by digital platforms, albeit in a more subtle way than letter mail. Traditional telecommunications carriers are not being substituted; they are carrying more traffic than ever, as platforms have multiplied the amount of data traffic and such traffic makes use of the telecommunications infrastructures, whether fixed or mobile. Their role is still fundamental and it is not expected to disappear with digitalization.

Nevertheless, digital platforms are increasingly providing telecommunications services to the public in competition with traditional telecom carriers. What is specific to these platforms is that they provide their telecommunications services on top of the infrastructure managed by the traditional carriers. This is why they are called over-the-top players (OTTs). OTTs are creating new larger communications networks on top of the traditional telecom networks, which means that the traditional telecom carriers are being platformed.

Platforms have enabled peer-to-peer (P2P) communications. Just as platforms have enabled the exchange of music and emails over the internet, platforms have also enabled the exchange of richer communications (calls, images, and video). Skype started the provision of voice-over-the-internet (VoIP) services in 2003. WhatsApp currently provides the most popular instant messaging service in the world, empowering the interaction of more than 2 billion users.

A common characteristic of all these platforms is their usage of the internet and the underlying infrastructure deployed by telecom carriers for the conveyance of their traffic. They merely empower customers to interact with each other through software. They create network effects in the data layer, not in the infrastructure layer. In fact, OTTs do not even participate in the routing of the calls or messages. On the contrary, customers contract their own access to internet services and the calls and messages are conveyed making use of these facilities and services. Since platforms do not have to invest in expensive infrastructure or incur any cost for its use, they can provide their service for a much lower cost than traditional telephone and SMS services.

Telecom carriers have not been able to exploit their position as the enablers of the digital revolution. Communications platforms such as Skype and WhatsApp have eroded a massive amount of value out of the telecommunications market, as they transformed the industry into a multi-sided market. Furthermore, platforms built on top of mobile telecom networks have become the gatekeepers of the "app economy," sidelining telecom carriers.

8.1 Skype and Voice Over IP

P2P exchange of data over the internet has been one of the key drivers of success of the digital platforms. It is the underlying technology supporting email. It was the underlying solution that supported Napster and other startups that disrupted the music industry. Voice communications were an obvious candidate for a similar disruption. Voice can be transformed into data packets and then transported over the internet, just like emails and music files, and for the same price: free. AT&T engineers who claimed that packet-switched telephony would not be possible have been proved wrong.

In August 2003, Nikolas Zennström from Sweden and Janus Friis from Denmark launched Skype, an application that allowed real-time voice communications between registered users. Skype was the first P2P service to allow the exchange of voice communications in real time, similar to a telephone service. The name Skype is short for Sky peer-to-peer. The founders originally intended to name the service Skyper, but the domain was not available so they settled for Skype.

Zennström and Friis were not newcomers to the P2P world. In 2000 they had developed Kazaa, a popular software for a P2P exchange of music files following the Napster model, but without a centralized server as Napster had. Files would directly be exchanged between users. The change in technological implementation did not escape legal attention. Kazaaa eventually closed, but the founders had understood the power of P2P networks in the internet. They just had to export the business model to a service without copyrights: voice was the next step.

Zennström and Friis were not newcomers in the telecommunications world either. They had both worked for the Danish telecom carrier Tele2 in the 1990s before creating Kazaa. They had actually proposed a VoIP project to their line of report. However, real-time voice communications require significant capacity to ensure the minimum quality of service. Broadband was not common in the 1990s, so that was not the right time to launch a VoIP service. This is why Zennström and Friis decided to start with Kazaa and the exchange of music files, as the exchange did not require real-time communication so it could be provided with lower-quality networks.

As broadband became increasingly popular in the early 2000s, VoIP was ready to go mass-market. Skype was the first popular VoIP service. Users would download the software in their personal computers, register a specific address to identify themselves, and then identify the address of the peers they wanted to call to. Skype would run the login server. Skype manages each customer's account, but is not involved in the actual transmission of the signal in each specific call. Each user runs its own ordinary node. Users with a good connection and computing power could act as supernodes, supporting communications between third parties.[1]

The differences between VoIP and the public switched telephone network (PSTN) service were relevant. Firstly, no telephone numbering was initially used, so communication was only possible between Skype users who were identified with a Skype address: it was not universal in the sense defined by Vail for the Bell System. This is why it was necessary to grow a large customer base. Secondly, no specific circuits were reserved for the communication. On the contrary, voice was cut in small packets, all with the destination address and sent to the public internet to reach the called party, where packets would be transformed back into voice. Consequently, the service was provided on a best-effort basis, as quality of service could not be ensured due to potential congestion in the public internet, jitter, and delays in the transmission. Thirdly, access to emergency services was not ensured.

But it was another difference which made Skype so popular: it was free. No fee was required to use Skype and users could talk for as long as they wanted at no cost. Skype was particularly attractive for international calls, as such calls were charged by telecommunications carriers at particularly high rates. This was a big hit in Europe, as all traditional telephone calls, even local ones, were subject to a per-minute charge.

Skype VoIP communications could be offered at no cost for the calling or called parties, as the communications would make use of the access to internet service already contracted by the calling and the called party with their regular provider (in most cases, a traditional telecommunications carrier). Skype did not deploy telecommunications infrastructure and not even pay interconnection fees for the use of telecommunications carriers' networks. Skype was free just like email or many other internet applications.

Skype is another digital platform built thanks to network effects. Skype allows individuals connected to the internet through different ISPs to have voice conversations. As in the early days of Vail's Bell System, network effects were key in the growth of Skype. The larger the pool of Skype users, the more useful the service was for the entire customer base. Skype grew very quickly. Just six months after it was launched, it had been downloaded 1.5 million times. Two years after its launch, Skype had 53 million registered users out of the 938 million of individuals connected to the internet. Skype went viral just like other closed networks, as early adopters convinced people around them to join the network so they could all benefit from the new service.

It is not by chance that the venture capital firm providing the seed capital to Skype is a familiar name: Draper, Fisher, Jurvetson, the same firm that had provided seed capital to Hotmail. Having understood the power of network effects created by digital platforms in communications industries, it seems now a pretty natural move from email (Hotmail) to voice (Skype). They acquired a 10 percent position in Skype.

In September 2005, just two years after its launch, Skype was acquired by eBay for $3.1 billion. eBay had already acquired PayPal in 2002. It had the ambition of creating a fully digital environment around its marketplace: communications, payments, etc.

However, the merger did not deliver the envisaged benefits, and eBay sold a controlling stake in Skype to a group of investors in September 2009 for $2.75 billion. Microsoft finally acquired Skype for $8.5 billion in May 2011. As of 2020, Skype had more than 300 million active users.

Eventually, Skype developed solutions to make calls to telephone numbers (Skype Out) and even to receive calls as a telephone number (Skype In). As these calls interconnect with the telephone networks, Skype has to pay interconnection fees to telephone carriers, so it charges users for such calls. In any case, fees tend to be low, particularly for international calls, as Skype uses the internet for the international transport of the call, and interconnection takes place at the local level, for a low fee. As a result, Skype can charge an international call to a telephone number as if it would be a local call. Over time, as broadband expanded, Skype supported not only messages and voice calls, but also video communications.

Skype services are not always easy to classify from a regulatory perspective. In most legislations around the world, communications services are defined as the transmission/

conveyance of information. This was the case in the US and the EU. Therefore, the regulatory nature of the service depends on whether Skype is actually transmitting – that is, transporting – the conversations.

However, Skype's VoIP service is fundamentally different from the public telephony services provided by traditional telecommunications carriers. Skype is not in charge of the transmission of the conversations. Skype merely provides software that allows a computer to interact with another computer making use of the same software. Conversations are transmitted over the public internet just like emails or any other packet of information. Traditional telecom carriers are actually transmitting the signals. As a result, Skype VoIP service was not originally considered a communications service according to telecommunications legislation in the US or the EU.

Skype has been required to notify or be licensed as a communications carrier in some jurisdictions, particularly as it has started to use telephone numbers and interoperate with the telephone network.[2] This is not a theoretical debate. Communications services are subject to a heavy regulatory framework across the world: compulsory interconnection, access to emergency services, call interception, specific consumer rights, specific privacy obligations, etc. In exchange, telephone service providers have the privilege to be granted telephone numbers by the national regulatory authorities.

Competition between traditional telecommunications carriers and Skype as well as other VoIP service providers remains a hot issue in the regulatory debate.

8.2 WhatsApp and OTTs

WhatsApp was incorporated in February 2009 by Jam Koum. A few months later, his former colleague in Yahoo, Brian Acton, provided $25,000 lent by Yahoo employees and became a co-founder. They were both in their 30s and well-seasoned in the Silicon Valley culture.

WhatsApp was an instant text messaging application. It was not a radically new service. Skype, for instance, had been providing instant messaging services for years. The key difference was that WhatsApp was developed specifically for the iPhone, being part of what we have called the third generation of the internet. Just as Skype was developed for the PC (second generation of the internet), WhatsApp was specifically designed to benefit the personalized and always-on nature of the smartphone. Once again, the medium is the message.

After some beta tests, the service was launched in November 2009. It was provided for free, being a substitute for the expensive short message services (SMS) provided by cellular carriers. The service went viral as early adopters invited their contacts to join the network. The selfish desire to make use of the free service with one's regular contacts helped to grow the network.

WhatsApp's fast growth was made possible by building on previously existing networks, as had already been observed in other cases. Of course, WhatsApp was made possible by the already universal internet. Furthermore, WhatsApp built upon the personal list of telephone numbers in the memory of the smartphone. Each user could invite the individuals who were in his or her smartphone's contact list. The service went viral so rapidly that a fee of $0.99 had to be introduced for new members to join the platform, to limit demand.

As usual, financing was provided by a well-known venture capital firm, Sequoia, which had already invested in several platform companies such as Google, YouTube, PayPal, LinkedIn, and Airbnb. Not only did Sequoia identify the power of network effects, but, more specifically, it also identified how WhatsApp could build on previously existing network effects: "WhatsApp is simple, secure, and fast. It does not ask you to spend time building up a new graph of your relationships; instead, it taps the one that's already there."[3]

Over time, WhatsApp was enriched to allow the exchange of pictures and voice messages, as well as VoIP services. WhatsApp quickly surpassed Skype as the leading communication platform despite being launched six years later.

WhatsApp works across operating systems (iOS and Android), across cellular carriers, and across countries. It is a digital platform that allows the interaction of large pools of users in a previously fragmented ecosystem.

Five years after the launch of the service, WhatsApp had 450 million users, but only 55 employees. Such a lean structure and the limited expenses – WhatsApp does not require infrastructure as it uses that of the traditional communications carriers – enabled WhatsApp not to rely on advertising. This was one of its main competitive advantages over other instant messaging platforms. WhatsApp was also built with limited access to users' data (no log in, no names, no profile; only the telephone number). Big data was not the business model. Even subscription fees have been on and off, being requested for some periods and in some countries, but not as a universal feature and a business model supporting the platform. The business model was to build the best communications service, grow the largest communication network, and then sell it. WhatsApp was sold to Facebook in February 2014 for $19 billion. As Mark Zuckerberg explained, "if you look at the number of things that have reached a billion people, they all end up being incredibly valuable and important things."[4]

Since the acquisition, WhatsApp has still not included advertising or subscription fees in the platform. However, Facebook has linked WhatsApp users' phone numbers with the identities of Facebook users, increasing its ability for targeted advertisement services. The European Commission imposed a fine of $122 million on Facebook in 2017 because, when Facebook notified the acquisition of WhatsApp in 2014, it had informed the Commission, both in the notification form and in a reply to a request of information from the Commission, that it would be unable to establish reliable automated matching between the accounts of Facebook users and those of WhatsApp users. However, in August 2016, WhatsApp announced updates to its terms of service and privacy policy, including the possibility to link WhatsApp users' phone numbers with Facebook users' identities. The Commission found that, contrary to Facebook's statements in the 2014 merger review process, the technical possibility of automatically matching Facebook and WhatsApp users' identities already existed in 2014 and that Facebook staff were aware of such a possibility.[5]

In December 2020, the US Federal Trade Commission filed an antitrust suit against Facebook, including the divestiture of WhatsApp (and also Instagram). "Facebook's acquisition and control of WhatsApp represents the neutralization of a significant threat to Facebook Blue's personal social networking monopoly, and the unlawful maintenance of that monopoly by means other than merits competition. This conduct deprives users of the benefits of competition from an independent WhatsApp (either on its own or acquired by a third party), which would have the ability and incentive to enter the U.S. personal social networking market. Moreover, WhatsApp's strong focus on the protection of user privacy would offer a distinctively valuable option for many users, and would provide an important form of product differentiation for WhatsApp as an independent competitive threat in personal social networking."[6]

Eventually, all large platforms developed their own communications apps. Facebook owns WhatsApp (with 2 billion users out of 4.5 billion individuals connected to the internet in 2020) and Messenger (1.3 billion users), Microsoft owns Skype (300 million users) and Hotmail (now Outlook). Tencent owns WeChat (1.2 billion users). Other popular platforms are QQ, Telegram, and Snapchat.

8.3 Platforms and Communications Services

Communications platforms present a fundamental regulatory challenge for the definition of the legal nature of the service. Should the services provided by platforms be classified as telecommunications services and subject to the regulatory obligations traditionally imposed on telecommunications carriers?

Telecommunications services, like most traditional network industry services, are subject to regulation as they are considered services of general interest (in EU terms) or common carriers (in US terms). Telecommunications operators are required to register with the local regulator and file reports about their activities. Thus, multiple obligations are imposed upon them based on different grounds: consumer protection (quality of service, invoicing, and so on), security (lawful interception, data retention, emergency services, etc.) and promotion of competition (number portability, special access regulation, etc.).

Communications platforms are exempt from most of the telecommunications regulations. OTTs, by definition, do not transmit information. Skype and WhatsApp do not own infrastructures. Instead, they curate platforms that run over their P2P software, enabling users to directly interact with each other over the public internet. Traditional telecommunications carriers convey the information over their infrastructures and access to internet services. As a result, many of the services provided by platforms do not qualify as telecommunications services.

However, as these services are increasingly substitutes for traditional telecommunications services, there are ever more convincing arguments to define such services as telecommunications services. This is what the EU authorities did when they adopted the new European Electronic Communications Code[7] in December 2018. There, the definition of telecommunications service has been expanded to include "number independent interpersonal communications services," a long name for OTT services.

In any case, many regulatory obligations are merely the legacy of another time and are probably no longer necessary (for example, public phones, phonebooks, and identification of the location of the calling party). Therefore, it does not seem sensible to expand these obligations to platforms. Some regulatory obligations are in fact defensive practices by telecommunications operators to protect telephony from competition.

Finally, there is a basic institutional challenge. Online platforms are global players, with little to no presence in most countries and it is not simple to effectively impose national regulations upon them. Only large countries (US and China) or organizations such as the European Union have the reach to effectively control the activity developed by digital platforms.

8.4 Telecom Carriers Are Platformed

"I knew it was over when I downloaded Skype. When the inventors of Kazaa are distributing for free a little program that you can use to talk to anybody else, and the quality is fantastic, and it's free – it's over. The world will change now inevitably." This was the reaction of the Chairman of the US Federal Communications Commission (FCC) Michael Powell when he used Skype for the first time in 2004.[8]

Telecom carriers have been disrupted by communications platforms. Initially, platforms provided a low-quality service that was not a substitute for more demanding customers. Network effects were limited, as the customer base of the platforms was initially small and the services were not interoperable with other communications services. Quality was also an issue. However, services were provided at no cost for the user, which made them particularly attractive for financially challenged users like students and migrant workers.

VoIP/instant messaging platforms exploited a loophole in the pricing structure and the regulation of the traditional telecommunications operators. Such operators had two coexisting pricing structures. On one hand, they traditionally charged by the minute for telephony services. Furthermore, prices were not cost-oriented and international services, charged at a high rate, subsidizing local services. In parallel, they charged for each SMS. On the other hand, they started to charge flat rates for access to internet services. VoIP/instant messaging services exploited the flat rates to provide for free services in competition with the per-minute pricing structure. This was particularly attractive for international voice services, charged at high rates.

As a result, a lot of value was eroded in the industry. Telecommunications carriers were slow to modify their pricing structures as the flat rate cannibalized their per-minute revenue. Platforms grew thanks to the distorted price structure. According to OVUM (a London-based research firm), the overall value erosion for the telecom industry by OTTs can be estimated at $386 billion between 2012 and 2018.[9]

As platforms have matured, their services have become a substitute for the services provided by telecom carriers. Telecom carriers have come to include telephony and SMS services into their flat rates. In this way, users have little or no incentive to bypass the telecommunications carrier in favor of the OTT for the provision of the service. Users, however, continue to use platforms as they are often more convenient and easier to use. Platforms such as WhatsApp concentrate, in a single app, different communications modes (messages, pictures, voice, and so on), making it more convenient than the competing services provided by traditional carriers.

Furthermore, platforms have reached the largest network effects. WhatsApp has 2 billion users, more than any telecom carrier in the world. Telecom carriers, by contrast, have been fragmented as technology has created parallel networks (fixed telephony, mobile telephony, cable, etc.) and regulation has fragmented the market by introducing competition among providers. While WhatsApp is basically present around the globe under a single trademark with a seamless service across handsets, networks, carriers, and countries, hundreds of telecom carriers compete for the provision of a fragmented service.

Telephony still achieves large network effects as regulation imposes interconnection both at a national level among the carriers in competition, and across borders. However, the regulatory regime is burdensome and comes at a price. The cost of regulation is very substantial, as standardization is slow and creates rigidities and interconnection and roaming have a high cost that is passed on to customers. Platforms reach a similar result at a fraction of that cost. The use of the internet and the Internet Protocol enable similar results at a much lower cost. It is cheaper to ensure coordination at the data layer (applications) than at the physical layer (infrastructure).

However, it is important to identify that the impact of digital platforms on telecommunications carriers is very different from the impact on postal or taxi services (see Chapter 9). In these cases, a new product or a new provider substitutes the traditional service provider, making it redundant.

Communications platforms do not make telecommunications carriers redundant. Traditional players are not substituted; their services are always necessary, as platforms do not develop alternative infrastructures. Communications platforms build upon the infrastructures that are already installed, operated, and maintained by traditional telecommunications carriers. Services such as Skype and WhatsApp can only run if users have their own access to the internet contracted with telecommunications carriers. Ultimately, the signal is transported by the traditional telecommunications carrier.

What is happening is that carriers are becoming just one side in a multi-sided market, which is now coordinated by a platform. For a start, platforms are replacing telecommunications carriers as the managers of the relationship with the final user of the service. Users increasingly rely on platforms for the management of their communications needs. They provide the interface for the user.

The pricing strategy of the traditional carriers is now also limited by platforms, forcing carriers to charge only a flat rate for connectivity. Carriers find it increasingly difficult to set their rate structure, as attempts to charge for a specific service (a call or a message) are made impossible because of the competition by the platforms. Traditional telecommunications operators have integrated both the management of the telecommunications infrastructure and the provision of the telecommunications services. The heavy investment in installing telecommunications networks was recovered over a prolonged period of time via small fees for the use of services; for instance, the fee per minute for making telephone calls. Liberalization did not affect this model, as the tariff structure was replicated at a whole-sale level in the special access regulation, and fees per minute were charged by incumbents to newcomers, who would then replicate this model in their retail pricing. As carriers are "platformed" they lose the power to define the tariff structure. They can merely set a flat rate for the connectivity service.

Furthermore, the connectivity service provided by a carrier is increasingly non-differentiable from the service provided by a competing carrier, whether it is fixed or mobile, or just a free Wi-Fi connection provided by the owner of a restaurant or a hotel, or even the municipality. Digital platforms need an underlying infrastructure and transportation service, but basically any functioning broadband service can support the platform. Telecommunications carriers are becoming mere managers of so-called "dumb pipes" and telecommunications services are becoming mere commodities.

This evolution is particularly apparent in Africa and other regions of the world. Facebook is subsidizing access to internet services to have access to Facebook and other apps and sites (no other social network or email service, however). When a user has access to this content from his/her mobile smartphone (the most common access technology in Africa), the data consumption by the telecommunications carrier providing the access to internet service is not taken into consideration. Facebook has an agreement with the carrier and is taking on the cost. In this way, the central position of the carrier is further eroded in favor of the position of the platform.

Finally, platforms might be in a position to select the underlying infrastructure and even the bandwidth necessary at each point in time. This is a reality for business users. Software-defined networking (SDN) can dynamically adapt capacity to demand by virtualizing infrastructure[10] and by providing capacity as a service, rather than as a fixed asset. The concept of SDN appeared in telecommunications as a solution to adapt the existing telecommunications infrastructure to the growing demands of big data. Not only are data flows growing at an exponential rate, but they are not as predictable and stable as traditional voice flows. Furthermore, data flows sustain operations of growing relevance that just cannot be suspended in case of congestion of the network, as was the case with voice traffic.

SDN decouples the physical infrastructure layer from the control layer and uses software to dynamically adapt the capacity in the physical layer to the existing demand. If a customer demands more capacity, it is provided in real time. If a customer demands little capacity at a given time, the vacant capacity is used to serve other customers. This is particularly useful for managing bandwidth in large data centers. In the same line, deep-packet inspection (DPI) makes it possible to prioritize traffic supporting critical applications over traffic that is not time-sensitive.

Consequently, and as traditional carriers are being "platformed" or put under the coordinating role of a platform, the role of the traditional carrier is increasingly limited to providing an undifferentiated infrastructure, and revenue is therefore eroded.

Still, the ecosystem in which online platforms live needs to be sustainable. Sustainability requires sufficient funding for the deployment and maintenance of infrastructures. This is particularly relevant as there is an increasing demand for bandwidth and carriers are requested to deploy optic fiber and 5G mobile networks. The deployment of such infrastructures requires heavy investment by carriers, which not only face more and more competition by other carriers, but are also being commoditized by digital platforms.

Telecommunications carriers are implementing different strategies to counter being platformed. On the one hand, some carriers are investing heavily in becoming platforms themselves, either through acquisitions (such as Verizon acquiring Yahoo and AOL) or by creating new platforms to make use of their financial resources and market knowledge. On the other hand, carriers are expanding into media in order to enrich the customer experience and differentiate their services from other carriers and from communications platforms. Media is being platformed as well, as we will see further on, so this strategy might be just delaying platformization.

8.5 IOS and Android: Apps Are Platformed

In March 2019, Daniel Ek, the co-founder and CEO of the leading music streaming app Spotify, announced in his blog the filing of an antitrust complaint before the European Commission. Ek condemned the abusive behavior of gatekeepers in the mobile internet, "companies like ours must operate in an ecosystem in which fair competition is not only encouraged, but guaranteed."[11]

Spotify was launched in Sweden in October 2008, just three months after Apple opened the iPhone up to apps developed by third parties. Spotify developed an app – a software that could be downloaded to smartphones – for the streaming of music. Users can still download and store songs, but they can also simply rely on on-demand access to the 50 million songs repository managed by Spotify. In 2010, Sean Parker, founder of Napster and first president of Facebook, acquired a 5 percent stake.

A free basic service was financed with advertisement. A premium service with no advertising and access to the entire music library was available for a fixed monthly fee of $9.99 a month. Or was it $12.99? It depends: if you contracted the service in Spotify's website the price was lower, but in case the service was contracted in the app making use of the smartphone, the price was higher, as a 30 percent commission had to be paid on top of the price. Who is charging such a commission for the use of the smartphone? Is it the mobile carrier supporting the service? No. Is it the manufacturer of the smartphone (Samsung, Huawei, etc.)? Not really. Who has developed the power to charge such a high fee on apps as popular as Spotify and Tinder to provide their digital services?

The 30 percent commission is paid by apps to the two companies that control the operating systems installed on 99 percent of the smartphones in the world: Apple (the owner of iOS) and Google (the owner of Android). These operating systems have a very different business model than the most common operating systems in PCs. Microsoft basically has a monopoly on the operating systems running on PCs and notebooks. A fee is charged for the use of the operating system, independently of the company producing the hardware. It is an open system that allows use of any software on top of it, including browsers to freely navigate the internet and have access to any website. This open model proved more successful than the closed model of Mac OS in Apple personal computers.[12]

When Apple launched the iPhone in 2008, a proprietary operating system, the iOS, was installed, just as Apple had proposed a proprietary operating system in the 1980s for the PC integrating hardware and software (as opposed to the Microsoft model of an operating system to run in any PC). With the iPhone, however, the move proved to be successful, probably because the iPhone had the best of the both worlds. It was a closed system fully integrating hardware and software, in the position to ensure the best consumer experience: easy to use, no interoperability problems, no virus, more privacy, and so on. At the same time, it integrated the best of the open systems, as it allowed third-party developers to have software installed in the phone, in the form of applications ("apps"). However, Apple would have full control of the apps – only apps approved by Apple could be installed on an iPhone.

Google developed a parallel operating system, Android, which would be licensed to the rest of the smartphone manufacturers. It is more openly managed than iOS, as manufacturers can make adaptations to Android for their handsets (called forks), but it followed the iOS model of pre-approved applications, unlike the fully open Microsoft model for PCs. iOS and Android were not the only operating systems developed for smartphones. Microsoft and other companies developed their own operating system, iOS and Android ended up being installed in 99 percent of smartphones: iOS in all Apple's iPhones and iPads (around 15 percent of devices worldwide), and Android in basically all other mobile devices.

The key feature both in iOS and Android is the so-called app store. A comprehensive analysis of the app store market was undertaken by the Dutch antitrust authority, the Autoriteit Consument & Markt, in 2019.[13] An app store is the distribution system for third-party app developers to interact with smartphone owners so they can download and use apps in their smartphones. Such an interaction is intermediated by the owners of the stores, who are the owners of the operating systems: Apple and Google. These are, again, digital platforms that facilitate third-party interaction in a multi-sided market: mobile phone users on one side, app developers on the other. This multi-sided market works similarly to the videogame console market. The more users an app store (or a console) has, the higher the incentive for developers to produce apps (or video games) for it. The more apps (or videogames) are available, the more interesting it is for users. Compared to mere transactional platforms, operating systems have been classified as "innovation platforms," as they "consist of common technological building blocks that the owner and the eco-system partners can share in order to create new complementary products and services."[14]

iOS is a fully closed app store, is only available on Apple devices (iPhones and iPads), and only iOS can be installed on such devices. Furthermore, in such devices, only the apps specifically approved by Apple can be installed and users can only have access to apps through the App Store operated by Apple. Android is more flexible, as it allows partners to develop their own app stores; and Samsung and Amazon have developed them (even if they are not heavily used).

The approval process, which exists both for iOS and Android, requires developers such as Spotify to respect the rules and conditions defined by the owner of the operating system. A specific review process is undertaken for each app. Furthermore, the review must be completed for each update of the software. In the review, it is determined whether the app works or it has crashes and bugs, whether it extracts data from the device without the owner's permission, and whether it respects some internal rules, such as those related to gambling and porn, all part of the strict policies implemented by iOS (not so strict in Android). This could be the reason why Android has more apps (2.5 million) than iOS (1.8 million). Complaints about the discretional application of rules are increasingly common.

Apple and Google also provide payment processing services in their app stores. Apps can get paid for their services directly through the app store, without the need to develop their own payment systems, require credit card details to each customer, and so on. While this is certainly a convenient feature, it has led to conflict, as Apple has imposed the obligation on developers not to use alternative payment solutions when selling goods and services to be consumed within the app. This is particularly relevant because both Apple and Google charge a 30 percent fee on payments made through the apps.

The difference between consuming goods and services within the app or outside the app is not always clear. Apple believes that Spotify is consumed within the app, as music is listened to on the smartphone. On the contrary, Uber services are consumed outside the app, so Uber does not pay the 30 percent cut as it charges it directly to the credit card provided by the user in the app. The distinction is not always so clear; Tinder's services are considered to be consumed inside the app, which would certainly not be considered satisfactory by most of the users.

Apple is particularly strict, as it does do not even allow apps to communicate to users so that they can directly subscribe to the service in the app's website from their PCs. Such messages cannot be included in the app and cannot even be displayed on the website operated by the app. If Apple finds that such an action is taking place, the app is expelled from the app store. As more apps simply refuse to pay the 30 percent cut, they do not use the payment processing service and payments can only be made on their website from a

PC. This is the case of some of the apps that have stronger brands, such as Netflix after 2018 and Spotify after 2016 (although subscriptions made before those dates still pay a 15 percent cut when renewed). Making payment outside the app is not the most convenient arrangement and it is confusing, as the apps are not even allowed to give instructions to customers to pay via the web. Customers have to figure it out themselves that payments are not available in the app and that they must change devices, go to their PC, browse through a website operated by the same company, and then make the payment there.

A 30 percent commission is certainly a high amount of money. It happens to be the same commission Spotify takes from the music rights-holders they intermediate (the labels that intermediate singers and writers). Apple claims that only a small fraction of apps (fewer than 20 percent) pay such a fee, as most apps either do not charge to customers or are considered to provide goods and services to be consumed outside the app, and are therefore exempted from the payment. In any case, app stores have proven to be very successful: total revenue of more than $26 billion was estimated for 2020 for app stores, two-thirds for iOS and one-third for Google.[15]

Spotify filed an antitrust complaint in 2019 on the grounds that apps should be free to decide their payment system and not be forced to use Apple's payment processing service. Furthermore, Spotify claims that apps should have no restrictions regarding communication with their customers. Finally, Spotify argues that app stores should not discriminate. Apple should not impose different conditions to apps providing similar services. For instance, there is no clear ground to exempt Amazon's Prime video app from the 30 per-cent cut, but to impose it on Spotify, as both services would be consumed within the app.

Even more relevantly, Spotify argues that Apple should not self-preference the streaming music service provided by Apple itself: Apple Music. The original iTunes service built for the iPod was substituted by Apple Music as streaming became more popular. Apple acquired the streaming startup Beats Electronics in 2014 for $3 billion. The new streaming service was launched in 2015. Apple Music enjoys important benefits over other streaming apps: it is preinstalled in all iPhones, it is interoperable with other Apple devices such as the Apple Watch and with Siri (Apple's virtual assistant), and it does not pay the 30 percent fee, so it can be contracted and paid for in the app without the need to pay the surcharge.

Altogether, this is a quite solid antitrust case.[16] Following the complaint by Spotify and a parallel one from an unnamed e-book distributor, the European Commission opened an investigation in June 2020. A similar case started in the US in 2011 as a result of a complaint from iPhone owners against Apple, as they considered that the 30 percent commission is a higher-than-competitive price. The case was dismissed by a ruling that iPhone owners could not file the complaint as the commission is charged to app developers. As of 2019, the Supreme Court confirmed in *Apple v. Pepper* that iPhone owners can sue. The case is still open for a final judgment. In the summer of 2020, Epic Games, the owner of the popular game Fortnite, launched an antitrust claim against Apple and Google for similar reasons.

This will not be the first case decided by the European Commission against an app store. In July 2018 the Commission imposed a €4.3 billion fine on Google for abuse of its dominant position in the management of Android.[17] The Commission decided that Google illegally tied the Google search app and Chrome, Google's browser, to Google's app store, reinforcing the dominant position in the search and mobile browser markets. Google provided portfolio-based revenue share payments to smartphone manufacturers conditional on the commitment not to pre-install competing general search engines, again to protect the dominant position in the search markets. Finally, the Commission found that Google had illegally conditioned the licensing of the app store and the Google search app to commitments by hardware manufacturers not to fork the Android software.

Apple and Google are the gatekeepers of the mobile internet. App developers decry the "tax" imposed on them as dishonest, similar to the taxes imposed by medieval German barons on vessels navigating the Rhine across their possessions, extracting value from an infrastructure they had not built.

Mobile carriers are completely out of these conflicts, and actually out of the mobile digital content value chain. Carriers invested for the deployment of the infrastructure supporting mobile broadband, but a large portion of the value derived from this investment was captured by Apple and Google, which built virtual networks on top of the telecom infrastructure networks. As the Dutch antitrust authority concluded: "Apple changed the hierarchy of the mobile landscape and shifted the power from the mobile carriers to the app stores and mobile OS."[18]

Notes

1 Baset, S. A. & Schulzrinne, H. (2004). An analysis of the skype peer-to-peer internet telephony protocol, *arXiv preprint cs/0412017*.

2 The Court of Justice of the European Union concluded that the SkypeOut service, a service that enables users to terminate calls in telephone numbers, is a telecommunications service subject to the obligation of notifications to the local regulator in the Judgment delivered in Case C-142/18 on June 5, 2019.

3 Carlson, N. (2014, Feb. 19). *Facebook is buying huge messaging app WhatsApp for 19! Billion*, *Business Insider*, retrieved from www.businessinsider.com/facebook-is-buying-whatsapp-2014–2?IR=T

4 Levy, S. (2020). *Facebook. The Inside Story*. Penguin, New York, p. 323.

5 Commission Decision of 17.5.2017 imposing fines under Article 14(1) of Council Regulation (EC) No. 139/2004 for the supply by an undertaking of incorrect or misleading information (Case No. M.8228–*Facebook/ WhatsApp*).

6 FTC, Complaint for injuctive and other equitable relief, December 2020, p. 38.

7 Directive 2018/1972, of December 11, 2019, establishing the European Electronic Communications Code.

8 Brafman, O. & Beckstrom, R. A. (2006). *The Starfish and the Spider. The Unstoppable Power of Leaderless Organizations*. Penguin, New York, p. 62.

9 See reference in Olson, P. (2015, Apr. 7). *Facebook's Phone Company: WhatsApp Goes to the Next Level with its voice Calling Service, Forbes,* retrieved from www.forbes.com/sites/parmyolson/2015/04/07/facebooks-whatsapp-voice-calling/#6fb4dc311388

10 Knieps, G. (2017). Internet of Things, future networks, and the economics of virtual networks, *Competition and Regulation in Network Industries*, 18(3–4), 240–255.

11 Ek, D. (2019). Consumers and Innovators Win on a Level Playing Field, retrieved from https://newsroom.spotify.com/2019–03-13/consumers-and-innovators-win-on-a-level-playing-field/

12 Isaacson, W. (2011). *Steve Jobs*. Abacus, London.

13 Autoriteit Consument & Markt (2019). *Market Study into Mobile App Stores*, Report, retrieved from www.acm.nl/sites/default/files/documents/market-study-into-mobile-app-stores.pdf

14 Cusumano, M. A., Gawer, A., & Yoffie, D. B. (2019). *The Business of Platforms. Strategy in the Age of Digital Competition, Innovation, and Power*. Harper Business, New York, p. 18.

15 Savitz, E. (2019, July 25). *App Stores Could Be Ripe for Regulation. Here's Who Benefits if Commissions Fall, Barrons,* retrieved from www.barrons.com/articles/app-store-fees-regulation-51564019432

16 Gerardin, D. & Katsifis, D. (2020). *The Antitrust Case against the Apple App Store*, retrieved from https://papers.ssrn.com/sol3/papers.cfm?abstract_id=3583029

17 Commission Decision of 18 July 2018 relating to a proceeding under Article 102 of the Treaty on the Functioning of the European Union and Article 54 of the EEA Agreement (Case AT.40099 – *Google Android*), OJ 2019 C 402, 28.11.2019.

18 Autoriteit Consument & Markt (2019). *Market Study into Mobile App Stores*. Report, p. 71.

Chapter 9

YouTube
Traditional Media Substituted and Platformed

The media industry has also been disrupted by digital platforms. What is specific about this industry is that disruption simultaneously takes the form of substitution, as with the postal industry, and platformization, as with telecoms.

Traditional media are being substituted as platforms are competing for the attention of the audience and succeeding in obtaining a larger share of advertising dollars. Individuals, particularly children and millennials, are spending more time on platforms such as Facebook and YouTube, and less time reading newspapers, listening to the radio, or watching traditional TV networks. User-generated content is cheap to produce (it is often free for the platform), audiences run into the billions, and advertisement is personalized and therefore more profitable. Consequently, a large share of adverting is migrating to the new media.

Nevertheless, audiences crave more curated content. Well researched and presented news are always of interest. Splendidly produced TV shows, series, movies, and live sporting events attract large audiences. Professionally produced content will always be relevant, just as telecom infrastructure will. They cannot be substituted.

The relevance of high-quality content does not mean that traditional content producers cannot be disrupted. They are being disrupted, but in a different way. On one hand, platforms are intermediating the access by the audience to the traditional media. Individuals are increasingly having access to articles and video through platforms. Professional content is aggregated and displayed as another option in YouTube, Facebook Newsfeed, Netflix, and so on.

On the other hand, platforms are intermediating the access by the advertisers to the advertising space displayed by traditional media. Platforms have a relevant (one could say non-transparent) role in the intermediation of digital advertisement, even when it is displayed in legacy media sites. Traditional media are increasingly relying on platforms, particularly Google, to sell their advertising space.

Platforms attract the attention of large audiences themselves and thus act as the gatekeepers for audiences and advertising money to reach the traditional media. The algorithms developed by the platforms decide which content is displayed, when and how often it is proposed, and how advertising money is distributed across the ecosystem.

9.1 YouTube and User-Generated Content Platforms

On April 23, 2005, "Me at the zoo," a video praising the coolness of the long trunks of elephants in the San Diego zoo, was uploaded in YouTube. It has been seen more than

53 million times and it has more than 1 million likes. The 18-second video was the first video uploaded in YouTube, as it was made available by the three founders of the platform: CEO Chad Hurley (son-in-law of Netscape founder James H. Clark), CTO Steve Chen, and Jawed Karim, who remained as an advisor while studying computer science at Stanford University. All were aged below 30 and had met as employees working for PayPal. With 888 million internet users worldwide at the time, and broadband becoming increasingly common in most developed countries, the world was ready for a video platform.

YouTube was certainly not the only startup with a proposition around audiovisual content. YouTube allowed individuals to upload, share, and see videos. YouTube solved the technical challenge of making it simple to upload and browse videos with different technical formats. Other start-ups were providing similar services, but they were not as successful as YouTube's.

What made YouTube successful was the community around the videos: making comments, expressing likes and dislikes, inviting users to share the video links with friends, adding the videos to their websites, etc. Building on their expertise from PayPal of how to build communities, YouTube founders made the platform go viral.

User-generated content was the original target of YouTube. The pitch for investors was "to become the primary outlet of user-generated content on the internet, and to allow anyone to upload, share and browse content." The first videos were just personal recordings that would have only been of interest to a very small audience (such as baby videos for grandparents).

The first video to go viral on YouTube and reach more than a million views was not generated by a user, but "Lazy Sunday," a hip hop video by professional comedians working for NBC's *Saturday Night Live* in December 2005 and then uploaded on YouTube. NBC filed a suit for the breach of intellectual property rights against YouTube for airing the video.

From the early days, YouTube had the intention of combining user-generated videos and more professional videos and to monetize both thanks to advertising. YouTube was presented to venture capital firms as a platform curating a sophisticated multi-sided market formed by content producers, viewers (eyeballs), and advertisers. The more eyeballs, the higher the revenue from advertisers, and the higher the fees paid to content producers, triggering the well-known virtuous cycle derived from indirect network effects.

This was the successful pitch made to Sequoia Capital, led by former PayPal's CFO Roelof Botha, who invested $3.5 million in YouTube. Funding was necessary, as YouTube was a centralized network. Videos were hosted on YouTube's servers, which required significant bandwidth to manage an increasing volume of data. By December of the same year, eight months after launch, eight million videos were being viewed per day; this number reached 100 million by July 2006.

Google acquired YouTube in November 2006 for $1.65 billion. YouTube was the perfect match for the search platform. YouTube was attracting large audiences and Google had already developed a successful business model to monetize the attention of such a large and growing audience. YouTube was also a natural addition to Google's business model. YouTube is a platform that curates a sophisticated multi-sided market formed by content producers, viewers, and advertisers.

As the medium is the message, YouTube, a new medium, created a new message on its own. Individuals developed new, creative, and unexpected ways to display themselves and their interests. As advertising revenue started to trickle down to content producers, individuals were empowered to professionalize their videos and make a living – in a few cases, a very comfortable living – out of this revenue stream. PewDiePie is the online name of Swedish Felix Kjellberg, the most popular YouTuber since 2013, who has more than 100 million subscribers to his YouTube channel of videogames commentary, jokes, and displays of his personality. It has been estimated that PewDiePie makes $8 million a month.[1]

YouTube has always combined user-generated content with more professional and traditional videos. Remember that the first video to reach 1 million views in YouTube was "Lazy Sunday," an NBC production. Music videos continue to be among the leading content in YouTube. Over time, YouTube has expanded into a paid video-streaming service. After the success of streaming services, like Netflix and Hulu, YouTube has expanded from being a non-transactional platform funded by advertising, to include also paid services, under the trademark YouTube Premium. This service allows access to regular videos without advertising, but also to original content developed for the platform.

YouTube's figures are impressive. On the content side, there are more than 2 billion videos on display. On the eyeballs side, YouTube reached 2 billion monthly users in 2019. On the advertisers' side, it has been estimated that YouTube revenue in 2019 was around $15 billion. Google has built a multi-platform business, as it provides a wide range of services that attract multibillion audiences. Advertising in Google sites, including YouTube, generated $134 billion in revenue in 2019.

9.2 Media Substituted by Platforms

YouTube and all other digital platforms are displacing traditional media. If we take the US as an example, individuals aged between 18 and 34 reduced traditional TV weekly consumption from 22 hours in 2012 to 11 hours in 2020,[2] a 50 percent reduction in just eight years. The only age group that increased traditional TV consumption were individuals over 65. Competition for the attention of the audience ("eyeballs") has intensified.

Eyeballs are migrating away from traditional TV to platforms. In 2019, the average adult in the US spent 6:31 hours per day browsing the internet via PCs, smartphones, and tablets. This is particularly the case with younger generations. The 16–21 age group spends 7:22 hours per day online. Among teens, slightly more than half of all screen use was dedicated to TV or videos, and 31 percent went to gaming. Video chatting, reading online, or creating content like art or music each accounted for 2 percent.[3]

Advertising money is migrating according to the attention focus of the audience. The revenue from internet advertising in the US ($124 billion in 2019, growing 15 percent) is much higher than that of TV advertising ($70 billion).[4] Global TV ad spending dropped 4 percent in 2019,[5] at a pace similar to the annual reductions in postal services during the past years.

One of the leading reasons for the success of platforms in the advertising market is the appearance of new content providers. Eyeballs are spending a growing proportion of their attention span on user-generated content. User-generated content is the driver of platforms such as YouTube and Facebook. This is a revolution, as the generators of content that attracts eyeballs not only have a low cost of production, but do not usually expect to recover their costs in the form of a payment by the platform (this is the case of Facebook users and most of the content producers in YouTube), or they accept incurring the risk of production and being paid only a fraction of the ad revenue generated by the content (YouTubers). By contrast, traditional media have to accept the fixed costs of producing content, often taking the risk of paying upfront and independently of future success in terms of audience and advertising revenue.

A deeper reason for the success of digital platforms in the advertising market is the indirect network effects. Media has always exploited indirect network effects. Network TV, newspapers, and radio broadcasters relied on massive audiences enticed by attractive content, whose attention was then sold to advertisers.[6] The larger the audience, the larger the payments of the advertisers. The larger the payments of the advertisers, the better the content that could be produced to attract audiences. Media has always been the ultimate model of a non-transactional multi-sided market.

Digital platforms have scaled-up to dwarf traditional media in terms of audience. The audience of the leading multi-platforms is measured in billions. Google, the most prominent example, attracts eyeballs with its search engine (90 percent of market share in most countries), YouTube (2 billion), Gmail (1.5 billion), and all kinds of minor products (maps, translations, etc.). The same strategy is followed by Facebook, which owns not only the ubiquitous social network (2.7 billion users), but also WhatsApp (2 billion) and Instagram (1 billion). Verizon Media owns Yahoo, AOL, the Huff Post, and TechCrunch and reaches an audience of 1 billion monthly active users. The combined audience of these multi-platforms runs deep into the billions.

By comparison, traditional media has always been fragmented. Television audiences have always been fragmented, as networks tend to be national, and various networks must share the attention of the audience in each country. The most popular TV event of the year in the US is the Super Bowl, which attracts an average of 100 million viewers for a few hours, twice as much as the second event – the semifinals leading to the Super Bowl. Newspapers such as the *New York Times* trail far behind, with around 6.5 million subscribers for its digital and paper versions.

Fragmentation has even been a regulatory goal in itself to protect pluralism. In most jurisdictions, specific rules limit ownership of media outlets, both to prevent concentration in a specific type of media, particularly TV networks, but also across media, by the creation of conglomerates in control of newspapers, radio stations, and TV broadcasters.[7] It is proving complex to extend this kind of regulation to digital platforms, despite concentration in digital media being far more acute than in the traditional media.[8]

Furthermore, indirect network effects are larger in platforms. Digital platforms have a larger number of advertisers than traditional media. Facebook has more than 8 million active advertisers. Platforms have made advertising more accessible, as the minimum investment is very small and the tools made available by advertisers are easy to use. In any case, the platforms still rely on large advertisers for most of their revenue. The largest 5–10 percent of advertisers make up more than 85 percent of Facebook revenue in the UK.[9]

Substitution of the traditional media is reinforced by algorithmic network effects. In 2020 the UK antitrust authority, the Competition & Markets Authority, published a comprehensive report on digital advertising that identifies the role of data.[10] Traditional media broadcasts content (and ads) to an unknown pool of readers and viewers. There is minimal segmentation of audiences by type of product (music channels are more attractive among youngsters, sport channels tend to be more popular among males, etc.) and geography (local papers, radio stations broadcasting to a city, TV stations broadcasting to a regional market, etc.). At the same time, advertisers in traditional media have limited tools for measuring the effectiveness of their campaigns. As the department-store mogul John Wanamaker famously stated: "Half the money I spend on advertising is wasted; the trouble is, I don't know which half."

On the contrary, data empowers platforms to sell more effective ads. Platforms have privileged access to a rich variety of data. They have data about users, either volunteered by themselves, observed by the platform or even inferred (such as income based on the location of the user's address), data related to searches, contextual data, and so on. Even more importantly, platforms have analytics data; in other words, data on how users respond to ads in the form of clicks, purchases, etc. Large platforms are in the position to harvest massive data from the platform's own consumer-facing products and they can complement this with data provided by third parties, such as content publishers, advertisers, and even data brokers selling data. For instance, Facebook partners can install the "like" button on their sites, as well as analytic software provided by Facebook and Google to scrutinize consumers' behavior. The platforms have massive amounts of information about individuals, even those who think they are not using their services.

Data has a dual function in digital advertising. On one hand, advertisers can better target their ads. Advertising can be contextualized, being displayed in the precise moment users are searching, reading, or commenting on an item. Advertising can also be personalized; that is, targeted based on personal information about the user. On the other hand – and equally if not more importantly – data are also used to ascertain the effectiveness of advertising, helping advertisers to optimize their investment. Data empower advertisers to measure attribution; that is, tracking what actions follow the display of an ad (clicks, purchases and so on). Thus, data are used to measure the effectiveness of ads. Finally, data are used to verify that there is no fraud (paying for ads that are not effectively displayed).

The more data a platform has, the better equipped it is to support advertisers in targeting their ads and the better it can help advertisers to ascertain the effectiveness of the campaigns. It has been calculated that "publishers earned around 70% less revenue when they were unable to sell personalized advertising but competed with others who could."[11]

The leading digital platforms benefit from the largest direct network effects, the largest indirect network effects, and largest algorithmic network effects. The combined network effects not only substitute traditional media, but are also concentrating the market. Thousands of newspapers, radio stations, and television networks around the world are being substituted by a very limited number of global digital platforms that are increasingly concentrating their advertising expenditure in the entire world.

9.3 Media Platformed in the Display of Content

Traditional media are affected by platforms in a more subtle way. They are not only being substituted, as platforms attract an increasing share of the attention of the audience, but they are also being "platformed." Traditional media is transitioning from being platforms managing their own multi-sided markets, to mere sides in a much larger multi-sided-market built on top of them, but managed by digital platforms.

Traditional media are being intermediated by platforms in two ways. First, an increasing share of the audience has access to traditional media through platforms, as audiences increasingly reach their sites though the platforms, because the platforms suggest their content to users. Second, traditional media are increasingly intermediated in the selling of advertisement space in their digital editions. Even when ads will be displayed in the websites and apps directly managed by traditional media, platforms will be receiving a cut, as intermediaries. Traditional media are increasingly intermediated on both the audience side and the advertisers' side of the multi-sided market.

Let us start with the intermediation on the audience side. Audiences are increasingly having access to content produced by traditional media through digital platforms. This is clearly the situation with news articles and podcasts, and is increasingly affecting video. Audiences have less and less direct access to the sites where content producers aggregate their content in the form of newspapers, magazines, or TV networks. Visiting the website of a newspaper or watching a TV network is becoming less common. Audiences are still enjoying the content, but they have access to it through the platforms. Platforms curate a selection of unbundled news pieces and videos produced by third parties, as they aggregate links, snippets, and highlight moments in TV. They make the selection of the most attractive content for each specific individual. Platforms become the gatekeepers.

The UK antitrust authority has calculated that only 43 percent of readers had direct access to the websites of the large newspapers in the country. Instead, most traffic to the sites was indirectly routed from Google (25 percent), Facebook (13 percent) and other platforms.[12] Large divergences exist as some media outlets have stronger brand recognition and attract audiences to their sites on their own. Any minor change in the platforms' algorithms is followed with apprehension by traditional media.

New media outlets have adapted to this new model. They do not have the intention of producing a bundled product with a balanced number of pieces so as to compose a newspaper, a magazine, a radio or a television station. Instead, they merely produce independent pieces of news, audio, or video, and they expect them to become viral as they are selected by digital platforms for their audiences. They work for the algorithm and they rely on digital platforms, particularly Facebook and Google/YouTube, to reach audiences. This strategy has been labeled as "networked strategy."[13] Platforms put the network on top of unbundled content. Traditional media are also starting to work for the algorithm, trying to attract audience to specific news pieces and videos, at the same time as producing traditional newspapers, both on paper and digitally.

It can get worse. Sometimes content produced by traditional media is just displayed by the platforms inside their products. Traffic is not directed to the original site where it was displayed, but it stays in Facebook and Google. This might happen on a voluntary basis (media uploading a video in YouTube for a 55 percent share of ad revenue in the platform) or against the will of the content producers, as with newspaper snippets in Google News (where no ads are displayed) and Facebook Newsfeed (which generates ad revenue that is not shared with content producers). The European Union adopted in 2019 the Copyright Directive[14] to force platforms to compensate publishers for the snippets.

Media was always a multi-sided market, as newspapers and radio and TV stations were platforms that connected advertisers and audiences. But digital platforms are taking the business model to the next level. They have built audiences that can be counted in the billions, with a potentially universal reach. Even more relevantly, they aggregate not only content generated by their users or by the platform itself, but also content generated by traditional media, content produced by as many content producers as possible, into a single platform. This is proving an unbeatable proposition for audiences, who spend countless hours "snacking" through content, but also with advertisers, as platforms multiply the audiences of any legacy media.

Only the very high-quality content produced by a low number of entities worldwide with very strong brands is in a position to escape platformization and maintain their direct relation with customers, mostly by relying on direct payments and only marginally on advertisement. This is the case with the highest-quality newspapers (the *New York Times*), magazines (*The Economist*), and audiovisual content producers (Disney, Hollywood majors).

For the others, content is becoming a commodity that is intermediated by the digital platforms. Traditional media compete for the attention of audiences that binge-watch an endless selection of user-generated content in the platform itself, but also newspaper articles, videos from traditional broadcasters, and pieces produced by new network media. Value is eroded, as cheaper content produced by unreliable sources often proves as attractive as high-quality journalism. Some "fake news" may attract a bigger audience than proper news, making it more attractive to advertisers. Content can no longer be commercialized at a premium as a bundled proposition. Traditional media is losing brand recognition and the direct relationship with the audience.

9.4 Media Platformed in the Relationship with Advertisers

Traditional media are also being intermediated by digital platforms in the relationship with advertisers when they sell the advertising space in their digital ventures. Google has built an invisible infrastructure that supports not only its own advertising business, but the digital advertising activity of traditional media, through a mesh of intermediaries that the UK antitrust authority, the Competition & Markets Authority (CMA) described perfectly in its report on digital advertising published in 2020.[15]

Publishers can directly interact with advertisers and sell their inventory. In reality, this is usually the case for the marketing of inventory by Google and Facebook: they do not rely on intermediaries. Google directly sells its search ads through sophisticated bids. In the UK, this method accounted for approximately 55 percent of the total digital advertising revenue in 2019. Google also sells display ads in YouTube and other ventures (6 percent of total ad revenue). Facebook sells its display inventory directly, which amounts to around 22 percent of the total revenue in the UK.[16]

Traditional media, by contrast, sell only a small percentage of the digital inventory directly to advertisers. They mostly rely on intermediaries to sell their inventory in what is called the open display segment of the market. This amounted to only 15 percent of the total digital advertising revenue in the UK in 2019 (£1.8 billion). Intermediaries play a relevant role and take a large piece of the pie for their intermediation. The UK antitrust authority has calculated that intermediaries capture at least 35 percent of the value, while alternative estimates raise this figure to 70 percent.[17] Not surprisingly, the main intermediaries in digital advertisement are, again, the digital platforms, particularly Google.

Publishers are hiring intermediaries to sell their ad space through real-time automated bidding for each ad as it is displayed. Every time a reader accesses the website of a newspaper, a request is automatically sent to various intermediaries hired by advertisers so they submit bids for a specific ad for that specific viewer. These intermediaries use data-fed algorithms and decide how much they are ready to pay for the ad space on behalf of an advertiser. They then send their bids to an intermediary hired by the newspaper (the ad server), who identifies the highest bidder and then contacts the ad server of the publisher (another intermediary) to receive the ad to be displayed. All this happens in a fraction of a second, millions of times every hour of every day.

Intermediaries are proliferating. Publishers are contracting ad server companies to manage their inventory and to automatically decide which ads to display at each moment, while advertisers hire other ad server companies to store the ads and deliver them to the publisher's ad server. In parallel, other intermediaries, the supply-side platforms and the demand-side platforms, interface in the bidding process to identify the winning option.

Surprisingly, Google is present throughout the supply chain with large market shares. It has a market share of more than 80 percent in the publishers and advertisers ad servers segments, and a share of more than 50 percent in both supply and demand platforms.[18] Basically, in most transactions, Google is contracting with itself, which creates an obvious conflict of interest and constant suspicion.

Digital platforms are intermediating traditional media even when they sell their own inventory. Google takes a large cut from the traditional media even when they are selling their own ad space in their digital editions.

9.5 Digital Platforms as Coordinators of the System

It is not always simple to identify the role of digital platforms in the media ecosystem and the legal nature of the services they provide. Traditional media were subject to a mature and detailed regulatory framework, based on the liability of the editor for the content that was published and broadcasted. It was not evident whether such a regulatory framework should be automatically extended to digital platforms, particularly as, in the early days, they had a weak position in the media ecosystem or they were not even recognized as actors in these markets. Over time, however, platforms have reached a position of power that has surpassed that of any of the traditional media corporations.

Specific regulation was adopted along the decades to rule traditional media. Media were often subject to licensing (particularly terrestrial TV), restrictions to concentration, obligations on intellectual property, liabilities in case content would breach defamation laws, protection of minors, restrictions on advertising, protection of independent content producers, as well as local content producers, among other regulations. In some jurisdictions, media were closely controlled by government, including content censorship. Such restrictions were defined at a national or even at a local level.

Digital platforms evolved from the hosting and transmitting of basic content such as emails, websites, and blogs. These activities were originally considered relatively harmless, as they had limited impact on society. Platform regulation was defined at an early stage, when network effects around platforms were not mature and the potential role of platforms in the media and advertising markets was not fully understood.

In the US, Section 230 of the Communications Decency Act, adopted in 1996, provides immunity from liability to providers and users of an "interactive computer service" who publish information provided by others. When NBC identified that YouTube was airing the video "Lazy Sundays," it asked YouTube to take down the video, as well as other 500 videos under their intellectual property rights. YouTube took down the videos, but NBC did not get the $1 billion they had requested in damages. YouTube was not liable for the content uploaded by a user; it just had to take it down at the request of the holders of the rights. The same provision was raised by eBay in the landmark case Tiffany v. eBay, adjudicated by the Second Circuit in 2010:[19] "For contributory trademark infringement liability…, a service provider must have more than a general knowledge or reason to know that its service is being used to sell counterfeit goods." Therefore, eBay was not liable for trademark infringement.

In the EU, providers of the most popular "information society services" were exempted from liability by the Directive on Electronic Commerce adopted in 2000.[20] The directive firstly allowed the providers of digital services to follow the regulatory regime of the country of establishment and not the regulation of the country where services were being used by consumers. Furthermore, service providers acting as intermediaries (those being a mere conduit that caught or hosted information) would be exempted from liability. They would have no general obligation to monitor the information they transmit or store. It was only expected that the platform would expeditiously remove illegal content upon obtaining actual knowledge of the illegality. This regulation was designed for ISPs and for platforms managing websites, blogs, and email services. It was subsequently extended to marketplaces such as eBay.

Video-sharing platforms were in this way subject to a privileged regulatory framework. This regulatory protection proved to be very effective against copyright infringements, YouTube being a good example. However, the exemption extended to the sector-specific obligations defined in the audiovisual legislation. In the EU, platforms were considered mere intermediaries, empowering content producers to interact with the audience. Digital platforms would not be subject to the obligations imposed on networks with "editorial responsibility" by the Audiovisual Media Services Directive.[21]

What seems clear at this stage, as platforms have matured, is the central role that the digital platforms play in the ecosystem created around them. Platforms may not create content, or play the role of an editor in a newspaper or TV network, but they have an increasingly active role and the possibility to shape the ecosystem around them, including traditional media. They have the power to determine the success of a content producer. They even have the power to exclude content providers from the platform. As the leading platforms concentrate and reach positions of market power, such powers are subjected to increasing levels of scrutiny.

Platforms shape their ecosystems as they decide how to match the different sides in the platform. Such matching is decided by algorithms that automatize the matching according to principles that are not always transparent and determine the success of some content and content providers against others. Let us take a closer look at the example of YouTube, following the analysis by Burgess and Green.[22]

YouTube originally ranked content according to objective criteria, such as the most viewed, the most favorited, the most responded, the most discussed, and so on. Rankings based on categories and across the whole platform were displayed. It was transparent. Viewers could identify which videos were more popular and decide whether or not they wanted to see them.

YouTube has evolved from passively providing transparent ranking for users to decide, to actively guiding viewers to specific content. In this way, YouTube has started to shape content in the platform.[23] The platform has developed automatized algorithms that personalize the videos proposed to each individual viewer.[24] The parameters introduced in the algorithms are not transparent. Given that YouTube is a commercial enterprise, and based on the understanding of its business model, it can be inferred that the platform has an incentive to attract the attention of the viewers for the longest possible time. In this way, the platform can sell more ads to advertisers. Based on the knowledge of each user, YouTube can guide them to content that they will like, reducing the element of chance and creating a bubble of safe and comfortable content for the viewer.

Monetizing eyeball time in the platform is the ultimate driver of the algorithm. Increasing time spent in the platform is a means, not an objective in itself. The real objective is to maximize advertising revenue; this is the objective that drives the type of content proposed to viewers and is the interest of the large advertisers that shapes the content proposed to viewers. Of course, large advertisers want to have illegal content such as terrorism and pedophilia excluded from the videos where their ads are displayed. However, large advertisers are also sensitive to potential boycotts as their ads are shown in videos that may be perfectly legal, but not in line with the sensitivity of pressure groups.

YouTube has developed specific tools to limit the visibility of perfectly legal but potentially risky content.[25] In 2017, following a bitter campaign by newspapers in the UK and the US, YouTube implemented a new policy. Only videos with more than 10,000 views would be eligible to receive a share of the advertising revenue. "This new threshold provides us enough information to determine the validity of the channel," explained YouTube. In the same way, some content is just permanently "demonetized," in the sense that it is allowed on the platform, but no ads are shown and no payments are made to the content producer, regardless of the number of viewers. Human reviewers have been introduced, as well as an appeals procedure against demonetization decisions. The most popular YouTuber, PewDiePie, was expelled from the Google Preferred program and his series in YouTube Premium was terminated, after anti-Semitic references were identified in some of his older videos. Restrictive policies have included all kinds of hate speech, but also conspiracy theory content and fake news, following so-called "targeted flagging" by users. The ultimate tool is platform exclusion, as happened in 2018 to Alex Jones, one of the most popular conspiracy theory promoters.

YouTube's curating policies demonstrate the central role of the platform to shape the ecosystem and to determine the nature and the identity of the services providers who will succeed in the platform. Most of the curating policies implemented by the leading platforms appear to be necessary and beneficial for the community. However, as platforms concentrate and gain market power, transparency in the curating policies of the platforms, particularly in the algorithms, and protection for the rights of the content providers seem to be necessary.

European authorities have decided to increase the regulatory obligations on the new category of "video-sharing platform providers." Amendments to the Audiovisual Media Services Directive adopted in 2018[26] have imposed new obligations on platforms such as YouTube and Facebook. Such platforms are not subject to the same regulatory obligations as TV networks and platforms that develop their own content or propose content on demand (Netflix). While the platforms like YouTube and Facebook have no editorial responsibility, they are obliged to take appropriate measures to protect minors, to protect the general public from content containing incitement to violence or hatred against minorities and from content which constitutes an offense such as terrorism and pedophilia.

They have to establish age verification systems for users and parental control systems, mechanisms for users to report or flag content, systems for explaining to users what effect has been given to the reporting, flagging and procedures to handle users complaints, among others.

Notes

1 Czarnecki, S. (2019, Aug. 6). *Study: PewDiePie is YouTube's highest Earner at $8m a Month, PR Week*, retrieved from www.prweek.com/article/1593191/study-pewdiepie-youtubes-highest-earner-8m-month

2 Lupis, J. C. (2020, Sep 14). *The State of Traditional TV: Updated with Q1 2020 Data, Marketing Charts*, retrieved from www.marketingcharts.com/featured-105414

3 Siegel, R. (2019, Oct. 29). *Tweens, teens and screens: The average time kids spend watching online videos has doubled in 4 year, The Washington Post*, retrieved from www.washingtonpost.com/technology/2019/10/29/survey-average-time-young-people-spend-watching-videos-mostly-youtube-has-doubled-since/

4 IAB (2020). Internet advertising revenue report, Full year 2019 and Q1 revenues, retrieved from www.iab.com/wp-content/uploads/2020/05/FY19-IAB-Internet-Ad-Revenue-Report_Final.pdf

5 Salkowitz, R. (2019, Dec 9). *New Data Shows Why TV Ad Spending Had Its Biggest Decline In A Decade, Forbes*, retrieved from www.forbes.com/sites/robsalkowitz/2019/12/09/heres-why-tv-ad-spending-fell-off-a-cliff-in-2019/#6611c3d368bd

6 Wu, T. (2016). *The Attention Merchants. From the Daily Newspaper to the Social Media, How Our Time and Attention is Harvested and Sold*. Atlantic Books, London.

7 Brogi, E. et al. (2020). *Monitoring media pluralism in the digital era: application of the Media Pluralism Monitor 2020 in the European Union, Albania & Turkey. Policy report*, Centre for Media Pluralism and Media Freedom, Florence, retrieved from https://cadmus.eui.eu//handle/1814/67828

8 Stasi, M. L. (2020). *Ensuring Pluralism in Social Media Markets: Some Suggestions*, Robert Schuman Centre for Advanced Studies Research Paper No. RSCAS 2020/05, retrieved from https://papers.ssrn.com/sol3/papers.cfm?abstract_id=3531794

9 Competition & Markets Authority (2020). *Online platforms and digital advertising*, Report, p. 61, retrieved from https://assets.publishing.service.gov.uk/media/5efc57ed3a6f4023d242ed56/Final_report_1_July_2020_.pdf

10 Competition & Markets Authority (2020). *Online platforms and digital advertising*, Report.

11 Ibid., p. 15.

12 Data for year 2019, as in Competition & Markets Authority (2020), *Online platforms and digital advertising*, p. 17.

13 Van Dijck, J., Poell, T. & De Waal, M. (2018). *The Platform Society. Public Values in a Connective World*. Oxford University Press, Oxford.

14 Directive (EU) 2019/790 of the European Parliament and of the Council of 17 April 2019 on copyright and related rights in the Digital Single Market and amending Directives 96/9/EC and 2001/29/EC, OJ 2019 L 130, 17.5.2019.

15 Competition & Markets Authority (2020). *Online platforms and digital advertising*. Report.

16 Ibid., p. 9.

17 Geradin, D. & Katsifis, D. (2018). *An EU Competition Law Analysis of Online Display Advertising in the Programmatic Age* (December 12, 2018), retrieved from SSRN: https://ssrn.com/abstract=3299931

18 Competition & Markets Authority (2020). *Online platforms and digital advertising*, Report, p. 18.

19 Tiffany Inc. v. eBay Inc., 600 F.3d 93 (2d Cir. 2010).

20 Directive 2000/31/EC of the European Parliament and of the Council of June 8, 2000 on certain legal aspects of information society services, in particular electronic commerce, in the Internal Market ("Directive on Electronic Commerce"), OJ 2000 L178/1, 17.7.2000.

21 Directive 2010/13/EU of the European Parliament and of the Council of March 10, 2010 on the coordination of certain provisions laid down by law, regulation or administrative action in Member States concerning the provision of audiovisual media services (Audiovisual Media Services Directive), OJ 2010 L 95/1, 15.4.2010.

22 Burgess, J. & Green. J. (2018). *YouTube*. Polity Press, Cambridge, 2nd edition.

23 Ibid., p. 99.

24 Davidson, J. et al. (2010). *The YouTube Video Recommendation System*, Proceedings of the Fourth ACM Conference on Recommender Systems. ACM, 2010, 293–296; Zhou, R., Khemmarat, S., & Gao, L. (2010). *The Impact of YouTube Recommendation System on Video Views*, Proceedings of the 10th ACM SIGCOMM conference on Internet measurement, 404–410.

25 Burgess, J. & Green. J. (2018). *YouTube*. Polity Press, Cambridge, 2nd edn, p. 149.

26 Directive (EU) 2018/1808 of the European Parliament and of the Council of November 14, 2018 amending Directive 2010/13/EU on the coordination of certain provisions laid down by law, regulation or administrative action in Member States concerning the provision of audiovisual media services (Audiovisual Media Services Directive) in view of changing market realities, OJ 2018 L 303, 28.11.2018.

Chapter 10

The Regulatory Challenges Posed by Communications Platforms

Regulation has played a fundamental role in the platformization of the communications industry, even if such a role has often not been made explicit and has even been ignored.

When digital platforms compete with traditional telecom and media companies, a level playing field must be ensured. Protecting traditional players against disruption should not be a policy goal. Competition on merits can lead to substitution if platforms are more efficient and deliver a better deal for consumers. Nonetheless, substitution should not be enabled or accelerated just because platforms enjoy regulatory advantages. Such advantages might derive from being excluded from the traditionally strict regulation of network industries. However, the level playing field should not be constructed by means of merely extending to platforms the rules designed for the tradicional players. Advantages might derive from benefits specifically granted to digital players from the outset of the digital revolution as with liability. It appears that the time has come to reconsider such privileges and achieve a new balance.

More nuanced is the vertical relationship between platforms and traditional network companies. Platforms often build their services on top of preexisting infrastructures and network effects. The conditions at which traditional companies make their assets and services available to platforms are often determined by regulation. Regulation must take into consideration the new balance of power among platforms, telecom carriers, and media companies. Regulation should not accelerate the commodification of telecom and media services to the advantage of platforms.

Finally, regulation should acknowledge the general interest in the communications industries. The traditional regulation of telecoms and media have often pursued the objective of fostering access to such services at affordable rates and protecting free speech and pluralism in public debate. The platform revolution is not diminishing the relevance of such public policy goals, but requires new instruments to make them effective.

10.1 Level Playing Field

Traditional communications service providers are subject to all kinds of regulatory obligations, as is usually the case in the network industries. However, platforms tend to be exempted from these obligations, even if the services they provide tend to increasingly compete with or even substitute the traditional communications services. This is the case with voice and instant messaging services, as well as of media services. A level playing field is necessary for the competition between the traditional communications networks and the new communications platforms.

A fundamental divergence exists in the regulation of liability between communications networks and communications platforms. While postal operators, telecom carriers, and media companies are considered liable for their services, digital platforms have been largely exempted from liability. Such divergence has benefited platforms in their competition with traditional communications networks.

Regulation has always considered communications corporations to be fully liable for carrying out their business ventures. Corporations pulled together assets and staff under a hierarchical structure. As hierarchical structures, there was no difficulty in considering the decision-makers in such structures liable for their decisions in the management of the structure. Communications corporations would be liable for the quality of their services, for fairness in the relationship with consumers, for the pricing of their products, and, in general, for any breach of legislation. Sector-specific regulation could be imposed on these corporations to attain public interest objectives: universalization of the service, rate regulation, and so on. Thus, communications corporations are heavily regulated, as is common in network industries.

The exception in the communications industries would be defined for the conveyance of messages beyond the control of the network managers. Postal and telecommunications carriers would not be liable for the content of the communications: letters and phone calls. Content would actually be protected from the control of the network managers, and the users' right to secrecy of communications is constitutionally enshrined. Legal interception of communications under exceptional circumstances would be regulated. These principles would be extended to the access to internet services provided by traditional carriers.

Media editors, by contrast, were made liable for the messages broadcasted by newspapers, radio, and TV stations. Such messages were no secret, of course, but were crafted for the widest circulation. Editors had access to the content to be broadcasted and, being hierarchical structures, editors had the possibility and the obligation to ensure that broadcasted content would meet legal and regulatory obligations.

As described above, at the turn of the century, platforms were exempted from liability in the provision of their services both in the US and in the EU (Chapter 9). Platforms were assimilated to postal and telecom carriers merely conveying messages over which they have no editorial control and were even hidden from the network managers. Platforms providing email and instant messaging service might be merely transporting messages produced by third parties with no liability, just like their traditional peers. However, there would be no fundamental reason to exempt them from obligations such as lawful interception, financial contribution to the universalization of basic communications services, quality of service, and so on. The European Electronic Communications Code has already extended its scope to communications platforms, imposing some obligations on platforms, even if not all the obligations that are imposed upon traditional carriers, as exposed in Chapter 8.

The debate is more heated around media platforms' liability. While editors of traditional media are liable for the content they broadcast, platforms such as Facebook and YouTube are exempted from liability, as long as they take down illegal content when required. Platforms claim that they are not producing nor exercising editorial control over content. They would merely act according to the Good Samaritan rule and take down obviously illegal content such as pedophilia and hate speech.

We already know that platforms such as Facebook and YouTube are not neutral intermediaries. Their algorithms actively determine the content to be displayed to their audience. They might not produce the content displayed in the platform. They might not actively select the content to be broadcasted as the editor of a newspaper or a radio and TV station. The algorithms developed by the platforms might not select content like a traditional editor, but they do determine the type of content to be displayed. They determine the parameters to be used by the algorithms. They can promote clickbait or they can promote reliable sources of content. They are in a position to control the results of the algorithm to ensure such results are in line with the platform's goals.

Thus, public authorities on both sides of the Atlantic are considering increasing the liability of platforms. It is not necessary to align the liability to that of a traditional editor; this would not be fair as platforms are not hand-picking content for their audiences like a traditional editor does. Nevertheless, platforms should be liable for what they do: for the promotion of certain types of content to their audiences.

10.2 Regulating Concentration

Another element of divergence in regulation is that regulation of traditional communications industries actively promotes fragmentation with the objective of counterbalancing the tendency towards concentration and monopolization due to network effects. Digital platforms, however, have been allowed to grow without effective limitations, even to the point of dwarfing the traditional communications giants.

Liberalization policies in the US and the EU first put an end to the national telecom monopolies and then actively promoted competition. Newcomers' access to scarce resources is guaranteed: telephone numbers, spectrum, and rights of way. In many countries, legislation poses limits on the amount of spectrum that can be assigned to a mobile carrier (anti-hoarding rules) in order to ensure that a minimum number of carriers can compete in the market. Interoperability of the telephone service (both fixed and mobile) is imposed upon carriers in the form of interconnection obligations, and in the case of carriers with market power, at non-discriminatory conditions, with no self-preferencing, at cost-oriented rates, which forces the largest carriers to share their network effects with competitors. Lock-in effects are regulated in the form of number portability obligations. These are just a few examples of how fragmentation is actively pursued by regulators in telecommunications.

In the media industry, fragmentation is a regulatory goal, not just to promote competition, but also to promote diversity, pluralism, and free speech. Again, obstacles were originally set in the US for spectrum hoarding in open-air television and radio.[1] Specific rules were adopted in most jurisdictions in order to prohibit ownership concentration. The FCC in the US regularly fine-tunes the media-ownership rules. Similar rules exist in most of the European countries.[2] In Spain, for example, mergers are not allowed if they would result in a TV network having a market share exceeding 27 percent of the audience in the whole country.[3]

It is clear at this stage that digital platforms are not subject to similar limitations. Communication platforms have expanded without limit. Search has concentrated around Google, which enjoys market dominance in search advertising. Social media has concentrated around Facebook, which was allowed to acquire Instagram, reaching a dominant position in display advertising. These two companies have grown to dominate not only the US market, but basically the global market, with the exception of China. Furthermore, they have grown cross-market to provide multiple communications services: Google expanded from search to email (Gmail), to media (YouTube), to advertisement intermediation (with the acquisition of DoubleClick), and to mobile operating systems (Android). Facebook expanded to communications by developing the Messenger service and by acquiring WhatsApp.

Platform concentration has matured without regulatory intervention, not even a robust antitrust enforcement. Antitrust scrutiny did not stop any of the large mergers that grew network effects through mergers and acquisitions. Facebook's acquisition of Instagram and WhatsApp and Google's acquisition of DoubleClick and YouTube did not originally raise objections by antitrust authorities, on either side of the Atlantic. Sector-specific regulation did not set any limit to organic growth. Authorities were too busy regulating competition in liberalized markets to notice the erosion of market power and the increasing insignificance of the traditional communications players.

A level playing field should balance the regulation on concentration to prevent unfair advantages for some market players. Some restrictions on growth in traditional industries could be reconsidered. Restrictions to the growth of platforms are increasingly requested.

Growth in traditional industries can take different forms. Mergers and acquisitions can help traditional players grow in scale and come closer to the largest platforms. Competition from digital platforms should be increasingly taken into consideration when antitrust authorities analyze mergers. Mergers are not the only way to grow network effects, however. Cooperation between traditional players can also provide scale and network effects to compete with platforms. For instance, this is the case of data pooling to grow algorithmic network effects. By sharing the data available to them, traditional players can grow enough data to compete with digital platforms. This is already happening in digital advertising, as traditional media pool their data to more effectively manage programmatic advertising. More and more deals of this kind will be made in the future, not only to pool data, but to pool audiences and advertising space for advertisers. Obviously, this poses antitrust concerns, but such concerns must be evaluated within the new environment created by the platforms.

At the same time, restrictions to concentration for digital platforms should be considered. Telecoms provide the ultimate example of the break-up of a monopoly. Horizontal and vertical unbundling is a common strategy in the network industries. The Federal Trade Commission has requested the divestiture of Facebook in December 2020. We will elaborate on this point in the final part of the book.

More sophisticated instruments are also available to fight concentration. A century of regulation of the network industries has shown that fighting network effects not only damages consumers as value is destroyed, but is often useless, as market players have too strong an incentive to grow to exploit the network effects. Incentives for concentration are diminished and even eliminated when dominant operators are forced to share the value of the network effects with other market players. This has proven to be the winning strategy in telephony. All carriers have to interconnect their networks in order to ensure the interoperability of the service. Furthermore, the dominant carrier must allow interconnection at cost-oriented prices, sharing with the rest of the market the value internalized by the largest network.

Communication platforms are not obliged to interoperate with other digital platforms or with traditional networks. Some platforms have successfully grown multibillion user pools, with WhatsApp as the leader, with 2 billion users. They are isolated networks, large enough to be successful and reluctant to share their value with other players, just like Mr. Vail and the Bell System one century ago (as described in Chapter 6).

However, things might be changing. The European Electronic Communications Code, adopted in 2018, empowers public authorities to impose upon platforms the obligation to make their services interoperable (Art.61(2)(c)). Such an obligation can only be imposed on platforms when "the geographic coverage and the number of end-users of the provider concerned represent a critical mass with a view to achieving the goal of ensuring end-to-end connectivity between end-users."

10.3 Platform to Telecoms Relations

Platformization creates a power struggle between the traditional operators and the digital platforms. This is particularly the case when platforms do not substitute traditional service providers, but continue to rely on their services, while substituting them as coordinators of the system. This struggle is highly relevant in the telecommunications sector, as the infrastructures supporting the internet – the basis for the operation of all digital networks – are run by the same telecommunications carriers that are being platformed (Chapter 8). The same effect has been identified in media (Chapter 9).

In the relationship between digital platforms and telecom carriers, the conditions under which platforms have access to the underlying telecommunications services are key. Platforms need access to telecommunications services and want such access at the most favorable conditions in terms of price and reliability. Telecommunications service providers need to ensure that, as their services are commoditized, they still get the revenue they need to deploy and maintain their infrastructures.

Regulation is key for the definition of the conditions of the relationships between telecommunications carriers and communications platforms. The long-running debate between carriers and platforms, as well as the ensuing regulation, has been confusingly termed as "net neutrality" regulation.

The expression "net neutrality" was launched in 2003 by the American academic Tim Wu.[4] Digital platforms insist on the need to impose regulation on telecommunications

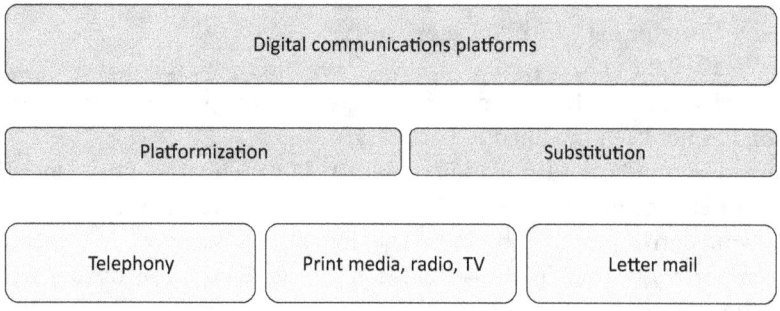

Figure 10.1

carriers to ensure they provide full and non-discriminatory access to their services to all applications and, more generally, to all content, particularly when carriers integrate vertically and not only manage the infrastructure but also their own applications (like telephony services and SMS) that compete with third parties that make use of their infrastructures.

Telecommunications carriers oppose such regulation, considering it an unnecessary restriction to the freedom to manage their assets. Firstly, they claim that network management is always necessary to avoid congestion in the network and to ensure that more demanding applications (such as real-time applications and VoIP) are correctly served, and that capacity, as a scarce resource, is not wasted on less demanding applications where time is not sensitive (such as email). Secondly, they claim that charging more for more demanding applications is a way to incentivize investment in the enhancement of the network, particularly the migration to fiber and 5G mobile networks. Thirdly, they claim that competition law is already sufficient to exclude potential exclusionary practices against competitors.

The debate has been particularly fierce in the US, where much of the conflict over net neutrality has focused on the classification of access to internet services under the Communications Act. Access to internet services were traditionally classified under Title I "information services." Consequently, they were exempted from the so called "common carrier" obligations to serve everyone under non-discriminatory conditions, defined in Tittle II "common carrier services," which applied to fixed telephony. However, in 2005 the Federal Communications Commission (FCC) adopted the so-called "Internet Policy Statement," outlining principles on how internet service providers were supposed to provide their services. In August 2008, when the FCC adopted a decision based on the Internet Policy Statement forcing Comcast not to block peer-to-peer connections, Comcast challenged it before the US Court of Appeals for the District of Columbia. In April 2010 the Court ruled that the FCC did not have the authority to regulate network management practices and vacated the order. In December 2010, the FCC adopted the Open Internet Order, imposing transparency, no blocking, and no unreasonable discrimination obligations on ISPs. The US Court of Appeals again struck down the main content of the order in 2014, as it was considered that the non-discrimination rules and the anti-blocking rules were common carrier obligations that could not be imposed on services under Title I "information services."

The FCC insisted by adopting the 2015 Open Internet Order, which reclassified access to internet services as Title II "common carrier services," imposing bans on blocking, throttling, and paid prioritization by ISPs. In June 2016, the US Court of Appeals for the DC Circuit upheld the legality of the order. A decade of debates and litigations was to conclude with a victory for net neutrality supporters.

However, in 2017 the FCC reversed the 2015 Open Internet Order and reclassified access to internet services as Title I "information services." Despite the legislative initiatives, the classification has not been modified in Congress.

Even more importantly, the abrogation of the 2015 Open Internet Order in 2017 did not lead to a major change in ISPs' behavior. There have been no major actions by platforms against anticompetitive practices by ISPs. The apocalyptic scenario described by the net neutrality supporters has not materialized, despite the abrogation of the net neutrality rules. The bitterness in the net neutrality debate has gradually diminished.

In the EU, legislation was only adopted in 2015, in the form of a Regulation to be applied in all the Member States.[5] The main obligation is defined in Article 3(3): "Providers of internet access services shall treat all traffic equally, when providing internet access services, without discrimination, restriction or interference, and irrespective of the sender and receiver, the content accessed or distributed, the applications or services used or provided, or the terminal equipment used." Reasonable traffic management is allowed, as long as it is transparent, non-discriminatory, and proportionate, and based on technical requirements, not in commercial considerations.

The experience over the past years is that the net neutrality rules have not been widely applied, as restrictive practices were actually not widespread. Also, in the EU the net neutrality debate has been reduced mostly to technicalities.[6] Some platforms have recognized the need to have some kind of network management (Google), while others (such as Netflix) have even signed specific agreements with telecommunications carriers to benefit from such management.

It is important to note how platforms understood from an early stage that they would benefit not only from securing access to the telecoms infrastructure, but also from imposing the conditions for the use of the infrastructure, thus restricting the ability to manage the technical conditions of the services and the pricing strategy of the communications carriers.

In the early 2000s, when the net neutrality debate sparked, telecommunications carriers were on the powerful side in the relation with platforms. Former monopolies had a strong market position and enjoyed healthy margins. New services like fixed and mobile broadband ensured the future of revenue growth. Platforms, on the other hand, were still in their formative years and many of the largest platforms did not exist yet. Even the most successful platforms at the time had a limited revenue as their business model based on advertising was in its infancy. Google, for instance, had an annual revenue of $1.4 billion in 2003, while Verizon reached $67 billion.

Twenty years later, however, the balance of power has shifted. Platforms have reached the revenue levels of the communications carriers, as their network effects have matured. Platforms' margins are healthier than those of carriers. Carriers face stronger competition from other carriers, and communications platforms have eroded a significant portion of the value of the industry.

In the new power equilibrium, regulating the conditions under which platforms can use the telecommunications infrastructures seems an unnecessary protection for the more powerful actors. Furthermore, it could reduce the availability of funding for the necessary maintenance and expansion of the communications infrastructure. On the contrary, calls have emerged, particularly in Europe, for "platform neutrality" rules.

10.4 Platform Neutrality

Platforms have become gatekeepers in a number of communications industries. Facebook and Google are the gatekeepers for audience access to the publishers' content through searches newsfeeds, and so on. Google is the gatekeeper for the publishers' access to advertisers. Google and Apple are the gatekeepers for app developers' access to their customers. There are increasing calls to regulate such gatekeeping positions.

All intermediation activities are prone to conflicts of interest, suspicion, and abuse. There is always suspicion that the intermediary is unduly benefiting the other side of the market. US advertisers were outraged when they discovered that ad agencies were receiving commission from publishers. There is the suspicion that the intermediary is benefiting competitors sending them more business, for instance because they pay a higher commission. An intermediary that works for more than one party and, at the same time works for supply and demand, has a fundamental conflict of interest that must be managed in the most transparent way.

Platforms are no different. As intermediaries active in many different markets, they are prone to the same conflicts of interest. Intermediated companies fear discrimination and being excluded from the market. The large scale of the intermediation, with millions of intermediated parties, increases the relevance of the conflicts of interest. The use of algorithms to automatize intermediation reduces transparency. Intermediated companies do not understand the logic behind the algorithmic decision-making and the sudden changes introduced to it, which can exclude companies from the market and even ruin them in a matter of hours.

Furthermore, there is always the suspicion that intermediaries tend to reinforce their position, provide the service for their own benefit, and extract excessive value out of the market. That has been the case, for instance, of excessive charges in payment cards, as described in Chapter 2. There is now an outcry from app developers, such as Spotify and Epic Games (producers of the game Fortnite), regarding the 30 percent fee charged by Apple's app store, as described in Chapter 8. This is also the case with content producers such as newspapers, music record labels, and book publishers in relation to the platforms intermediating their services (Chapter 9). And it was certainly the case of freelancers, musicians, and writers against the traditional intermediaries. This is increasingly happening with large media platforms as well. Content producers complain about value appropriation by the platforms. The high margins of Facebook and Google demonstrate that they are not sharing a fair portion of their profits with the intermediated parties: YouTubers and professional content producers.

Furthermore, the largest platforms have grown to achieve such market power that they are in a position to drive the market in the direction of their choice, defining the type of content they prefer at each point of time. Platforms originally preferred user-generated content over professionally produced content, such as family pictures over news in Facebook's newsfeed, or amateur videos or professional music videos in YouTube. This type of content was cheaper and often even free for the platforms. As large advertisers grew wary of user-generated content, fake news, and so on, platforms evolved to prefer safer professional content. It is the interest of the platform to determine whether one type of content is privileged and "goes viral" over another. Content producers are totally dependent on platforms' changes in strategy.

The concern is particularly acute when platforms vertically integrate and intermediate their own services provided in competition with other companies. This is becoming an issue around all the large platforms. The European Commission imposed a fine on Google for abusing its dominant position in the search market for self-preferencing the Google Shopping service against comparison services provided by competitors (Chapter 3). Similar claims have been made by other companies providing specialized search services, such as travel (Chapter 12). Apple has been accused of self-preferencing its own music service in the management of the app store (Chapter 8), while Amazon has been accused of self-preferencing its own products against third parties in the Amazon Marketplace (Chapter 3).

Platformization might be the unavoidable evolution of the communications industries due to the power of network effects. Traditional communications might have to adapt to being intermediated by digital platforms. In any case, it seems clear that digital platforms, particularly those with market power, have a special responsibility to be fair in their relations with the underlying service providers. This is the scope of the intervention of the European Union in Regulation 2019/1150 on promoting fairness and transparency for business users of online intermediation services.[7] We will analyze this Regulation in Chapter 22.

10.5 General Interest

Platformization is affecting reliability and other public values such as pluralism, public debate in democratic societies, equity, affordability, and universality. The traditional content editors (newspapers, radio and TV stations) in charge of ensuring all these values are no longer in a position to act as gatekeepers, as they are losing that status in favor of digital platforms, the new coordinators of the media markets.

Traditional players are struggling to compete with new network media that product media specifically to be intermediated by platforms and to become viral. Traditional editing activities come at a cost, which makes it challenging to compete with content producers that ignore the relevance of objectivity, fact checking, etc., as they care only about "eyeballs," Furthermore, eye-catching headlines, whether true or fake, spread faster than rigorous description of reality. Finally, as news pieces and content in general is being unbundled, traditional editors are no longer in a position to ensure a balanced and comprehensive perspective.

Digital platforms have so far refused to assume editorial responsibility. They claim that they are mere intermediaries, with no liability for the content they display. However, it is becoming clear that platforms play an important role in the new media ecosystem. As intermediaries, platforms are not responsible for the creation of content generated by third parties, but they are responsible for selecting and displaying content in their platforms and even more so for actively feeding the news to their users

Firstly, platforms are responsible for the criteria used for the selection of content to be displayed to users. It has become increasingly clear that all platforms moderate.[8] Algorithms used to connect content and eyeballs are not neutral. They are designed to meet the objectives of the platform manager; that is, profit. Platforms design self-learning algorithms using artificial intelligence to reach specific objectives, and they control the results they deliver.

Platforms set objectives based on their business incentives. As the current business model of the leading platforms is based on advertising, the objective defined for the algorithms would be to maximize the time spent by eyeballs in the platform, given that more ads can be displayed the longer the attention is retained. Initially, platforms did not know whether attention would be attracted through user-generated videos of babies and kitten, through news pieces from traditional media, or through clickbait content produced cheaply by new network media. The more time devoted to it, the higher it would get in the priority of the algorithm.

Traditional media has fiercely counterattacked, raising doubts about the role of the platforms. Doubts are being raised about the reliability of the content promoted by the platforms (fake news), about the impact of content in the democratic process, about the liability of the platforms for content they promote (hate content), and about the benefit of advertisers to see their brand connected with such content. Public authorities are increasingly concerned with the role of the platforms in the media industry, particularly in Europe, where initiatives against fake news and hate speech have been undertaken both at the EU and national level. The European Commission adopted a Code of Practice on Disinformation in 2018.

Platforms are reacting. More resources are being used for the curation of content in all major platforms, particularly in Google/YouTube and Facebook. Algorithms are starting to consider values other than monetizing attention. Human resources are being deployed to take more nuanced decisions. Certain types of content are being radically excluded from the major platforms.

The ultimate challenge, however, is whether platforms will be in a position to promote a sustainable ecosystem that enhances values of general interest, such as diversity in audio-visual content, reliability of informative content, and participation in a transparent and balanced public debate, necessary for the democratic process.

The lesson that can be extracted from recent history is that platformization leads to commodification of content. High-quality and reliable content, which is expensive to produce, might not be viable under the new multi-sided, advertising-funded market structure. Given the current algorithms, high-cost quality content cannot compete with low-cost content, attracting at least as much attention and the same or even more advertising revenue.

Platforms do not seem to have the economic incentive nor the legal obligation to ensure public values such as reliability, diversity, or pluralism in the public debate that they increasingly manage and shape. There are growing calls for public intervention to introduce public values into the equation; that is, in the case of platforms, into the algorithms.

As the leading platforms become the gatekeepers and coordinators of the media ecosystem, the public debate should increasingly focus on how to ensure that content has fair access to the platforms. It may be time to guarantee content producers fair, transparent, and non-discriminatory access to the platforms. We think that corresponding regulation is needed.

Notes

1   Starr, P. (2005). *The Creation of Media: Political Origins of Modern Communications*. Basic Books, New York.
2   Bárd, P., & Bayer, J. (2016). *A Comparative Analysis of Media Freedom and Pluralism in the EU Member States*. Study for the LIBE Committee of the European Parliament, retrieved from www.europarl.europa.eu/thinktank/en/document.html?reference=IPOL_STU%282016%29571376
3   Article 36 Act 7/2010, on Audio-visual Communication.
4   Wu, T. (2003). Network neutrality, broadband discrimination. *Journal of Telecommunications and High Technology* Law, 2, 141, retrieved from SSRN: https://ssrn.com/abstract=388863.
5   Regulation (EU) 2015/2120 of the European Parliament and of the Council of 25 November 2015 laying down measures concerning open internet access and amending Directive 2002/22/EC on universal service and users' rights relating to electronic communications networks and services and Regulation (EU) No 531/2012 on roaming on public mobile communications networks within the Union, OJ 2015 L 310, 26.11.2015, p. 1.

6 BEREC, *Report on the implementation of Regulation (EU) 2015/2120 and BEREC Net neutrality Guidelines*, BoR (19)117, October 2019.

7 Regulation (EU) 2019/1150 of 20 June 2019 on promoting fairness and transparency for business users of online intermediation services, OJ 2019 L 186, 11.7.2019.

8 Gillespie, T. (2018). *Custodians of the Internet. Platforms, Content Moderation, and the Hidden Decision That Shape Social Media.* Yale University Press, New Haven, CT.

Part III

Platforms in the Transport Industries

Transport provides vivid examples of the transformation of the network industries, as digital platforms build a new network – a digital network – on top of the traditional transport infrastructures and services.

Over the past two centuries, most transport modes have evolved into network industries by connecting originally scattered assets and exploiting them as coordinated large-scale networks. Highly efficient networks have emerged in maritime transport, with huge container ships connecting a global network of mega-ports, aviation, with hub-and spoke intercontinental networks, and railways, regional, long-distance, and high-speed. However, transport modes still work mostly in silos, with little coordination between them. Coordination is left to passengers and shippers, with the support of expensive and often unreliable intermediaries.

Platforms have the potential to create network effects inside a transport mode by facilitating the interaction between a fragmented supply and an even more fragmented demand. In aviation, platforms such as the Sabre and Amadeus booking systems provide good examples, both of which are discussed in greater depth below.

Furthermore, platforms can also build new networks in previously fragmented markets such as road transportation. This is the case of Uber in urban mobility and BlaBlaCar in long-distance services.

The ultimate frontier for transport platforms is to transform all the transport industries into a multi-sided market, a network of networks with digital platforms coordinating transport modes to provide a seamless door-to-door experience for passengers and shippers. Massive new direct, indirect, and algorithmic network effects enabled by platforms are indeed transforming transportation.

Consequently, traditional transport providers have started to be disrupted. Some are being substituted, sometimes in a quite drastic way (such as taxis), sometimes at a slower pace (railways, for instances). Nevertheless, the most transformative disruption is the more subtle evolution of transport into a global multi-sided market. Traditional services become mere commodities, intermediated by platforms. Traditional players are increasingly working for an invisible algorithm that determines their fate.

The regulatory implications of such transformation are evident. Transport is a general interest service. Over decades, a convoluted mesh of rules and procedures has been built to protect the general interest, but existing regulations were not designed for the new market structure. The regulations sometimes obstruct the efficiencies generated by platforms and sometimes benefit platforms, as legacy players are artificially fragmented. However, the existing framework cannot meet the traditional general interest objectives in a completely new environment.

Chapter 11

Scale and Networks in Transportation

Transportation has always been a highly fragmented industry. Different transport modes compete for the transport of passengers and goods, and there are even various degrees of fragmentation within each transport mode. Fragmentation is extreme in road transport (taxis and trucks, for example), but it also exists in maritime transport, aviation, and railways, as there is a high number of operators, usually limited in their geographical scope or market segment.

Over the past two centuries, efficiency gains have been pursued in each transport mode, mostly by increasing scale on the supply side: larger vehicles (larger vessels, airplanes, and trains) and larger infrastructures to support such vehicles (larger ports and airports). Scale has been built and managed by ever larger operators (airlines, shipping companies, railway undertakings), which have assumed the role of coordinating the service in terms of routes, schedules, price structures, quality, and so on, all for the benefit of passengers and shippers. Monopolies are the ultimate example of such a strategy to build scale at the supply side. Globalization took scale to the next level. Economies of scale have triggered cost reductions that have attracted more traffic, igniting a virtuous cycle that transformed economies and societies.

However, this trend might well be reaching its limits. Larger vessels are creating operative problems in ports and do not provide the necessary flexibility to face downturns in demand. The largest airplanes, such as the Airbus 380, have proven to be commercial failures. Liberalization put an end to monopolies and introduced new horizontal and vertical fragmentation in the different transport industries. At the end of the day, scaling-up in each transport mode has not solved the structural challenge of coordinating the different transport modes and providing door-to-door solutions, both for passengers and shippers.

Digitalization is providing an alternative model to build scale and improve coordination. Platforms are building scale on the demand side by pooling together large numbers of passengers and shippers. They are also building scale at the supply side, but not through the integration of activities under a single hierarchical structure integrating all assets. Platforms merely aggregate the supply provided by other parties, mostly traditional transport service providers. Finally, platforms facilitate the interaction of demand and supply at a massive scale with their sophisticated algorithms. Platforms can coordinate different transport suppliers in different transport modes into multi-sided markets, facilitating multimodal door-to-door solutions for passengers and shippers. The new and powerful network effects are triggering new virtuous cycles with a transformative power that has equaled the historic transformation of transport over the last two centuries.

11.1 The Adventure of Traveling

In 1873, the French writer Jules Verne published *Around the World in Eighty Days*. The main character, Phileas Fogg, circumnavigated the world in 79 days, making use of the most popular transport modes in his time: railways and steamers (also elephants for a short ride). Such an impressive mark was the result of investment in infrastructures: the Transcontinental Railroad in the US was completed in 1869, and the Suez Canal was inaugurated in 1870. In any case, traveling around the world required frequent changes in transport mode, different service providers, uncertainty with bookings, and the spirit of adventure that made Verne's novel so popular.

In the twenty-first century, travelers must still navigate different uncoordinated transport modes (aircrafts, railways, private cars, etc.), provided by different companies, under different conditions and passenger rights schemes, with different booking and contracting methods, thanks to the intervention of expensive intermediaries and the like. Even urban mobility is a mesh of uncoordinated transport modes. The challenge of providing seamless door-to-door transport, either for passengers or for goods, remains unmet.

All transport modes have evolved over time, albeit to different degrees, to reach more efficient results in terms of capacity, speed, reliability, safety, environmental performance, and, in particular, costs. Such evolutions have taken different forms in each transport mode, according to each mode's own characteristics, and sometimes convoluted traditions.

Scale has been the process by which efficiency has been increased in many transport modes. This strategy has proven successful in the densest routes. For example, maritime transport between large ports has become more efficient thanks to large container ships, and railways provide efficient mass-transit solutions, both urban and long-distance. Network effects have been relevant to feed larger and larger infrastructure and service providers.

However, scale has its physical limits: for example, large ships cannot use the Panama channel, the A380 cannot land at many airports, and most roads cannot handle very large trucks. Furthermore, scale in supply has not brought efficiencies on routes with scarce demand, which is the case of first/last mile transportation.

Despite the amount of evolution since Phileas Fogg circumnavigated the world in 1873, passengers (or shippers) still have to coordinate different transport modes, supplied by different transport service providers. The alternative is to contract the coordination of the necessary services with expensive intermediaries (travel agents, forwarders). There is an unmet demand for a seamless door-to-door transport experience.

11.2 Scale Up!

A fine example of scaling up in transportation can be found in maritime transport and the history of Maersk, the market leader. Captain Peter Mærsk-Møller bought his first steamship, the SS *Laura*, in 1886, pioneering steam shipping in Denmark. Forty years later, under the leadership of one of his sons, the Maersk corporation launched its first line service, between the Far East and the American West Coast. The Maersk family led the transformation of maritime transport from tramp services into line shipping, the provision of transport services along prearranged routes and schedules. Maersk led the maritime industry to become a network industry, with maritime companies accumulating cargo from different shippers thanks to standardized services connecting trade hubs (ports).

Maersk also ended up leading the containerization of general cargo. General cargo is increasingly transported in standardized shipping containers. As shipping became a standardized activity, scale became the key competitive advantage. In 1996, Maersk built ships with a capacity of 6,000 containers. Today, more than 90 percent of the non-bulk cargo worldwide is transported in container ships. The largest container ships can carry more than 20,000 containers. Economies of scale are utterly relevant, as costs are dramatically reduced with larger vessels: a ship with 10,000 containers uses half as much fuel per container as a ship with 3,000 containers, and the size of the crew is similar; therefore, total costs can be 30 percent lower.[1]

Larger ships required concentration. Maersk initiated an acquisitions spree around the turn of the century in a race to build scale. In parallel, Maersk started to operate terminal facilities in the reduced number of larger container ports across the world ready to operate such large vessels. As of 2020, more than 50 percent of the world's liner fleet market share was controlled by four shipping companies. Maersk's strategy to build scale has been successful; it is the largest shipping company in the world, with a market share of more than 15 percent. Maritime services are a good example of the potential of economies of scale to increase efficiency and rationalize transport operations.

Economies of scale have defined industrial trends over the past two centuries. Industrialization transformed economic activity in the nineteenth and twentieth centuries with larger production centers to benefit from economies of scale. National, and later global, markets allowed the concentration of activity in even larger corporations, exhausting all the potential economies of scale.

Transport companies have sought to keep up with the overall trend in industry by scaling up. Economies of scale reduced cost and generated concentration in the provision of transport services. Concentration has made coordination simpler and more efficient, as a single corporation has internalized the coordination of an increasingly larger set of services, even at a global scale.

11.3 Transport networks

Railways provide a good example of how network effects play in infrastructure industries. It is not by chance that transport, as well as telecoms and energy, are considered "network industries."

National railway networks have evolved from the early isolated, non-standardized railway lines.[2] Standardization was the precondition for the creation of a national railway network. An early example was the so-called "battle of the gauges" in the United Kingdom. The Liverpool and Manchester Railway was built with a narrow gauge of 1, 435 mm, while the Great Western Railway engineered by Lord Brummel was built with a "broad gauge" of 2, 134 mm. Gloucester was the first town at which a narrow gauge and a broad gauge converged, and the break of gauge obviously hindered train traffic flow from one line to the another: passengers and goods had to transfer from one train to the other. Finally, "an Act for regulating the gauge of railways" was enacted by Parliament in 1846, imposing the "narrow gauge" as a standard. The "narrow gauge" would become the international standard – but not everywhere.

Another example was the standardization of time measurement. In 1844, the clock of the cathedral in Exeter (England) was still set according to the sun, as it had been done for centuries. As Jules Verne's Phileas Fogg had experienced when circumnavigating the world, accelerated traveling speed challenged assumptions about time. Given that Exeter was a few hundred kilometers west of London, the time set in the clock in the cathedral showed a ten-minute difference from the clock in Greenwich Royal Observatory, which set the time in London. For centuries, no one realized or cared. However, as railways started to connect London and Exeter with prearranged line services, it was necessary to set a common time. This was not a minor issue: poor timing coordination actually led to collisions resulting in deaths. For a while, local times coexisted with "railway time" and clocks often displayed three hands: two would display the local hour and minute, and a third one would display the "railway time." Eventually, time in all of England and around the world followed the time set by the Greenwich Royal Observatory, known as Greenwich Mean Time (GMT).

Railways brought standardization and rationalization not only on gauge and timing, but in general to the overall management of transportation. Railways require heavy investments in dedicated infrastructure (the tracks). Not only railway infrastructure has a high cost. It is also a sunk cost, as infrastructure cannot be used for other purposes in case of market exit. Railway infrastructure is highly standardized. Rigid rules define all kinds of issues such as track gauge, standards for electrification, signaling, maximum speeds, maximum freight supported by tracks, rolling stock operable over the tracks, training of the drivers, and so on. The operation of the system is highly rationalized, as services are usually organized as prearranged regular services.

Railway infrastructure dramatically increased transport capacity against traditional transport modes, creating new economies of scale. A high number of passengers and a high volume of freight could be transported along railway infrastructure. Only high volumes of passengers and freight would pay back the high cost of building the infrastructure. At the same time, however, scale was the competitive advantage, as the concentration of passengers and freight in a single infrastructure would dramatically reduce the cost of transportation for each single passenger or piece of cargo. Railways brought economies of scale in land transportation. As prices were reduced, new demand for transport was created, initiating a virtuous cycle that transformed the world.

Network effects increase when infrastructure is arranged with a hub-and-spoke structure. Infrastructure supports the transport of passengers and goods to a central hub, where the flow is sorted and redirected to its final destination. This is the classical organization of intercontinental air transport, as passengers fly in small planes from their cities to the hub airport where they can take a larger airplane to travel to a paired hub airport in another continent. This model also works for maritime transport. In this way, it is not necessary to build dedicated infrastructure for point-to-point transportation; the maximum economies of scale are reached in the infrastructure, and the maximum positive network effects are reached.

Infrastructure in transport, but also in telecoms, energy, and water, creates network effects, which is why they were referred to as "network industries." The number of users of an infrastructure has an impact on each of the users. It will usually be a positive effect, as fixed costs can be distributed across a larger pool of users. However, it can also have negative effects, as the capacity of the infrastructure can become exhausted and congestion can damage the user's interest, as early morning commuters can confirm.

Some common traits can be derived from infrastructures analyzed as networks. Firstly, the network effects that define infrastructures are usually of the so-called direct type. All users of a transport infrastructure are basically making the same use of the infrastructure and share the same type of impact derived from the use of the infrastructure by other users, either positive (cost reduction) or negative (congestion). Secondly, network effects require investment in the infrastructure, which is usually a risky investment, as it requires large volumes of capital to be deployed before the entry into operations of the infrastructure, it is a sunk cost, and it requires a long period (sometimes even decades) to be recovered in the form of small recurring payments for the use of the infrastructure. Thirdly, the manager of the infrastructure becomes the market coordinator, as he or she has the ability to define routes, frequencies, schedules, type of vehicles, and so on. The management of the network effects will largely determine the success of the operation of the infrastructure. The network manager must incentivize the use of the infrastructure by the largest number of users, which is obtained by sharing economies of scale with users in the form of low prices. However, prices cannot be too low, as they may attract too many users, thus creating congestion. The creation of network effects and, subsequently, the appropriate distribution of the benefits derived from such effects, are the key roles of a network manager.

Network industries, as front runners of industrialization, were built on economies of scale on the supply side. Transport is no exception. Infrastructure generated large capacity for transportation, even if at a large sunk cost. The larger the pool of users of the infrastructure, the lower the cost for each of them, in an evident network effect. The lower the cost of the transport service, the larger the demand, triggering a virtuous cycle that reinforced the infrastructure manager. Infrastructure managers became the organizers and coordinators of the system. As the twentieth century progressed, large infrastructure managers succeeded, often with monopoly rights: railway monopolists, large hub ports supporting large shipping companies, large hub airports supporting hub-and-spoke air companies, large integrated urban transportation systems and so on.

11.4 Legal Monopolies

Legal monopolies are common in transport infrastructures. With network effects being so obvious, it was only logical that the maximum positive network effects would be reached if all passenger and freight would use the same infrastructure. The fixed cost of the infrastructure would be distributed among the largest possible number of users. Granting a legal monopoly for the exploitation of the infrastructure would ensure that everyone uses the infrastructure to the maximum efficiency in terms of economies of scale and network effects. Transport infrastructure was often considered to be a natural monopoly

In most countries, public authorities took the lead in the deployment and exploitation of transport infrastructures. At that time, only the state was in a position to mobilize such unprecedented volumes of capital. Once established, the exploitation by the public authorities of the infrastructure, and often of the transport services over such infrastructure, was supposed to ensure fairness in the provision of the service. This was the case of railways, large ports, airports and airlines, etc.

In a few countries, capitalistic structures such as corporations, trusts, and large banks were created and developed specifically to meet the challenge of the financing of transport infrastructures. Railroads triggered many of these developments.[3] Competing infrastructures were deployed, such as parallel railroads in the US (just as competing telecommunications and electricity infrastructures). In the nineteenth and the twentieth centuries, financiers such as J. P. Morgan built fortunes by rationalizing competing infrastructures into single corporations, or at least cartels, in order to reduce competition, increase scale, and ensure the recovery of investments (plus a good margin). This was the case of railroads across the US, but also of the AT&T's telecommunications network and Edison's electricity network.[4]

All around the world, transport infrastructures ended up being exploited as monopolies. This was the case of railways, roads, ports, airports, and so on. In most cases the monopolies were state-owned, while in other cases, particularly in the US, they would sometimes be private monopolies subject to regulation. Monopolies would facilitate the financing of construction and ensure the full exploitation of the network effects.

11.5 The Limits of Supply Side Economies of Scale

After two centuries of scaling up, both through market mechanisms and regulatory intervention, supply-side economies of scale might have reached their limits. Scale in transport has limitations that are being confronted by large market players. For instance, growth in container ships seems to be stopping. The largest ships can only operate in the largest ports and cannot use certain routes, such as the Panama or the Suez canals. Furthermore, such large ships are difficult to operate and prone to accidents in ports. In aviation, scale also has its limits: the Airbus A380, with capacity for up to 868 passengers, has turned out to be a failure.

Furthermore, economies of scale are not present in all transport modes. For example, road transport offers limited possibilities in terms of larger vehicles, as road standards do not allow very large trucks. This is why the truck industry is highly fragmented, even in large and mature markets such as the US or Europe. Trucks represent more than two-thirds of the value of freight that is moved within the US – twice as much as rail, air, waterways, and pipes combined[5]–but truck operations are highly fragmented, with thousands of active companies. The most obvious example of fragmentation is the taxi industry. Thousands of independent taxicabs circulate within the world's large cities. No coordination exists. Taxi drivers are not aware of the location of potential passengers, so they cruise the city looking for potential passengers who might wave their arm to hail a taxi. Hours of empty runs are wasted, unnecessarily consuming fuel, while increasing the cost of operations. Passengers are frustrated, particularly at peak times (when raining or after a show or sport event) with long waiting times and uncertainty.

Regulatory intervention in the form of legal monopolies, the guarantee to exhaust economies of scale, has proven to have downsides. Monopolies, either state-owned monopolies or privately regulated monopolies, have little incentive to curb costs, to innovate in the provision of services, or to improve efficiency generally. Over the past decades, economists have challenged the assumption that all transport infrastructures and activities are natural monopolies.

Competition has proven to be possible and sustainable in transport modes that were monopolized for decades. This has been the case in air transportation. New airlines have entered the market, which have been stricter with costs and have innovated thanks to new business models, such as point-to-point low cost services, to the benefit of passengers. The industry has been horizontally fragmented, as newcomers have been allowed to enter the market and more companies could provide services.

In other transport modes, natural monopolies have come to be accepted in certain segments, which leaves room for competition in other segments of the industry. This seems to be the case in railways. Regulators in the European Union have accepted national monopolies for the management of the railway infrastructure, but they are opening the provision of transport services over such an infrastructure to competition, both in passengers and freight transport. The industry has been vertically fragmented, as separation of infrastructure management and transport service provision has been mandated.[6]

Overall, regulatory intervention during the past decades has fragmented the transport industries, both horizontally and vertically, and scale is no longer a goal in itself. Regulation has sought to build a more nuanced combination of efficiencies derived from scale with efficiencies derived from competition.

Finally, a more structural limitation of supply side economies of scale can be identified. Network effects emerge in dense routes that pool together large volumes of passengers and goods. However, first/last mile transport rarely generates enough volume to create supply economies of scale. Scale creates benefits when, for example, passengers traveling between large cities pool together in a high-speed train with more than 500 seats, but not when moving from a home in the suburbs to the local train station.

Flexibility is required to meet the requests of first/last mile transport. Road transportation is often necessary to reach hubs. Passengers often need a dedicated vehicle (usually their own car) to reach a high-speed train station (or even the local train station to commute to the high-speed station) or airport. The same problem exists for freight transportation. Large container ships or large railway freight trains can be efficient for the transport of large volumes of goods, but when goods have to be picked up in the factory or delivered to the commercial center (or, even worse, to the purchaser's home in the case of e-commerce), smaller trucks are necessary, as volumes do not justify large-scale transportation.

Scale on the supply side does not provide a solution for the coordination of first/last mile transport with large-scale transport in dense segments. This is the main limitation of transportation in the twenty-first century. Large transport providers are focused on improving the efficiency of their operations, but they only cover a segment of the overall transport needs. Concentration of scale in hubs creates a structural problem of congestion around the hub. It is a challenge to park the car in a high-speed train station, or even in a local commuting train station, not to mention an airport. It is a challenge for a truck to get in and out of a large port. The transition from first/last mile transport to backbone transport, particularly if intermodal, is slow, expensive, unreliable, and uncomfortable. It is often an adventure, just as it was for Phileas Fogg.

Coordination and inter-modality are so poor that users often avoid it all together. Once passengers have their own car, they do not merely use it to reach the hub (the train station), but they often use it to travel door to door. They use it for daily commutes, creating congestion in urban areas. They use it for long-distance trips, raising the cost of transportation and damaging the environment. The same occurs with freight transportation in medium distances. Once freight is loaded in a truck, it might simply head to its final destination rather than to the next charging station for transshipment.

The limits of our current transport system, particular the limits of scale in the supply of transport services, might well have been reached. The transport system, as it is designed, neither meets the passengers' nor the shippers' demand of door-to-door seamless transportation.

11.6 From Supply to Demand Side Economies of Scale

Digitalization offers an opportunity to transform transport by creating new network effects, reducing costs, and igniting new virtuous cycles. New network effects are not triggered by further scaling up the suppliers of transport services. On the contrary, they are being triggered by the smartest aggregation of the demand side, in order to better interact with supply.

Digitalization and platforms in multi-sided markets are a game changer. Platforms have the potential to overcome some of the traditional limitations of transport infrastructure, as they generate new network effects. Such effects do not derive from scale on the supply side – these are pretty much exhausted – and instead derive from scale on the demand side. Pooling large groups of users, and coordinating them thanks to technology with transport service providers (traditional large service providers, traditional fragmented service providers, and even new small non-professional "sharing economy providers") creates new cost reductions and new virtuous cycles.

Online platforms can generate network effects in transport activities that were traditionally too fragmented to reach scale. This explains the success of platforms in land transportation such as Uber and BlaBlaCar. These platforms pool large numbers of independent drivers with even larger numbers of potential passengers. Platforms then coordinate the interaction between drivers and passengers. Benefits are derived from scale. However, such benefits do not derive from the concentration of scale in the supply side.

The benefits do not derive from building the largest concentrated and coordinated transport operation. Platforms do not invest in deploying infrastructure or in building the largest vehicles or the largest fleet of vehicles. On the contrary, benefits derive from the ability to coordinate previously fragmented and isolated service providers. Such coordination is possible thanks to the evolution of both technological hardware (internet and smartphones) and software (algorithms/artificial intelligence). Investment is necessary to grow the large pools of users (both drivers and passengers) and to develop especially the software technology in order to efficiently coordinate the interactions among them. However, the investment required is much lower than it is to deploy infrastructures.

Online platforms can also generate new network effects by coordinating the services of the traditional infrastructure network managers, in such a way that a network of networks is being made available for passengers. This is the model behind the concept of "Mobility-as-a-service" (MaaS) and platforms such as Whim.

Traditional transport network industries can become just one side in a multi-sided market established by a digital platform. Online platforms can aggregate a pool of transport service providers, both traditional network operators and traditional independent service providers, and even new service providers under the umbrella of another platform such as Uber, and efficiently empower the interaction of such service providers with large pools of passengers. The coordination role of the platform can generate efficiencies by better coordinating the available transport modes. Intermodality can be improved. Coordination can also ease congestion, thus reducing pollution, as flows of passengers can dynamically be distributed across transport modes.

New network effects derive from growing scale on the demand side, pooling together passengers and distributing them more efficiently across the transport infrastructures as well as across the available transport services providers.

Notes

1 Levison, M. (2016). *The Box. How the Shipping Container Made the World Smaller and the World Economy Bigger*. Princeton University Press, Princeton and Oxford, p. 369.
2 Finger, M. & Montero, J. (2020). *Handbook on Railway Regulation: Concepts and Practice*. Edward Elgar, Cheltenham.
3 Eli, J. W. (2001). *Railroads & American Law*. University Press of Kansas, Lawrence described how railroads triggered the creation of many of the institutions of capitalist societies.
4 Strouse, J. (2000). *Morgan. American Financier*. Perennial, New York.
5 US Department of Transportation, Bureau of Transportation Statistics and Federal Highway Administration, Freight Analysis Framework, version 4.5, 2019, retrieved from https://data.bts.gov/stories/s/Moving-Goods-in-the-United-States/bcyt-rqmu
6 Finger, M. & Montero, J. (2020). *Handbook on Railway Regulation: Concepts and Practice*. Edward Elgar, Cheltenham.

Chapter 12

Amadeus, Sabre, and Air Transport

An early precedent of a digital platform can be identified in the aviation industry in the form of two global distribution systems: Amadeus and Sabre. In the 1980s, both started using digital technologies to match supply and demand in the aviation industry, facilitating interactions between airlines and travel agents for the booking of flights. They later expanded to rail tickets, hotels, and so on at the supply side, as well as online travel agencies in the demand side. Amadeus, the global market leader, is currently one of the largest digital platforms in Europe.

As these platforms have been active for more than 30 years, they provide a good example of the kind of conflicts that exist around digital platforms. Intermediaries, particularly if they grow to participate in a large share of transactions in a market, are prone to raise suspicion among the intermediated parties, calling for maximum transparency as well as for fair relationships with the ecosystem around them.

Amadeus and Sabre are also particularly interesting because, even in the 1980s, they were subject to the first digital platform regulation we have identified, both in the US and in Europe. The regulation, after review in 2009, remains in place in Europe and it advanced many of the regulatory solutions under debate today for the larger digital platforms: fairness in platform to business relations, non-discrimination, search neutrality, multi-homing, and liability.

12.1 Global Distribution Systems

We have been told for years that the internet is disintermediating human transactions. However, if a young and ambitious engineer based in Minsk (Belarus) wants to fly tomorrow to San Francisco to launch a start-up, it is probable he will use more intermediaries than he is aware of.

If our engineer tries to find the best deal on the web or on an app of the Belarus national airline, he will be disappointed to find out that Belavia, the country's national airline, does not fly either to San Francisco or to US. If he tries American Airlines, and then Delta and finally Continental Airlines, he will find out that the US companies have no direct flight from Belarus to the US. They do not even provide the option of connecting two of their flights to complete the trip. He should not be surprised. Aviation is a highly fragmented industry. Even the largest airline in the world, American Airlines, does not carry more than 5 percent of the total annual passengers in the world.

Our internet-savvy engineer will probably use an online travel agency such as Expedia, or being European, eDreams, to identify that the best traveling option is provided by Turkish Airlines, with a stopover in Istanbul. If he decides to buy the ticket, he will be already using an intermediary, but will also, inadvertently, be using a second one, probably Amadeus. Online travel agencies use Amadeus to gain access to travel data, to compare fares, and to book tickets. Two intermediaries, not bad for a disintermediated digital world! If the engineer decided to buy his ticket in a traditional travel agency, the agent would probably also use Amadeus to identify the best travel option and book the ticket.

Amadeus, based in Spain, is the global leader (with a 40 percent market share) among a small group of digital platforms called global distribution systems (GDS) that intermediate in multi-sided markets. On the supply side, airlines provide full details about their flights and the available fares and allow them to be booked through the platform. On the demand side are both online travel agents and traditional agents, as well as corporations that directly manage their travel needs. Individual users usually do not interact directly with Amadeus, which is why our Belarusian engineer has probably never heard of Amadeus, even though he relies on the services provided by the platform in most of his traveling.

Had our engineer acquired his ticket through a traditional travel agent, the travel agent would have had a contract with Amadeus, having paid a set-up fee and a monthly fee to use the platform. He would introduce in his terminal the request for information to Amadeus about a flight from Mink to San Francisco. The screen would display the most attractive options for the trip. Agents typically do not go beyond the first screen, just as in most Google searches, so Amadeus ranking is key in the final decision as to which airline to contract. The agent and the airline both pay Amadeus a commission for every ticket sold. If our friend buys his ticket at an online travel agent, the process is basically the same: as the user introduces a request at the travel agent's, it is redirected to Amadeus, which provides the relevant information to be displayed by the online agent in his site. As the transaction is completed, Amadeus receives a commission from the online agent and the airline. The difference is that the process is automated and the agent's costs are lower than with a physical agent, so the price to be paid by the user is usually lower.

Amadeus was created in 1987 as a joint venture among four European airlines – Air France, Iberia, Lufthansa, and SAS – to create a European alternative to US-based Sabre. Amadeus and Sabre are the largest GDSs in the world. In a typical European arrangement, Amadeus' headquarters are in Spain, the product development department in France, and the main data center is in Germany.

Amadeus was one of the first platforms in the transport industry. Amadeus displays direct network effects: the more travel agents used it, the more it became the standard, so agents would not need to learn to use other platforms. It displays more relevant indirect network effects: the more airlines use it to feed the supply, the more interesting it is for travel agents, and the more agents use it, the more useful it is for airlines. Over time Amadeus has created also significant algorithmic network effects. The more searches are made in the platform, the more the platform learns about customers' profiles, travel patterns, and so on, and the better the information options that can be provided to users.

What is specifically interesting about Amadeus is how it overcame the "chicken-and-egg" challenge faced by all platforms during their initial stage. When Amadeus was created, US-based Sabre was already well established in the market. Sabre had started operating in 1964 as the first computerized booking system in the world, developed for American Airlines by IBM. SABRE evolved to include not only American Airlines flights, but flights of all the major US carriers and beyond. Amadeus was created by the leading airlines in Europe as a strategy to have more control over the distribution of their own products. They did not want to rely on an intermediary. Collaboration between competitors controlling a large share of supply ensured a critical mass on the supply side. Such a critical mass was more than enough for travel agents, the demand side, to join the platform. This is a strategy that can be exported to other industries where traditional players are afraid of being "platformed."[1] However, vertical integration of suppliers into platforms poses challenges of its own.

12.2 Conflicts in the Aviation Ecosystem

Amadeus has built a complex ecosystem around it: a mesh of 500 airlines, 100 rail operators, 32 insurance companies, 770,000 hotels, and hundreds of thousands of traditional travel agents, as well as online travel agencies, serving all of the world's air travelers. Amadeus rests at the heart of it, intermediating millions of transactions per day.

The role of intermediaries is often misrepresented, when not despised or secretly envied, and Amadeus is no exception. Describing the conflicts in the ecosystem that Amadeus triggered provides an illuminating perspective of the kind of conflicts that intermediaries, and digital intermediaries in particular, raise in most markets.

An early conflict existed, as Sabre complained that the European airlines controlling Amadeus would not provide fare data that was accurate and up-to-date, at least not to the same degree as the data they provided to Amadeus. Because data were so critical for a computer reservation system to compete effectively, the refusal to provide such data would exclude Sabre from the European market. Such a complaint even triggered the first formal referral of the Department of Justice of the US to the Competition Directorate of the European Commission in 1997.[2]

The vertical integration of European airlines as owners of the leading global distribution system was also a source of conflict with other airlines. Exclusion was feared by smaller airlines entering the liberalized aviation market. If they were not allowed to make their services available in Amadeus, they would not be able to effectively enter the market.[3] Furthermore, they feared that once in Amadeus, they could be discriminated in the display of search results. Amadeus could rank the services provided by the parent airlines first and leave for a second or third screen the services provided by competitors. Airlines called for search neutrality.

These concerns triggered a dual response. On one hand, regulation was adopted to ensure fairness and neutrality in the intermediation services of the global distribution systems, both in Europe and in the US. We believe this represents the first example of platform regulation in the transport industry, and we will develop it further in the next section. On the other hand, first Sabre and later Amadeus decided to vertically unbundle: airlines sold their shares and they no longer own either Amadeus or Sabre.

Despite regulation and unbundling, conflicts persist. Not all airlines have the same interest in participating in Amadeus. Airlines provide the possibility to directly book services with them, both for passengers and to travel agents. However, only the largest airlines can count on their direct distribution systems to sell a significant portion of their seats.

They have a large market share, a strong position in their home markets, and customers know they can receive good service directly from them. Large airlines are increasingly confident that they can by-pass Amadeus. Smaller airlines, however, rely on intermediaries for a larger part of their sales and perceive that a weaker indirect distribution channel would threaten their position in the market.

In 2015, German airline Lufthansa introduced a €16 distribution surcharge on tickets sold through Amadeus and other global distribution systems. Travel agents could contract Lufthansa directly, bypassing Amadeus. This seems to be a strategy to reduce the dependency on Amadeus or, in other words, not to be "platformed" by Amadeus. Ryanair, the largest European low-cost airline, went further and terminated its contract with Amadeus in 2017.

In response, in 2018 the trade association uniting global distribution systems and online travel agents filed an antitrust complaint against Lufthansa's surcharge, claiming that Lufthansa was abusing its dominant position in the German and Austrian aviation markets to distort the downstream distribution market. In 2019, the European Travel Agents & Tour Operators Association (ECTAA) filed another antirust complaint before the European Commission for discriminatory practices against IATA, the international organization of airlines, for similar practices.

In parallel, also in 2018, Lufthansa filed an antitrust complaint against Amadeus and Sabre before the European Commission. Lufthansa claimed that Amadeus' and Sabre's agreements with airlines and travel agents restrict the ability to use alternative suppliers of ticket distribution services. This may make it harder for suppliers of new ticket distribution services to enter the market, as well as increase distribution costs for airlines, which are ultimately passed on to the ticket prices paid by consumers.

To complicate things further, other conflicts exist between online travel agents (OTAs) and Google. Our Belarus engineer can directly connect to an online agent's site to search for a quote, but he might just use Google to search for a Minsk–San Francisco flight. The UK antitrust authority has identified that at least 40 percent of traffic to online travel agencies derives from Google (and it is common to have as much as 60 percent).[4] Google's search algorithms traditionally provided online agencies' propositions among the first organic (non-paid) ranked results. In parallel, online agencies pay Google to display their proposals as advertisements. Online agents are among Google's best customers: Expedia spent more than $5 billion on Google advertising in 2019, before COVID-19 hit. Online travel agencies often spend one-third of their revenue on digital advertisement.

In 2019 Google launched Google Travel, its own specialized service to search, compare, and book flights, accommodation, and other travel services. Online agencies complained that the search algorithms were modified to show ads first (as before), then Google Travel results, in a visually rich box reserved for Google Travel, and only later the organic results showing the OTAs proposition. In 2020, Expedia's CEO accused Google of trying to disintermediate them.

A formal antitrust complaint before the European Commission was filed in 2020, accusing Google of self-preferencing in travel search results. Recall that Google was fined €2.4 billion in 2017 for self-preferencing its price comparison service (Google Shopping) when delivering search results. To no one's surprise, online travel agents, fully familiar with the neutrality regulation imposed on searches in Amadeus, have also asked to expand the search neutrality regulation imposed on Amadeus to general searches in Google.

The European Commission had still not decided this mesh of complaints and counter-complaints two years after they were filed.

12.3 The First Platform Regulation?

As early as the 1980s, global distribution systems were subject to regulation, both in the US and in Europe. These are the first precedents of digital platform regulation we have identified. Many of the issues covered in these regulations are similar to the challenges posed today by the largest digital platforms: access to the platform, self-preferencing and search neutrality, data sharing, and so on.

In the US, global distribution systems were originally regulated in 1984 by the Civil Aeronautics Board and the rules were readopted in 1992. The rules blocked practices that would cause consumers and their travel agents to receive misleading information and would distort airline competition in favor of the airlines in control of Sabre. However, as Sabre became a public company in 2000, no longer controlled by American Airlines, the Civil Aeronautics Board decided not to readopt the rules, and so they were repealed in 2004.[5]

In Europe, Regulation No 2299/89[6] and later Regulation No 80/2009[7] imposed obligations on all companies managing computer reservation systems in aviation, not only on Amadeus as the market leader with market power. Regulation No 80/2009 is still in place, even if Amadeus became an independent public company in 1999. Most obligations resonate with the current debate on platform regulation.

Firstly, fairness and transparency are imposed on the vertical relationship between platforms and businesses providing the intermediated services. (1) The regulation explicitly prohibits platforms to impose exclusivity on intermediated parties. Airlines can sell their services on other platforms; in modern parlance, they can multi-home (in parallel a similar clause is imposed in relationship to travel agents). (2) The Regulation prohibits contractual clauses that exclude the direct commercialization of services by the airlines through their own digital systems. (3) Platforms will provide information about changes to their distribution facilities and loading/processing procedures to all intermediated carriers.

Secondly, non-discrimination is also imposed on the platforms. (1) Platforms are obliged to load and process data provided by participating carriers with equal care and timeliness. No discrimination is allowed. (2) Distribution facilities are separated, at least by means of software and in a clear and verifiable manner, from any carrier's private inventory, management, and marketing facilities. (3) Search results will be presented in a neutral form, ranked according non-discriminatory criteria, respecting the specific rules defined in the annex of the regulation.

Thirdly, self-preferencing is subject to specific rules. (1) Platforms must disclose direct and indirect capital holding by an airline or a rail company in the platform, or of the platform in one of such companies. (2) An independent audit must be made every four years, detailing the ownership structure and governance model. (3) An airline with shares in a platform (as originally in Amadeus) cannot discriminate against another platform (such as Sabre) by, for instance, not providing data. (4) The platform cannot discriminate in favor of parent carriers, particularly in the display of search results.

Fourthly, on data sharing, the regulation requires that (1) intermediated service providers ensure that the data they submit to the platform are accurate and provided as determined in the annex of the regulation; (2) all data generated by the platform can be made available by the platform, but if it does, it has to offer it with equal timeliness and on a non-discriminatory basis to all participating carriers; and (3) provisions are introduced to protect the privacy of the passenger.

Lastly, the regulation even a provision on liability for what we could refer to as "illegal content." In case an airline is subject to an operating ban imposed by the European authorities, platforms can display offers, but the ban must be "clearly and specifically identified in the display."

This first regulation of a transport platform seems a very balanced approach to the regulation of a complex ecosystem, with obligations imposed on transport service providers to share data, but also with effective obligations to control the platform.

Notes

1   Sismanidou, A., Palacios, M., & Tafur, J. (2008). New developments in Global Distribution Systems (GDSs) for the airline industry: first-mover mechanisms that enabled incumbent firms to maintain a leading position, *II International Conference on Industrial Engineering and Industrial Management*, pp. 727–734.
2   Rill, J., Wilson, C., & Bauers, S. (2000). The Amadeus Global Travel Distribution Case. In S. J. Evenett, A. Lehman, and B. Steil (eds), *Antitrust Goes Global, What Future for Transatlantic Cooperation?* Brookings Institution Press, Washington, DC.
3   Monti, G. & Augenhofer, S. (1998). *Consumer Choice and Fair Competition on the Digital Single Market in the Areas of Air Transportation and Accommodation*, Research for the IMCO Committee of the European Parliament, retrieved from www.europarl.europa.eu/RegData/etudes/STUD/2018/626082/IPOL_STU(2018)626082_EN.pdf
4   CMA (2020). *Online platforms and digital advertising market study*, Annex P, Specialised Search, p. 6.
5   Ravich, T. M. (2004). Deregulation of the Airline Computer Reservation Systems (CRS) Industry. *Journal of Air Law and Commerce*, 69, p. 387.
6   Council Regulation (EEC) No. 2299/89 of July 24, 1989 on a code of conduct for computerized reservation systems, OJ L 220, 29.07.1989.
7   Regulation (EC) No. 80/2009 of the European Parliament and of the Council of January 14, 2009 on a Code of Conduct for computerised reservation systems and repealing Council Regulation (EEC) No 2299/89, OJ L 35, 4.2.2009. Comments in Peeters, M. (2009). The New EC Regulation on Computer Reservation Systems, *Air and Space Law*, 34(4–5).

Chapter 13

Uber and Urban Mobility

Uber's history is the best example of how network effects are the cornerstone for digital platforms. As graphically described by early investors, the scope was to "put a network on top" of physical assets. This was the business model Travis Kalanick had mastered in his previous business ventures. Furthermore, Uber explicitly views itself as a network: "Our massive, efficient, and intelligent network consists of tens of millions of drivers, consumers, restaurants, shippers, carriers, and dockless (also called free-floating) e-bikes and e-scooters, as well as underlying data, technology, and shared infrastructure. Our network becomes smarter with every trip."[1]

Uber is also an illustrative example of the regulatory challenges posed by digital platforms. At an early stage, Uber was challenged by the San Francisco and California Authorities as the service did not fit the rigid template of transport regulation. It was not obvious how to classify the service provided by Uber: was it a digital service, an intermediation service, a taxi service? The same questions would eventually emerge in all the jurisdictions where Uber started operating.

Finally, Uber is a leading example of how disruptive platforms can be. Within just a few years, even months, the value of taxi medallions in the leading world cities simply collapsed, taxi companies went bankrupt, and a new transport mode emerged out of nowhere. Uber did not intermediate taxi services, but transport services provided by limos first, and then by non-professional drivers. Uber substituted taxi companies as an alternative underlying service was used at the supply side of the multi-sided market.

13.1 A Wild Ride

In late 1997, a group of friends studying in UCLA developed the idea of a software that would allow the exchange of music files among peers. It would "crawl" the computers of the participants looking for multimedia files, particularly music files, displaying the list of available files. Once identified and following a request from the interested participant in the platform, the other party would transfer the file. It was a hit among the UCLA students. Travis Kalanick was one of the students behind the platform, which would be named Scour. Scour did not allow the automatic downloading of the file from any of its participants. That was the value that Napster added in its winning proposition in June 1999. Scour adopted the Napster solution and it grew to a relevant size (more than 7 million downloads), attracting millions of dollars in investment, but in July 2000 it received a copyright infringement lawsuit. The company declared Chapter 11 protection and by the end of 2000 its assets had been liquidated.

However, Kalanick had identified a powerful business model: a digital platform connecting third parties to share their idle assets. He would build a new business around the same concept, but the second enterprise would be a law-abiding platform. In 2000, he developed software that enabled internet companies to store their content close to their users, so it would be downloaded faster and cheaper, under the name Red Swoosh. It was not a new idea. Content delivery networks (CDNs) were already a multimillion business with market leaders such as Akamai. Kalanick's proposition was different, building on the platform concept: "Red Swoosh solution is not dependent on massive investments in thousands of servers or dozens of data centers deployed close to the edge throughout the world. Instead, Red Swoosh leverages the massive unused capacity of the desktop to share and deliver inexpensively, effectively, and legally."[2] Red Swoosh was sold to Akamai for $18.7 million in 2007.

In the summer of 2008, Garrett Camp, a native of Calgary (Canada) living in the San Francisco Area, had the idea of an app that would request a ride in a limousine – a service he was often using himself after selling his own start-up to eBay for $75 million. He even reserved the domain name ubercab.com on August 8, 2008. The first code was developed by Camp's Mexican friends back from college in Calgary. Camp bought some limos and even leased a garage to store them. At this point, Kalanick stepped in. Building on his past experience, he proposed building the company as a platform, allowing existing limo drivers and passengers to interact through the app. In this way, the company could grow without large investments in vehicles. Scour had grown without producing music. Red Swoosh had grown without investing in its own data centers. Uber would grow without owning vehicles.

In May 2010, Uber launched on a small scale in San Francisco. Around that period, the seed round to finance the start-up was completed. The first-round investors, Lower Case Capital and some angel investors, including Napster's founder Shawn Fanning, provided $1.3 million, evaluating the start up at $5.3 million.[3] The main seed investors, as well as the founders, would become billionaires.

Kalanick joined Uber as CEO in November 2010, just after the first cease-and-desist order was received. It was the beginning of a fast but rough drive.

13.2 "A Network Layer on Top of It"

In September 2010, the month before Kalanick joined as CEO, Uber managed 427 rides. In the following months, the service exploded. Early riders were students and young tech professionals. They were attracted by novelty, but also by simplicity in the hailing process through the smartphone as well as the sense of control provided by the ability to follow the progress of the vehicle on a map when traveling around the city, particularly when approaching to pick up the rider. Attracting drivers was more of a challenge. Licensed limo drivers were scarce, they already had their own customers and they were not always tech-savvy. Uber provided the drivers with an iPhone, which was a valuable offer at that early stage. As drivers joined the platform, they realized that they were able to sell more rides and reduce the waiting periods with no activity. Limo drivers became interested, but they were always behind demand. This was good for the drivers in the platform, as they were busy with the rides requested by a larger and larger pool of riders, but it reduced the quality of the experience for the riders, as limited supply increased waiting times.

Expansion into other cities took place at an early stage: Seattle, Chicago, Boston, New York, and Washington DC were launched within a matter of months. Launching became more standardized. Drivers were given iPhones and economic incentives, while local celebrities and influencers were used to raise awareness among riders. Legal actions from taxi associations and local authorities were the most effective free advertising for the company. The platform ignited in all the cities, proving that the business model was a success.

A new round of investment began in 2011 and attracted more capital. Benchmark Capital led an $11 million fund-rising drive, becoming the largest investor in the company.[4] The company perfectly understood the business model. It had already invested in eBay and OpenTable (a platform connecting restaurants and diners): "We had an internal thesis that other industries might benefit from a network layer on top of them."[5]

"A network layer on top of it" is the secret behind Uber's success. The platform managed by Uber creates a new offer by pooling together a large number of independent drivers, coordinating them, and connecting them with a large pool of riders. The new offer is far superior to the service traditionally provided by taxi drivers.

Road transport has been traditionally served by independent vehicle drivers or small fleets. This is the case with taxis, but also of long-distance freight services provided with trucks. The taxi industry is an extremely fragmented industry. Taxi drivers navigate around the cities looking for passengers in an inefficient manner, spending much of their time on empty runs. Taxi dispatchers make it possible to call a taxi and identify the location of the rider. However, the process is cumbersome (making a phone call, identifying the caller, the location, and so on) and slow (the vehicle might not be close enough to the location of the passenger and 15/20 minute waits are the rule in many cities). There is also uncertainty; when vehicles are scarce, the waiting time might be longer or vehicles might even not be available at all.

Furthermore, the taxi industry is heavily regulated all around the world, and not always to the benefit of the passengers. Licenses are usually required in order to enter the market, and the number of licenses is capped. In some cities, like San Francisco, the cap is particularly stringent, but basically everywhere the industry is not in a position to meet peaks in demand. The number of licenses is not designed to meet such peaks, but to ensure minimum revenue to existing licensees. Secondary markets have emerged in most cities in the world, allowing licensees to sell their licenses for hefty prices, even as much as $1 million in New York City at some point. Such high prices trigger a vicious cycle, as new licensees put pressure on local governments to increase regulated tariffs to be able to recover the investment in the license, but when a local government increases tariffs, the price of the licenses in the secondary market goes up again.

The Uber model is more efficient. Firstly, the app reduces friction in the contracting process. No phone call to a dispatcher is necessary. A few clicks on a smartphone are sufficient for the passenger to identify himself/herself, locate the pick-up place, and request the service. Payment is also simplified, as the platform takes care of the payment on behalf of the driver, mostly through payment cards.

However, it would be a gross misunderstanding of Uber's business model to conclude that the reason behind the company's success is the app and its simplicity. More important is the role of the platform as the coordinator of the supply side. Cars can now work as a network and no longer as independent units. When Uber receives a request from a rider, such a request is matched with the most efficient supply available. The matching is managed automatically thanks to artificial intelligent tools – the famous algorithms – which factor in different parameters, such as how close the vehicle is to the pick-up location, but also traffic conditions, whether the vehicle is driving in the direction of the pick-up point or in the opposite direction, whether the driver has served the rider in the past (with positive or negative comments), etc.

Coordination on the supply side reduces waiting times for the passengers, thus improving the experience. More importantly, efficient coordination reduces empty runs for the drivers, thereby contributing to the reduction of fuel burn and emissions. The shorter the ride to the pick-up point is, the shorter the empty run will be and, therefore, the lower the cost of the provision of the service. Coordination improves response times in peak hours. Drivers can head directly to the next rider with short waiting times in between rides. They can charge more rides in the same time.

The real revolution derives from indirect network effects. As large pools of drivers interact with large pools of passengers, the efficiency of the service multiplies. When large pools of passengers are available, drivers can link one service with the next one, without the need to navigate looking for riders, thus reducing the cost of providing the service. As the cost reduction is passed to passengers, new demand is created. Passengers benefit from better service as high numbers of vehicles ensure that a vehicle will be close to the pick-up place for an ever lower price. A virtuous cycle is created. The industry is transformed. This is not a theoretical explanation of how Uber works. This is the way Uber described itself in the 2019 Form S-1 when it went public.[6]

The network effects really materialized when Uber started to work with non-professional drivers. Large pools of drivers could not be built with limo drivers, as originally envisaged by Uber. The sedans were cool among young techies in large metropolitan areas, but the service could not scale to serve a larger share of the population and exhaust the potential indirect network effects.

It was not Uber's idea to work with non-professional drivers. It was Lyft, the competitor that entered the San Francisco market in May 2012, that opened a platform to non-professional drivers – basically anyone with a vehicle and a driver license. Lyft had been working on carpooling platforms for universities and corporations. The peer-to-peer model had arrived to urban transportation. Uber could only follow competition in order not to miss the growing opportunity. Non-professional drivers were accepted in Uber's platform in mid-2013 and the new service launched in more than 70 cities before the end of 2013.

Uber's proposition was transformed from a luxurious "Everyone's private driver" to a new company slogan: "Evolving the way the world moves." Uber went international, serving hundreds of cities around the globe. In parallel, riders started to use the platform more and more often. If riders initially used the app twice a month, three years later they were using it 15 times a month on average.[7]

Non-professional drivers transformed the company. An unlimited pool of drivers could meet the rapidly growing demand. In 2017, more than 2 million drivers were working for

Uber, adding 50,000 drivers per month.[8] Furthermore, it was easier to adapt supply to peaks in demand. Professional drivers would work a limited but stable number of hours, making it difficult to meet peaks in demand and reduce supply at off-peak hours. Non-professional drivers were happy to use the platform for just a few hours a week, when demand was strong and revenue was more fluid, in order to receive extra revenue on top of their regular salary. In this way, it was easier to meet demand at week-end nights. Surge pricing (the increase of rates when supply is short) was another mechanism to ensure equilibrium between supply and demand, even if it was more controversial and not always understood by passengers.

The main evolution of the platform was UberPool, a service that allows passengers to share a vehicle for all or part of a ride, as the platform coordinates different requests for transportation in the same area or direction. Large pools of riders make it possible to match different travel requests to be served by a single driver. Rides take a little longer, but the 30 percent cost reduction per rider was achieved by distributing costs among more than one passenger. Lower prices generate new demand. A further twist was introduced in late 2017, when a further 25 percent price reduction was offered by Uber (Express Pool) if the rider was willing to compromise on pick-up and drop-off points, by walking to more efficient locations in order to streamline routes. This is the power of indirect network effects.

Eventually, Uber expanded the platform to include different services. Different types of vehicles would be supported (regular cars, limos, SUVs even motorcycles in some cities), as well as different cargo, not only passengers but also food (Uber Eats) and small parcels (Uber Rush) and finally even mass-transit services in different US cities.

The ride was over for Travis Kalanick in 2017.[9] Just as Mr. Vail was ousted from AT&T a century earlier, Kalanick was ousted as CEO by investors eager to capitalize on their investment. Benchmark Capital and other shareholders were not ready to keep investing in the creation of network effects. A new CEO was appointed to reduce investment, make the company profitable, and aim for an IPO. The company went public in 2019. In ten years the company has grown from zero to a value of $75 billion, an impressive figure but lower than the $120 billion expected by some investors. Travis Kalanick's shares were worth more than $5 billion. Benchmark Capital's initial investment of $11 million had transformed into $6.5 billion. Despite all this, Uber has never made a profit.

13.3 Regulatory Bumps

On October 20, 2010, Uber received its first cease-and-desist order. That was just five months after the launch of a modest operation in San Francisco. Remember that Uber managed only 427 rides in the month before the order, in September 2010.

There were three issues under discussion. The first was whether Uber was a technology company or a transport company. While transportation is heavily regulated by local or state authorities, digital services tend to be lightly regulated, typically at a federal (national) or even supranational level (as with the European Union). They also tend to benefit from privileges and exemptions defined at the early stage of development of the internet to promote innovation and foster the growth of the digital economy.

Uber tried to build on legislation and case-law from both sides of the Atlantic that exempted digital platforms from liability. In the United States, Section 230 of the

Communications Decency Act of 1996 provides immunity from liability for providers and users of an "interactive computer service" who publish information provided by others. Such immunity was originally designed for internet service providers (ISPs) and then extended to search engines and social network platforms. In the landmark case Tiffany v. eBay, the platform was considered non-liable for the sale of counterfeit items by third parties using the platform.[10] In the European Union, the E-Commerce Directive[11] provides protection against restrictions to the provision of so-called information society services by Member States, and even an exemption of liability for content they host. The Court of Justice of the European Union (CJEU) decided in favor of eBay, exempting it from liability for counterfeited items sold through the platform.[12]

The second question was whether Uber was providing a transportation service or whether it was merely a digital intermediary facilitating the contracting of services provided by a third party. For Kalanick, the answer was clear. Like all of his previous companies, Uber was based on the platform model. Uber did not own cars or employ drivers; it was only a platform that mediated between drivers and riders. For taxi drivers and public authorities, the situation was not that clear.

The third question was whether Uber was offering taxi services or other types of transport services. Taxi services were regulated by local authorities, in this case the San Francisco Metro Transit Authority. Drivers required a license (medallion) capped in number (around 1, 500 for the whole city) that allowed them to pick up passengers if hailed on the street. Regulated tariffs would apply, and they would be displayed by a certified meter. Black car services were regulated at a state level by the Public Utilities Commission of California. While there was no defined cap on the number of service providers, black cabs had to be pre-arranged and were not allowed to pick up passengers on the street. No price regulation existed for these services.

Uber blurred the distinction between taxicabs and prearranged services, not only in San Francisco, but all around the world. The prohibition on picking up riders on the street was no longer relevant, as the app made it possible to seamlessly prearrange the service with the touch of the screen of a smartphone. Smartphones could also be used as taxi meters. Black cars and later non-professional drivers coordinated by Uber became direct competitors for taxi services. The blurring lines between taxicabs and black car services explain why the first cease-and-desist order was jointly submitted by the San Francisco Metro Transit Authority and the Public Utilities Commission of California.

After years of operations all around the world, the same questions have been raised as Uber has started activities in a new country. Contradictory rulings have been adopted by public administrations and courts. Sometimes the platform business model is not understood. This was the case, for instance, with the England Employment Tribunal, which stated that "The notion that Uber in London is a mosaic of 30,000 small businesses linked by a common "platform" is to our minds faintly ridiculous."[13] However, as the dust settles, some clarity has emerged.

As for the first question – whether Uber provides a communications/digital service – the debate is pretty much solved. Authorities on both sides of the Atlantic have ruled that Uber is not merely a technology company that is exempted from the rules governing transport. This was the position of the PUC in California back in 2013: "Specifically, we reject the argument that [Transport Network Companies] are simply providers of IP-enabled services and therefore exempt from our jurisdiction. We find this argument to be factually and legally flawed and, therefore, do not accept that the method by which information is communicated, or the transportation service arranged, changes the underlying nature of the transportation service being offered."[14] In the EU, the Court of Justice decided in the same direction in 2017: "it must be classified not as 'an information society service' [...], but as 'a service in the field of transport'."[15]

As for the second question – whether platforms are intermediating in the provision of services or providing under their own name and liability of the transportation service – there are different answers. State rules in the US do not provide a clear-cut answer to this question. Uber continues to provide an intermediation service, so passengers directly contract with drivers. The debate has been focused on the legal classification of the relationship between Uber and the drivers. It has been said that drivers are under an employment relationship with Uber. Uber has countered by arguing that drivers are self-contractors who provide services to passengers in their own name, only contracting with Uber an intermediation service to connect with passengers.

The Court of Justice of the European Union made a clearer analysis in the Elite Taxi/Uber judgment.[16] The Court confirmed that transport platforms are intermediaries and not the providers of the underlying transport service. The Court repeatedly differentiated between the non-collective urban transport service and the service offered by Uber. The Court confirmed that "passengers are transported by non-professional drivers using their own vehicle," while the platform provides a different service, which the Court has repeatedly qualified as "intermediation service." The Court of Justice of the European Union stated that Uber "is more than an intermediation service consisting of connecting [...] a non-professional driver using his or her own vehicle with a person who wishes to make an urban journey." Thus, the platform is not limited to the transfer of information between the driver and the passenger. The Court confirmed that "the provider of that intermediation service simultaneously offers urban transport services, which it renders accessible [...] and whose general operation it organizes for the benefit of persons who wish to accept that offer in order to make an urban journey." The platform allows each individual driver to benefit from coordination and the effects of operating as a network: shorter waiting time, less driving time, consequent cost reduction, etc. This is precisely the added value offered by the platform. The Court concluded by stating that the mediation of platforms in multilateral markets transforms the market by facilitating that a previously fragmented demand functions as a structured network.

In any case, this powerful effect of the service provided by the platform does not mean that the platform is the provider of the transport service. The transport service provider remains the driver and the platform provides an intermediation service. An intermediation service offers great added value since it allows each driver to benefit from the network effects. In this way, the Court of Justice of the European Union has validated this new model of industrial organization carried out by platforms in multilateral markets. It recognizes that transport platforms provide an intermediation service, but not the underlying transport service, which remains the ownership (and responsibility) of each individual provider (in the case of Uber, each driver).

As for the third question – whether there is a difference between taxicabs and platform-mediated services – US authorities have understood that the element that differentiates taxicab services from other transportation services is that taxicabs are hailed on the street, so passengers need protection as they have no information on the identity of the service providers, the conditions of the service to be provided, etc. Transport platforms, by contrast, allow passengers to identify the service provider and be aware of the service conditions (including price) before asking for the service. As a result, the regulatory protections for the taxicab user do not necessarily have to be extended to these different services. The US District Court of the Northern District of California confirmed this approach in a 2017 judgment: "In a street-hail situation, a passenger is (as a general matter) more likely to be in a vulnerable position compared to a passenger who prearranges a ride. That is, in a street-hail situation, the passenger typically has an immediate need for transportation services and therefore, in a more vulnerable position, lacks the assurances that come with a pre-arranged ride. Thus, there is a conceivable basis for a differential approach to regulation, including, e.g., closer regulation of rates and imposition of requirements for taxis."[17] In this way, the debate was closed, at least in California. Other jurisdictions are following a similar path, such as Chicago, Washington DC, and London, where traditional black cab services are differentiated from so-called private hire vehicles.

The largest US states have defined a new category of prearranged transportation services providers, managed by transport platforms and provided by non-professional services: transport network companies (TNCs). TNCs are not taxicab services or black car/limo services. It is important to underline how the same name recognizes the relevance of the network effect. The California Public Utilities Commission set the example in 2013, defining a TNC as: "an organization [...] that provides prearranged transportation services for compensation using an online-enabled application (app) or platform to connect passengers with drivers using their personal vehicles." Services are provided by drivers using their personal vehicles. Platforms connect drivers with passengers using an online app. The platform must have a license and meet a series of obligations to ensure safety: the platform has to contract insurance, conduct criminal background checks for each driver, and conduct vehicle inspections before starting operations and then once a year.

TNC regulations have been adopted by a high number of US states, including Illinois, Massachusetts, New York, and Texas, with interesting twists in some jurisdictions. In California, licensed platforms cannot own the vehicles. In New York, licensed platforms cannot mediate taxicab services. In Massachusetts, TNCs are obliged to pay to state authorities $0.20 for each ride. This revenue is then distributed among state and local authorities for infrastructure funding: 10 cents go to municipalities for transport infrastructure funding, 5 cents go to a state transportation fund, and 5 cents go to fund transition to competition for taxis.[18]

In conclusion, there is growing consensus around the role of transport platforms as intermediaries facilitating the contracting of transport services between drivers and passengers connected by the platform. However, such an intermediation services are not limited to connecting drivers and passengers, as they create network effects that transform the underlying services and creates a new offer of services. Consequently, transport platforms are increasingly subject to regulatory obligations such as licensing (even if not capped in number), safety (drivers and vehicles test, insurance), and even taxes.

13.4 Substitution of taxi

The price of taxi licenses (medallions) has collapsed all around the world. In New York City, it has been reported that the price of a medallion in 2013 was around $1.3 million, but that one was sold in 2020 for as low as $130,000.[19] That is a reduction in value of 90 percent. The reason is clear: the number of daily taxi rides is going down, while the number of rides managed by platforms reached the equivalent number of taxi rides in New York City already in 2016. In 2019 Uber completed more than 500,000 rides, while yellow cabs completed fewer than 250,000.[20]

Notwithstanding, figures for New York City confirm that taxi services remain popular. There is plenty of room for taxicab services in dense urban areas like Manhattan. However, Uber and other platforms (Lyft, Via, Juno or Gett) have demonstrated that there was an unmet demand in New York City for urban transportation services. From 2014 to 2018, ride-hailing apps grew from zero to 15 million trips per month, while taxi usage has only declined by around 5 million trips per month. Uber alone is now bigger than yellow and green taxis combined, first achieving that milestone in November 2017.[21]

Service has improved in boroughs other than Manhattan, which were traditionally underserved, as well as during the night or in peak times, when available supply was not sufficient to meet demand.

Taxis are suffering the same substitution effect that postal services and the music industry were already experiencing due to email and Napster, respectively. The largest taxi company in San Francisco filed for bankruptcy in 2016.[22] Taxicab services, hailed on the street, will always have a role, just as letter mail services have a role after the introduction of email, but it is a matter of time before non-connected taxi services become secondary in urban areas, as the leadership role is being taken by digital platforms.

It can be argued that a level playing field has not always been ensured for the competition between taxicabs and digital platforms. This was particularly the case when non-professional drivers entered the market. Some similarities can be identified between Napster's services in breach of intellectual property rights and Uber's reliance on non-licensed drivers.

However, the competitive advantage that transport platforms have over taxicab services does not rely on the regulatory framework. The key advantage is that platforms "put a network layer on top"of individual vehicles. Transport platforms create a new offer, as they coordinate the provision of services by individual service providers. Efficiency is generated by the smart matching of large pools of drivers and large pools of passengers. Transport platforms do not own the vehicles and they do not hire the drivers. They merely mobilize and coordinate large numbers of independent drivers, efficiently matching them with passengers. Such smart matching reduces the cost of the provision of the service in a dramatic proportion, particularly when pooling together more than one passenger in a vehicle by matching passengers traveling in the same direction with a single driver. The most fragmented of the transport industries, taxicabs, is being substituted by a new offer of networked, more efficient services, organized by transport platforms.

The relationship of Uber with collective transportation is more nuanced. A clear substitution trend cannot be identified between Uber services and urban collective transportation services (bus, railway, and subway). Uber services, either based on black cars or non-professional drivers, have a price that is clearly higher than that of collective transportation services, so substitution is necessarily limited. Even in the more developed markets in the US, ridesharing usage for commuting is limited. Only 20 percent of ridesharing trips are for commuting. It is more common to use ridesharing for recreation/social purposes (above 55 percent) and shopping errands (18 percent).[23]

Studies undertaken on the potential substitution effects confirm that ride-hailing services have not created substantial substitution effects in relation to collective urban transportation. According to surveys, almost 70 percent of ride-hailing platform users would have used a private car for the last ride they took with a platform, and only around 15 percent would have used a public bus or a train.[24] A similar result emerges when analyzing ridership trends as Uber launched services in different US cities.[25]

It has been said that ride-hailing services could complement collective transportation. Collective transformation combined with ride-hailing and other alternative transport modes offers a reliable alternative to the usage and ownership of a private car. As the International Association of Public Transport (UITP) has stated repeatedly: "it is the offer of an integrated combination of sustainable urban mobility services that most effectively challenges the flexibility and convenience of the private car. A broader mix of mobility services is the answer to ever more complex and intense mobility needs. [...] car-based services and especially car-sharing are the obvious services that complement public transport as they offer the benefits linked to car usage without the need to own the car."[26]

Transport platforms complement mass-transit as they help to overcome the three most relevant weak points of collective transportation. Firstly, as collective transportation can only serve high-density areas, platforms allow passengers to move from their premises to stations. Uber has often provided data on how a significant share of their services initiate or terminate at a subway or train station, both in Europe and in the US.[27] Secondly, platforms complement mass-transit when it is not active, particularly at night time. Ride-hailing is particularly used between 8 p.m. and 4 a.m., when mass-transit systems reduce or even suspend operations. Thirdly, platforms provide the necessary flexibility to make specific trips for which collective transportation might not be suitable (those involving heavy luggage, disabled passengers, or requiring urgency, etc.).

The most relevant study on the subject so far, undertaken for the American Public Transportation Association (APTA) and by the Shared-Use Mobility Center (SUMC) confirms that "people who take greater advantage of shared modes report lower household vehicle ownership and decreased spending on transportation." In parallel, "the more people use shared modes, the more likely they are to use public transit."[28]

The main downside of Uber, from a transport policy perspective, might be congestion in dense urban areas like New York City. The number of vehicles providing transport services has increased since Uber has enter the market. In New York City there are more than 80,000 vehicles working with platforms, compared to approximately 13,500 taxicabs.[29] Some drivers working with Uber work only a limited number of hours, as opposed to the intensive use of taxicabs. Vehicles working with platforms tend to provide services across a wider geographical area than taxicabs. In New York City, they are more present in the outer boroughs than taxicabs, which are concentrated in Manhattan and around airports. Transport platforms reduce the use of private vehicles and the need to drive around looking for parking, as well as the need for parking itself. In any case, it is not obvious that these benefits compensate the large increase in the number of vehicles on the streets, particularly in dense urban areas. Such congestion, and the consequent reduction in average speed, has been identified in Manhattan.[30] We have not even mentioned air pollution. Some studies suggest that even if only 15 percent of ride-hailing services used collective transportation had ride-hailing not been available, a high percentage of passengers would have not been made the trip (22 percent) or would have made it by foot

(17 percent) or bicycle (7 percent).[31] According to these results, ride-hailing is increasing road traffic and, consequently, congestion and pollution.

Uber is disrupting urban transportation worldwide. Uber and other ride-hailing platforms such as Lyft, Via, and Juno in the US, Didi in China, Ola in India, and Cabify in Latin America, were initially fringe services for a specific market niche. They were not in a position to compete directly with well-established transport providers such as taxicabs, black cars, or collective urban transportation. However, as network effects grew, the services became more attractive. A substitution effect can already be identified for taxicab services.

Notes

1 Uber Form S-1 for the IPO, April 11, 2019.
2 Red Swoosh investor prospectus, March 2001, in Lashinsky, A. (2017). *Wild Ride. Inside Uber's Quest for World Domination.* Penguin, New York, p. 45.
3 Stone, B. (2017). *The Upstarts. How Uber, Airbnb and the Killer Companies of the New Silicon Valley Are Changing the World.* Transworld Publishers, London, p. 61.
4 Ibid., p. 126.
5 Bill Gurley, Partner in Benchmark Capital, in Lashinsky, A. (2017). *Wild Ride. Inside Uber's Quest for World Domination.* Penguin, New York, p. 102.
6 Uber Form S-1 for the IPO, April 11, 2019.
7 Lashinsky, A. (2017). *Wild Ride. Inside Uber's Quest for World Domination.* Penguin, New York, p. 153.
8 Galloway, S. (2017). *The Four. The Hidden DNA of Amazon, Facebook and Google.* Portfolio/Penguin, New York, p. 215.
9 Isaac, M. (2019). *Superpumped. The Battle for Uber.* W.W. Norton & Company, New York.
10 Tiffany (NJ) Inc. and Tiffany and Company v. eBay, Inc. 600 F.3d 93 (2d Cir. 2010).
11 Directive 2000/31/EC of the European Parliament and of the Council of June 8, 2000 on certain legal aspects of information society services, in particular electronic commerce, in the Internal Market ("Directive on Electronic Commerce"), OJ 2000 178, 17.7.2000.
12 Judgment of July 12, 2011, *L'Oréal v. eBay*, C-324/09, ECLI:EU:C:2011:474.
13 *Aslam & Others v. Uber, Employment Tribunals*, Case Nos 2202550/2015 & Others, October 28, 2016.
14 Decision Adopting Rules and Regulations to Protect Public Safety while Allowing New Entrants to the Transportation Industry, Decision 13–09-045 of 19.9.2013.
15 Judgment of December 20, 2017, *Elite Taxi/Uber*, C-434/15, ECLI:EU:C:2017:981.
16 Ibid.
17 *Desoto Cab Company v. Michael Picker et al.*, Case No 15-cv-04375_EMC, January 12, 2017.
18 Act regulating transportation network companies, adopted in the One Hundred and Eighty-Ninth General Court, 31.7.2016.
19 McEnery, T. (2020, May 11). NYC taxi rescue plan calls for revaluing all medallions at $250,000, *New York Post*, retrieved from https://nypost.com/2020/05/11/nyc-taxi-rescue-plan-calls-for-medallions-to-be-250000/
20 Pesce, N. L. (2019, Aug. 9). This chart shows how Uber rides sped past NYC yellow cabs in just six years, *Market Watch*, retrieved from www.marketwatch.com/story/this-chart-shows-how-uber-rides-sped-past-nyc-yellow-cabs-in-just-six-years-2019–08-09
21 Schneider, T. (2018). Analyzing 1.1 billion NYC taxi and Uber trips, with a vengeance, retrieved from https://toddwschneider.com/posts/analyzing-1–1-billion-nyc-taxi-and-uber-trips-with-a-vengeance/#update-2017
22 Corrigan, T. (2016, Jan. 24). San Francisco's biggest taxi operator seeks bankruptcy protection, *The Wall Street Journal*, retrieved from www.wsj.com/articles/san-franciscos-biggest-taxi-operator-seeks-bankruptcy-protection-1453677177

23 SUMC (2016). *Shared Mobility and the Transformation of Public Transit*, Research Analysis for the American Public Transportation Association. TCRP Report 188 Pre-Publication Draft–Subject to Revision, p. 11, retrieved from www.tcrponline.org/PDFDocuments/tcrp_rpt_188.pdf

24 SUMC (2016). *Shared Mobility and the Transformation of Public Transit*, Research Analysis for the American Public Transportation Association. TCRP Report 188 Pre-Publication Draft–Subject to Revision, p. 15 retrieved from www.tcrponline.org/PDFDocuments/tcrp_rpt_188.pdf

25 Hall, J.D., Palsson, C., & Price, J. (2017). Is Uber a Substitute or Complement for Public Transit? Working Paper 585, University of Toronto, Department of Economics.

26 UITP (2016). *Public transport at the heart of the integrated urban mobility solution*, Policy Brief, April 2016, retrieved from www.uitp.org/sites/default/files/cck-focus-papers-files/Public%20transport%20at%20the%20heart%20of%20the%20integrated%20urban%20mobility%20solution.pdf (08–02-2017).

27 Smith, A. & Salzberg, A. (2016, May 18).*Uber + Public Transit: Changing SoCal's Car Culture*, *Medium*, retrieved from https://medium.com/uber-under-the-hood/uber-public-transit-changing-southern-californias-car-culture-540b2021091#.vi82ie5z3

28 SUMC (2016). *Shared Mobility and the Transformation of Public Transit*, Research Analysis for the American Public Transportation Association. TCRP Report 188 Pre-Publication Draft–Subject to Revision, pp. 3 and 7, retrieved from www.tcrponline.org/PDFDocuments/tcrp_rpt_188.pdf

29 DeBord, M. (2019, Apr. 1). Congestion pricing could mark the beginning of the end of New York's famous yellow taxis, *Business Insider*, retrieved from www.businessinsider.com/congestion-pricing-end-of-new-yorks-famous-yellow-taxis-2019-4

30 Shaller, B. (2017). Empty seats, full streets. fixing Manhattan's traffic problem, Schaller Consulting, retrieved from http://schallerconsult.com/rideservices/emptyseats.pdf

31 Clewlow, R. & Mishra, G. S. (2017). *Disruptive Transportation. The Adoption, Utilization, and Impacts of Ride-Hailing in the United States*, Institute of Transportation Studies, University of California Davis, Research Report UCD-ITS-RR-17–07.

Chapter 14

BlaBlaCar and Long-Distance Mobility

BlaBlaCar is one of the most successful European digital platforms. It connects drivers with passengers for long-distance carpooling.[1] BlaBlaCar has grown a community with the right distribution of benefits derived from indirect network effects. BlaBlaCar has a market share of more than 90 percent in all the large European countries, starting with France, where it was created.

BlaBlaCar is a good example of the so-called sharing economy; that is, the provision of services by non-professional drivers making use of their personal assets. It is to transport what peer-to-peer is in telecoms and prosumers are in the electricity industry. Platforms are central in the sharing economy, as they reduce transaction costs to such a low level as to put them below the small value of sharing a personal service, enabling exchanges that were previously impossible.

BlaBlaCar has disrupted long-distance traveling in Europe. It has partially substituted traditional service providers such as railways and coaches by attracting significant shares of passengers who previously traveled by train (12 percent in France) and coach (20 percent in Spain). BlaBlaCar has increased the use of the private vehicle against the use of collective transportation means.

14.1 Carpooling Digital Platforms

On Christmas Eve 2003, Frédéric Mazzella, a French entrepreneur living in Paris, was unable to find a train ticket to travel back home to the region of Vendée. He thought about carpooling, a solution he had often used when studying computer science in Stanford, where he benefited from HOV (high occupancy vehicle) lanes.[2] Mazzella looked online for available seats in cars traveling to his home region, but could not find any. He finally traveled home with his sister, but the idea of using the internet as a platform to connect drivers and passengers took root.

Mazzella's idea grew slowly. His carpooling platform was only incorporated in 2006. Initially, it tried to replicate the "Car Sharing Club Exchanges" sponsored in 1942 by the US Office of Civilian Defense in order to save gas and rubber that were necessary for the war effort, but this time online.[3] Such clubs were established in working centers, churches, and schools. An interested individual would fill out a card that would be displayed in public areas, for drivers and passengers to seek a match. Instructions made it clear that prices would be agreed by the participants and that the exchange would take no liability in the provision of the service. Exchanges were supported thanks to an aggressive campaign with slogans such as *"When you ride alone, you ride with Hitler. Join a Car-Sharing Club Today."* Carpooling increased in popularity again in the 1970s in the US and in Europe, as the energy crisis increased gas prices, and again in the 1990s, as environmentally conscious drivers and passengers started to share drives. Carpooling became particularly popular in the US in the large metropolitan areas that established HOV lanes. Drivers picked up passengers at bus stops (or dedicated locations) to meet the occupancy requirements and use the fast lanes, a practice that was termed "slugging."

It was only in 2010, when smartphones became widely available and a mobile app was developed, that the platform attracted the interest of investors, and really ignited. That was the beginning of the success story of BlaBlaCar, as the platform was renamed for the international expansion.

Long-distance carpooling through digital platforms has been mostly a European success story. BlaBlaCar approached 90 million users in 2019, mostly in Europe. It claims to carry 70 million passenger per year.[4] BlaBlaCar is the market leader for carpooling in all the large European countries, with markets shares above 90 percent in France, Spain, Germany, Italy, and Eastern Europe.[5]

Carpooling has reached a significant portion of the overall long-distance transportation market in the more mature European markets. A total of 8 million carpooling trips were made in France in 2015, for a total of 6 billion passenger-kilometers (pkm). To put it into perspective, this represents 12 percent of long-distance person/km traveled by train in France in the same year, and 2.72 percent of all the long-distance person/km traveled in France.[6] In economic terms, it can be estimated that, in 2015, the fees paid by carpooling passengers to drivers in France amounted to EUR 210 million.[7]

The key to the success of carpooling in Europe has been the accumulation of significant indirect network effects, as large pools of drivers and riders joined the platform. Large pools of drivers and riders are necessary to ensure the availability of trips for passengers and of passengers for drivers. The larger the volumes of drivers, the greater the chances of finding a driver who is making the desired trip at the right time. The larger the number of passengers, the greater the chances that a driver will find passengers for his/her ride.

At the early stages of BlaBlaCar, only trips between large urban areas were available on the platform and such trips tended to be concentrated on weekends and holidays. As the pool of users became larger, trips between large metropolitan areas and smaller towns became more common. Finally, as the number of registered users reached the millions, trips between smaller towns also became available.

The same evolution can be identified for times of departure. The time of departure is an important factor when choosing one driver over another.[8] However, availability of the service diverges significantly among routes. Dense routes among large cities offer a wide range of options, often more than collective transportation (which has few night services, frequent strikes, few cross-border services, etc.). Routes connecting rural areas, on the contrary, provide fewer options or even none at all. The day of the week is also relevant. Carpooling services tend to be concentrated over Fridays and Sundays and, in general, over holiday periods.

Since low prices and flexibility in departure times are the key parameters for using carpooling, as the number of users increases, the service improves, generating a further increase in the number of users. Such a virtuous cycle is the defining trait of the indirect network effect.

Carpooling exists in the United States but is far less popular than in Europe, particularly for long-distance trips. In the US, companies like Zimride and rdvouz.com are active in the long-distance carpooling market, but far from the volumes of users and mediated rides of BlaBlaCar in Europe. BlaBlaCar has also launched operations in Brazil, Mexico, and India. The volume of these operations has not yet reached the levels of success of the European operations. In the US, population density in suburban areas is low and public transportation from homes to pick-up areas is mostly non-existent. This might be one of the reasons why long-distance carpooling has not taken off in the US as much as in Europe.

A further factor in the success of carpooling in Europe is that BlaBlaCar has identified the right balance in the distribution of the indirect network effects created by the platform, in terms of pricing, both for drivers and passengers, and commissions for the platform. The benefits derived from the indirect network externality have to be adequately distributed in order for the platform to succeed. It is not enough to generate a large pool of users and therefore to generate large network effects. Furthermore, it is necessary to evenly distribute the benefits of the indirect network effects across the ecosystem in order to maintain and grow the large pool of users.

Economic reasons are the main driver for the growth of long-distance carpooling. Seventy percent of carpooling users chose the service for this reason.[9] Carpooling prices are clearly below those of railway travel (particularly high-speed services) and usually below those of bus services.

Since price drives carpooling usage, it is interesting to identify how prices are set. BlaBlaCar recommends that drivers set the price by dividing the cost of gasoline and tolls by three. Prices rarely diverge substantially from this reference point. In France, it appears that the average price per passenger is EUR 0.06 per kilometer.[10] There are several reasons for the low prices of carpooling services. Firstly, drivers tend to share only variable costs (gasoline and tolls) with passengers; fixed costs, which amount to two-thirds of the total cost in France,[11] are often ignored. This is possible because the driver is traveling anyway and not merely transporting third parties to the destination. Secondly, national regulation on carpooling often introduces a limit on fees that can be charged, as the provision of the service for a profit is often prohibited.[12] Thirdly, carpooling has some regulatory advantages over traditional public transportation: the driver pays no income taxes when charging passengers, no time limits are imposed on drivers, no public service obligations are imposed, no tolls are imposed for the use of many roads (while the main cost borne by railway undertakings is the access charge for the use of the railway infrastructure), etc.

Benefits derived from the indirect network effect must be distributed not only among both sides of the multi-sided market (drivers and passengers), but also to the platform itself. Only in this way does the platform have an incentive to invest in the growth of the pools of users and to continue providing the matching service.

When BlaBlaCar launches its services in a new country, no commission is charged; the service is provided for free, both to drivers and to passengers. This is a strategy to grow the pool of users and recognition that the value for users might be limited in the initial stage of the service, as a small pool of users generates small network effects. As the service becomes more popular, network effects kick in and the service becomes more competitive. At this stage, BlaBlaCar starts charging for its matching service. BlaBlaCar charges a commission of approximately 18 percent of the price paid for the carpooling service. The commission is charged to passengers. The commission is charged to the passenger and not to the driver, contrary to the charging model of other transport platforms such as Uber, which charges a similar commission, but to the drivers.

Every platform must find the right equilibrium. In the case of BlaBlaCar, the price charged by the driver is low (it only covers some variable costs), so it would not be sensible to charge the commission to the driver. Passengers, on the contrary, who are benefiting from a price that is lower than the alternatives (trains, buses, etc.) can more easily foot the bill.

14.2 The Sharing Economy

Back on Christmas Eve of 2013, Frédéric Mazzella had to travel back home with his sister. Many drivers would probably have been happy to share a seat in their own private cars for a fee with him. However, the cost of looking for them, negotiating the conditions (pick up time and place, fee to be paid, etc.), and checking whether the driver would be safe were too large to make such transaction possible. The transaction costs were actually higher than the benefit derived from sharing the ride. Thus, sharing drives was traditionally limited to relatives, friends, and work colleagues. Transaction costs were lower among closely related individuals, but the chances of finding a match were also very low.

BlaBlaCar, like all successful platforms, reduces the costs of sharing. Parties interested in sharing a ride identify themselves via the platform, identify the route to be driven, their location, the sharing conditions, etc. Platforms match drivers offering a specific ride with passengers interested in that specific ride. Algorithms automate the matching process[13] at a low cost. Communication between drivers and passengers takes place though the app at no cost for the users. Online platforms reduce the cost of sharing to below the benefit derived from sharing

Mazzella was always interested in building trust among participants. Trust is particularly necessary in carpooling, as passengers must rely on the driving ability of an unknown person, and risks are high. Trust is facilitated by the new cultural values created by the social networks, as well as by specific instruments provided by the platform such as identification of the users and the provision of personal information about them. For instance, BlaBlaCar users must provide information about how chatty they are (their *blabla level*). Users rate themselves and provide evaluations, "likes," etc. Artificial intelligence tools help to manage ratings by excluding fraudulent or non-relevant ratings, thus generating algorithmic network effects, even if less relevant than in other platforms.

BlaBlaCar has achieved a high level of trust among users. In a paper that Mazzella drafted with the sharing economy guru Arun Sundararajan,[14] the results of a large survey among BlaBlaCar users confirmed that trust among the users of the platform was higher than trust in work colleagues and neighbors, and close to trust in relatives.[15]

BlaBlaCar is the leading example of the so-called "sharing economy" in the transportation industry. Sharing, defined as using an asset jointly, either at the same time or in turns, is as old as humanity. However, over the past decade, a new socio-economic model has emerged around the notion of sharing both assets and services. It is the "sharing economy," also called "collaborative consumption."[16]

From the perspective of supply, individuals are empowered by online platforms such as BlaBlaCar to provide transport services to other individuals, making use of their own idle resources, such as empty seats in their cars when making a trip. Non-professional drivers are in a position to compete with traditional transport companies such as railway and bus companies.

From the demand perspective, there is a general trend towards choosing access over ownership. Individual ownership of assets is replaced with the possibility to use assets without owning them. Savvy consumers are increasingly aware of the cost of owning assets in terms of maintenance, repairs, insurance, storage, etc., as well as of the externalities in the form of congestion, environmental damage, etc. The reduction in the rate of private vehicle ownership in the most developed societies is a good example, as is the even more significant trend among young people to either not obtain a driver's license or to delay taking their license exam.[17] At the same time, increasing economic hardship is generating a new demand for low-cost services. Good examples are air transportation and the proliferation of low-cost bus services. Carpooling is becoming a new transport mode in competition with other public transportation modes such as railways and buses.

14.3 Substitution of Mass-Transit Services

BlaBlaCar is disrupting the long-distance transport markets in Europe, from Portugal to Russia. BlaBlaCar was originally a fringe service that was popular among students and young professionals, with limited routes and timings, unable to meet the standard demand of transportation and therefore to compete with railways and buses. However, as the pool of users grew, indirect network effects kicked in; routes and timings improved to surpass the offerings of public transportation in certain occasions (night services, rural services, etc.). The low prices of carpooling are also a threat to established transport companies. As a result, public authorities in France, the country where BlaBlaCar is most mature, have estimated that BlaBlaCar represented around 12 percent of the passenger-km of long-distance railway person/km in 2015, and might represent 20 percent in 2030 (including passengers and drivers).[18]

The key question is where carpooling users come from. Are they drivers leaving their car at home to travel with someone else? Are they passengers who previously used trains and buses? Are they new travelers, who would not be traveling if carpooling was not available?

French users were asked what type of transport mode they would have used had carpooling not been available.[19] A majority of passengers, 69 percent, declared that they would have traveled by train. Some of them would have traveled with a private car (16 percent) and only a few would have not traveled at all (12 percent). This confirms that BlaBlaCar passengers are mostly migrating from mass-transit low-emission and efficient trains to private cars. The picture is different for drivers: most of them (67 percent) would have continued traveling with their own car had carpooling not been available. Only 8 percent would not be traveling. What is more interesting is that 26 percent of them would have used public transportation had BlaBlaCar not been available.

French researchers have quantified the number of passengers that railways are losing to BlaBlaCar. For 2015, the French railway monopolist SNCF lost 6 percent of its long-distance domestic passengers,[20] and it has been estimated that in 2030 it could lose 8.5 percent of passengers.[21] Parallel research has quantified that each vehicle-km traveled by carpooling reduced the use of the train by two person/km.[22] It can be estimated that the impact on revenue is even larger, as BlaBlaCar mostly attracts passengers paying full-price tickets at peak time and is less popular at off-peak times, when railway companies offer discounted prices.

Managers of coach services in Spain claim a similar impact upon their long-distance operations. Bus services are popular in Spain among low-income passengers, while rail services, particularly high-speed services, tend to be more popular among high-income passengers. While no significant impact on the number of passengers has been raised by the national railway company, CONFEBUS, the Spanish trade association of bus companies claims that carpooling has caused a 20 percent reduction in the number of long-distance bus passengers.

The position of bus services in Spain is particularly weak, as they are managed under exclusive rights granted for each route by the public authorities. Monopoly rights allow bus companies to extract rent from profitable routes between large cities and fund the provision of services to small towns and villages, which is not profitable, with these rents. If a new transport mode, carpooling, detracts passengers and revenue from profitable routes, the financial position of the bus companies will not be sustainable and the regulatory model will soon collapse.

In conclusion, the existing evidence shows that a large volume of passengers are migrating from long-distance mass-transit services to carpooling. This is the main customer base for carpooling services, with a much-reduced volume of drivers becoming passengers in someone else's car. Carpooling is disrupting the traditional public transportation markets and the existing regulatory framework.

Notes

1 Montero, J. (2019). Regulating Transport Platforms: The Case of Carpooling in Europe. In M. Finger & M. Audouin (eds.), *The Governance of Smart Transportation Systems*. Springer, Cham, 13–35.
2 Retrieved from www.blablacar.fr/a-propos/success-story
3 Chan, N. D. & Shaheen, S. A. (2012). Ridesharing in North America: Past, present and future. *Transport Review*, 32:1, 93–112.
4 Dillet, R. (2020, Feb 6). *BlaBlaCar's Revenue Grew by 71% in 2019, TechCrunch*, retrieved from https://techcrunch.com/2020/02/06/blablacars-revenue-grew-by-71-in-2019/?guccounter=1&guce_referrer=aHR0cHM6Ly93d3cuZ29vZ2xlLmNvbS8&guce_referrer_sig=AQAAAKXjAcpgr3j1UamqMtV2pBSY8xv0ZXUo2q_ESEqwXuPWegN7KPd9hvbt1_sSkr4MOeHgARIy2YdOGlmXShDnWabzhl8DHZZRkNMnfH34Ux_u-ilDkvHO6MVXv8bE0XzJCWPgh6eiQ1ABgKbpG6FTnjdc6Uaorvb2YqLLncIQeIjv
5 BlaBlaCar grew in Germany through the acquisition in April 2015 of its local competitor, carpooling.com, which had 6 million registered users. BlaBlaCar grew in Eastern Europe through the acquisition in March 2015 of Budapest based AutoHop and in January 2016 of Jizdomat, active in the Czech Republic and Slovakia.
6 CGDD (2016). Covoiturage longue distance: état des lieux et potentiel de croissance, *Etudes & Documents du Commissariat Général au Développement Durable*, n° 146, May 2016. These figures are coherent with the figures in other studies.
7 This is the result of multiplying the 3.5 billion pkm by the average price per km (EUR 0.06).
8 ADEME/6T Bureau de recherche (2015). *Enquête auprès des utilisateurs du covoiturage longue distance*, September 2015.
9 MAIF (2009). *Usages et attitudes des utilisateurs du site Internet covoiturage*. December 2009, retrieved from www.maif.fr/content/pdf/particuliers/auto-moto/covoiturage/maif-etude-covoiturage-12–2009.pdf. More altruistic reasons were also considered: protection of the environment (12 percent) or providing a service (4 percent).

10 CGDD (2016), Covoiturage lingue distance: état des lieux et potentiel de croissance, *Etudes & Documents du Commissariat Général au Développement Durable*, n° 146, May 2016, p. 5.

11 GART/UTP (2014). *Covoiturage et transports collectives: concurrence ou complementarite sur les deplacements longue distance?*, July 2014, p. 42.

12 In France, carpooling is allowed as long as the driver makes the trip for him/herself and charges no fee, but merely divides the costs of the trip. (Article L. 3131–1 Loi de transition énergétique of 2015). The reference for costs is the annual publication by the tax authorities of the maximum deduction in the income tax for the use of the private vehicle. Similar rules apply in many European countries.

13 McAfee, A. & Brynjolsson, E. (2017). *Machine, Platform Crowd: Harnessing Our Digital Future*. W. W. Norton & Company, New York.

14 Mazzella, F. & Sundarajan, A. (2016). *Entering the Trust Age*, retrieved from www.blablacar.com/wp-content/uploads/2016/05/entering-the-trust-age.pdf

15 A survey of 18, 289 BlaBlaCar users in 11 countries found that 88 percent of respondents had high or very high trust in a member with a full profile in the platform. Only 58 percent of respondents declared the same level of trust in colleagues, and 42 percent in their neighbors. Trust in relatives reached 94 percent.

16 There is abundant literature on the sharing economy, such as Botsman, R. & Rogers, R. (2010). *What's Mine Is Yours: How Collaborative Consumption Is Changing the Way We Live.* HarperCollins, New York; and Sundararajan, A. (2016). *The Sharing Economy: The End of Employment and the Rise of Crowd-Based Capitalism.* MIT Press, Cambridge (MA).

17 Beck (2016, Jan. 22). The decline of the driver's license, *The Atlantic*, retrieved from www.theatlantic.com/technology/archive/2016/01/the-decline-of-the-driverslicense/425169/

18 Minisere de l'Environment, de l'Energie et de la Mer (2016). Projections de la demande de transport sur le long terme, p. 83.

19 ADEME/6T Bureau de recherche (2015). Enquête auprès des utilisateurs du covoiturage longue distance, September 2015.

20 CGDD (2016). Covoiturage longue distance: état des lieux et potentiel de croissance, *Etudes & Documents du Commissariat Général au Développement Durable*, n° 146, May 2016.

21 Minisere de l'Environment, de l'Energie et de la Mer (2016). Projections de la demande de transport sur le long terme, p. 83.

22 ADEME/6T Bureau de recherche (2015). Enquête auprès des utilisateurs du covoiturage longue distance, September 2015, p. 74.

Chapter 15

Mobility-as-a-Service

The Network of Networks

Mobility-as-a-service (MaaS) is the integration of various forms of transport services into a single mobility service accessible on demand.[1] The integration can take different forms, but digital technologies and platforms certainly play a fundamental role.

Whim is a platform that was launched in Finland and is active in other European countries, which integrates different transport services in an app under a common ticketing and pricing framework, including a flat monthly rate. The creation of a multi-sided market for the interaction between passengers and different mobility companies could trigger indirect network effects, as well as algorithmic network effects if AI is used to automate the interaction between the parties. This is not the case at the moment, but seems to be the natural evolution of Whim.

Platforms such as Whim, Uber, and Google are the candidates to build a digital network on top of the existing urban transport networks: a network of networks providing door-to-door solutions to passengers, fully exploiting the different network effects.

15.1 What Is MaaS?

In 2012, Sampo Hietanen, a Finnish civil engineer, developed the idea of MaaS when he was leading the non-profit association Intelligent Transport Systems Finland (ITS Finland). The idea was the integration of various forms of transport services into a single mobility service accessible on demand. A single app, with a single payment, would allow access to different transport modes: collective transportation, car rental, taxi, car-sharing, bike-sharing, and ride-hailing, with the aim of substituting the ownership of private cars.[2]

From theory to practice, Hietanen incorporated MaaS Global in 2015 and became its CEO. In February 2016 a seed round was completed, raising €2.2 million from Transdev (a French transport company), Veho (a British consumer electronics company), and Tekes (the Finnish Funding Agency for Technology and Innovation). The service was launched in Helsinki in 2017 under the name Whim. In August 2017, a second round of investment was closed, raising €10 million from Toyota. The international expansion started with Birmingham in 2018, to be followed by Antwerp and Vienna.

The Whim app is an evolving exercise. In 2020 it offered four options to travel around Helsinki. In exchange for a monthly fee of €499, Whim Unlimited provides unlimited rides in the local mass-transit system (provided by the local public company); unlimited car rental services (provided by Sixt); the unlimited taxi rides originally proposed were later substituted by a cap of 80 taxi rides in a 5 km radius, available only on the days that car rental is not used (provided by local taxi company Taksi Helsinki); and unlimited use of local shared bikes. "Whim urban 30" provides unlimited access to local mass transit, shared bikes and a limited number of taxi ride for a monthly fee of €59.70. A similar offer includes car rental for weekends. "Whim to Go" is a pay-per-ride option with no monthly fee, which allows access to local mass-transit, taxi rides, car rental shared bikes, and e-scooters at regular prices (no surcharge) with the convenience of contracting and paying through the app.

MaaS would be the next stage of an existing trend of combining various transport services into a single ticket or payment solution. Mass-transit authorities have been coordinating different services (subway, tram, buses, etc.) under a single ticketing or payment system for decades. The next step is to also include transport services provided by private operators (taxi, car rental, shared mobility services, etc.), particularly to overcome the "first/last mile" challenge. There is a high number of pilot programs offering combined solutions with mass-transit and shared mobility services. For example, Uber is working with local transit systems in Atlanta (MARTA), Dallas (DART), and Portland,[3] among others, to combine services. FreeNow, the German car-sharing operator, has reached an agreement with the Italian high-speed railway operator Italo to sell combined tickets.[4]

MaaS programs include traditional mass-transit services, plus private services such as taxi and shared mobility services, to meet all the mobility needs of different passenger profiles.

15.2 Network Effects

While MaaS is a convenient booking app, it is not clear at this stage whether Whim by MaaS Global will trigger the indirect network effects that ignite a platform. Convenience is a common trait for mobile apps, but it is not the main reason behind their success. In order for platforms to succeed, they must trigger network effects in a way that reduces costs, and such cost reduction must be properly distributed across the ecosystem.

MaaS certainly reduces transaction costs, as most local transport services can be contracted in a single app, either for each service or for a flat monthly fee. MaaS reduces the hassle of constant micropayments for each ride. The journey planning tool provided by Whim is also useful and reduces search costs.

It is too early to tell whether the incentives for all the members of the ecosystem have been properly defined. In particular, it is well known that flat rates for users are often non-sustainable when the platform must pay a fixed fee for every use of the service to the service provider. This was particularly the case of taxis and car rental services in the Whim offer. The flat rate proposed by Whim might face the same challenge as ClassPass.

ClassPass was launched in New York in 2013. It became popular, as it allowed users to select an unlimited number of classes in different gyms for a flat monthly rate. The flat rate attracted heavy users who attended a high number of classes. They would always pay a flat rate, but the platform had to pay gym owners by the class. ClassPass was spending heavily on customer acquisition, but it had the resources because investors had provided substantial capital. After a certain point, heavy users were not allowed to contract the flat rate, and at a later stage, such a rate was dropped in favor of a pay-as-you-go pricing model. On the other side, when the number of users was low, gyms benefited from extra revenue to fill the empty spots in their classes – an interesting proposition. However, users started to drop their monthly payments to their favorite gym in favor of contracting directly with ClassPass, as it allowed more flexibility and attending different gyms. Gym services were commoditized and providers faced more competition. The most popular gyms can survive without the platform, but regular gyms cannot, even if this new business model puts strong pressure on their finances.

The flat rate proposed by Whim might face the same challenges as ClassPass. Heavy users of the more expensive services (taxi and rental cars) might pose an excessive financial burden on the platform. The unlimited taxi rides initially provided were later substituted with a cap of 80 rides per month.

However, it is not obvious how Whim by MaaS Global, at this stage, triggers cost reductions derived from network effects. There is not yet an algorithm that optimizes the interaction of passengers and mobility companies. Whim does not automatically match passengers and service providers. From the demand perspective, the platform does not identify and actively propose the best traveling option for passengers. On the supply side, service providers are not benefiting from new passengers in off-peak times or avoiding empty runs.

At this stage, the value in Whim's proposition is to substitute private vehicles with a more convenient aggregation of collective and shared mobility services. By giving a flat rate and access to a diversified set of mobility options, individuals can more easily drop their private vehicles. This is high on the political agenda of most cities, so MaaS often has the active support of the local government where it is active.

In any case, MaaS has the potential to evolve into a far more effective proposition, creating powerful network effects as it is transformed into the network of networks.

15.3 The Network of Networks

The concept of mobility-as-a-service goes further than just the convenience of a booking app and a flat rate. Transport platforms can create a multi-sided market where travelers on one side and mobility service providers on the other can interact in a coordinated way, thus transforming the previously isolated transport modes into a fully coordinated transport system: a network of networks.

Some urban transport modes already operate as a network. This is the case with urban railway services, public bus services, and new shared mobility services. They concentrate a high number of travelers in a coordinated way. Railway and bus passengers travel together in the same vehicle, reaching high levels of efficiency in terms of cost and occupation of congested public space. New shared mobility services, including ride-hailing, car-sharing, and bike-sharing, go a step further, introducing a network on top of previously isolated vehicles, increasing the efficiency in the provision of the service.

However, each network operates with little coordination with the rest of the transport networks. Local authorities try to coordinate different transport modes by creating hubs so that passengers can transfer from one network to another; they try to coordinate schedules and they try to develop common ticketing systems. However, such efforts have had limited success so far.

Technology empowers better coordination of the different transport modes, creating a network on top of the pre-existing networks. The Internet of Things (IoT) has made it possible to gather information about the location and conditions of the vehicles (trains, buses, cars, bikes, and so on). Artificial intelligence has made it possible to analyze such information and to determine the optimum coordination of the transport mode.

Platforms play a central role. Firstly, the IoT empowers the network of networks. Users can be connected though their smartphones. And now vehicles can also be connected through specific sensors that provide information on the location of the vehicles, the number of passengers in the vehicle, the speed, etc. Furthermore, it is possible to incorporate sensors into the roads in order to obtain information about the evolution of traffic: congested streets, incidents in the service, and so on. The IoT makes it possible to digitalize the real world, mirroring it and creating a digital proxy of it – a digital twin.

Secondly, software is in the position to organize the information in a sensible way. Algorithms and AI can identify the patterns and the opportunities for the coordination of the different transport modes. Intermodality can be increased as there is more information about each transport mode and such information can be organized in favor of the passenger. It is possible to identify the best combination of services for a traveler to move from Point A to Point B in a seamless and effortless way. In this way, passengers can optimize the coordination of mass-transit services with first/last mile mobility.

Travelers already have access to good sources of information about travel options. Waze (owned by Google) provides real-time information about traffic conditions and the best routes to drive around the city or region. Google Maps also provides information about available traveling modes, including mass-transit, shared-mobility and the best intermodality options to reach a destination.

A similar, if not more powerful effect can be identified on the supply side. It is not only that travelers can passively identify the best mobility alternative. It is also that the providers of mobility services can actively adapt their services to the passengers' needs. As demand can be anticipated in a more precise way, suppliers can adapt their services in terms of schedule, frequencies, and coordination with other transport modes.

Most mobility managers are aware of the need to digitalize their operations, to install sensors into their infrastructures and vehicles, to gather data about their operations, and to improve the efficiency of their systems. They perceive the possibility to increase the efficiency in their operations and to increase consumer satisfaction.

However, it is not sufficient to digitalize the mobility networks. The deepest transformation will come from the interactions between passengers and the different transport modes that digitalization makes possible. Of course, digitalization will transform the direct relationship between each mobility service provider and the customers. However, the transformation will not finish there.

Digital platforms have the ability to facilitate the interaction between different parties, between the different sides in multi-sided markets. Digital platforms are in a position to receive data from all the parties thanks to the internet (not only from computers and smartphones, but from all kinds of sensors). Platforms, then, have the capability to make sense out of all these data through their algorithms, their AI systems. Thanks to these capabilities, they can optimize the interaction between all the parties. Such interactions are not limited to passengers or to a specific mobility service provider. The most powerful role of platforms is to facilitate interaction among passengers and all the service providers at the same time, making them work as a single system, thus creating a network of networks.

However, the role of digital platforms in transport could be even more relevant. Platforms, being at the center of the system and having the AI capability, can become the organizers of the mobility system. They will inform travelers of the best traveling options. They will inform mobility providers about the demand for their services. They will not only inform the parties, but also determine the available mobility options. They will have the power to nudge passengers to use a specific mobility option. Through their suggestions, they can steer the flow of travelers from one transport mode to another, from one route to another, even from one moment of the day to another.

On the supply side, platforms might substitute traditional transport managers as coordinators of each transport mode. Railway or bus companies not only provide the mobility service, but they also organize it, as they decide the routes and schedules. They decide when a specific train or bus is operated, when the maximum capacity is in operation, etc. Traditional network operators had control of the direct network effects, which they had created by supplying the infrastructure/service that allowed the aggregation of traffic (passengers/freight). They decided how traffic would be managed (schedules, frequencies, quality of service, etc.) and how the benefits of the direct network effects would be distributed between the passengers and the service provider, in the form of rates for passengers and benefits for the service provider.

Platforms create larger network effects by adding to the traditional direct network effects in each network, the indirect network effects derived from the interaction of all the different networks and the algorithmic network effects that empower machine learning. As large pools of travelers make their traveling decisions through the platform, the system as a whole can increase its efficiency, distributing traffic flows better across the existing transport modes. The network of networks brings new efficiencies that will far exceed the benefits created by the isolated networks.

As platforms are creating these new network effects, they have the power to coordinate the system and the power to nudge travelers into a specific network and route. Transport service providers will increasingly adapt their services to the flow of traffic directed by the platform. They will always have the role of transporting passengers and freight, but will gradually lose their direct relationship with the passenger and, more importantly, their role of coordinator of the service. As described in the following sections, transport services providers might get "platformed."

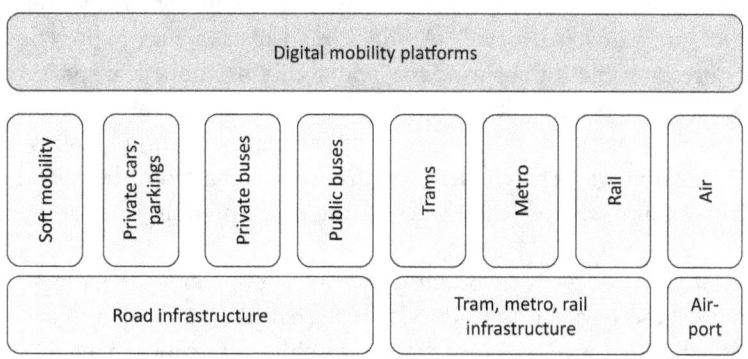

Figure 15.1

15.4 The Race to Be the Network on Top

Who will win the race to become the leading mobility platform? Whim by MaaS Global is currently the front runner in the aggregation of urban transport modes under a single digital platform with the ambition of meeting all the mobility needs of a regular individual. Undoubtedly, it will be a valuable experience showing the path for future developments. However, Whim is not alone in the race to build the leading mobility platform.[5]

Mobility platforms such as Uber and, on a smaller scale, Didi in China and BlaBlaCar in Europe, already enjoy large numbers of satisfied users who repeatedly return to their favorite mobility app to meet their mobility needs. We have already described how Uber has expanded into other services (Uber Eats and Uber Rush) and also how Uber is partnering with mass-transit systems to coordinate their offers. Mobility platforms are evolving to aggregate new transport modes into their offerings. MaaS would only be a natural evolution of their business model, which is to facilitate the contracting of transport services provided by third parties. However, operative challenges are very relevant. It is not simple to coordinate different transport modes and to ensure the seamless navigation from one to another. Pricing is also a challenge.

The lead could also be taken by large multipurpose platforms such as Google. Google has the largest pool of users and the best data and is already active in the organization of mobility information. Google Maps provides accurate information about travel times using different transport modes, including private cars, mass-transit, and walking. It also includes information about Uber services. Google Travel allows users to search and book the best traveling options. In 2013 Google acquired Waze for $966 million. Waze pools together a community of drivers feeding traffic information into the platform. Google also has the advantage of having a large set of data about passengers (location, calendar, preferences, and routines, etc.) at its disposal. In view of this, Google is in a good position to lead a mobility platform aggregating different mobility solutions.

Public authorities managing urban transport systems might vertically integrate and try to aggregate other transport services into a platform. Mass-transit is the backbone of the urban mobility system in most major metropolitan areas, particularly in Europe and increasingly in Asia. It is easier to complement the already coordinated set of mass-transit services (buses, railway, subway) with secondary services such as taxis, rental cars, and shared mobility solutions (ride-hailing, car-sharing, and bike-sharing) provided by third parties, under a single platform. "Transport operators will benefit by creating a larger market via the integrated platform."[6] A widespread idea in Europe is that local public authorities are responsible for ensuring the availability of public transport to all citizens. European local transport authorities perceive themselves as the natural managers, at least of the local mobility platform. Managers of public transportation systems have already identified this possibility: "The challenge is that something from the outside world will act faster than us, and become the leader of the urban mobility market. This threat is very real."[7]

Regardless of who succeeds in building the most popular aggregating platform, the management of the ecosystem will always be a challenge, as the number of players is high, they have different incentives (particularly public authorities managing mass-transit systems), and the operational challenges are relevant. However, transport is one of the largest markets in the world, so the prize in this race is definitely worth the effort.

Notes

1 MaaS Alliance (2017). *Guidelines & Recommendations to create the foundations for a thriving MaaS Ecosystem*, p. 2, retrieved from https://maas-alliance.eu/wp-content/uploads/sites/7/2017/09/MaaS-WhitePaper_final_040917–2.pdf
2 Heikkilä, S. (2014). *Mobility as a Service–A Proposal for Action for the Public Administration*, Case Helsinki, Master's Thesis, Aalto University.
3 Uber (2017, July 23), *Uber & MARTA: Connecting The Last Mile*, retrieved from https://newsroom.uber.com/us-georgia/uber-marta-connecting-the-last-mile/
4 Ferpress (2014, Feb. 19). *Car2go e Italo: parte da Milano l'alleanza tra il car sharing e il treno di NTV*, retrieved from www.ferpress.it/car2go-e-italo-parte-da-milano-lalleanza-tra-il-car-sharing-e-il-treno-di-ntv/
5 Holmberg, P., Collado, M., Sarasini, S., & Williander, M. (2016). *Mobility as a Service: Describing the Framework*, Viktoria Swedish ICT AB, January 15, 2016.
6 Kamargianni, M., Matyas, M., Li, W., & Schäfer, A. (2015). *Feasibility Study for "Mobility as a Service" concept in London*. UCL Energy Institute for the Department for Transport.
7 Caroline Cerfontaine, UITP Combined Mobility Expert, in UITP (2014). 5th UITP Combined Mobility Workshop: Barriers, *Branding, and Business Models for Integrated Mobility*, retrieved from www.uitp.org/barriers-branding-and-business-models-integrated-mobility

Chapter 16

Transport Providers are Platformed

Transport platforms do not aim to deploy their own infrastructures and vehicles, thus substituting the existing transport providers. On the contrary, they are positioning themselves on top of them, to coordinate the existing transport networks and services to build a seamless door-to-door proposition for passengers. Transport markets are being "platformized"; that is, transformed into multi-sided markets intermediated by a digital platform. This is the case of urban transportation with Mobility-as-a-Service, but also in long-distance transport both of passengers and goods.

However, it is already well known that, as service providers are "platformed," they lose their direct relationship with the passengers and their services are commodified. Platforms extract an increasing share of the value in the industry and have the potential to become system coordinators, substituting service providers in that role. The stronger the position of the platform, the weaker the position of the underlying service providers.

This poses a dilemma to traditional players, as they might be tempted not only not to work with platforms, but to delay digitalization in order to obstruct platformization, or at least block data sharing to make platformization more difficult.

As transport is vested with general interest, platformization requires that funding is available for the construction and maintenance of infrastructure and that the operation of the transport system meets general interest objectives such as a wide geographical coverage, affordability, and sustainability. Traditional regulatory tools might have to be redefined for the new market structure. The general interest will have to be included in the algorithm.

16.1 Platformed

Mobility platforms do not have the intention or ability to substitute the current providers of transport services. Platforms can build digital twins, but do not have the financial means to duplicate the physical transport infrastructure at a global or even at a local level (roads, railways, ports and airports). Platforms do not even have plans, at least currently, to massively acquire vehicles (trains, buses, cars, vessels, and aircrafts) in order to substitute the fleets of the existing transport companies. We do not think transport infrastructures and assets will be substituted as is happening, at least partially, to postal services or taxicabs.

While transport infrastructure and assets will not be substituted, they will be "platformed." As described in the case of telecommunications operators and content producers, they will always be necessary, but their role in the system will gradually diminish.

The assets and services will be firstly commoditized; later, their role as system coordinators will be taken over by the platforms.

The first and most evident effect when a platform grows in an industry is that users (in this case passengers and shippers) start to interface with the platform for contracting the service, and not with the traditional service provider. Users of Whim by MaaS Global substitute the direct relationship with the mass-transit operator, the taxi operator or the rent-a-car operator, with a single interface: the app run by the platform. Services are contracted and paid for through the app. Passengers might be formally contracting with the traditional transport companies, but through an intermediary that controls the relationship with the customer.

The relevance of the direct contact with the customer is fully understood by all players. Transport providers are fully aware of the risks generated by delegating the relationships with their customers to an intermediary. A relevant effect is the difficulty of implementing innovative pricing policies like discounts for heavy users, yield management techniques, and so on. Intermediaries are also an obstacle to gathering data from customers, as intermediaries might be the only entities that have full visibility over the identities of the passengers, their traveling habits, and other relevant data. All players are increasingly aware of the relevance of these issues.

The second effect is that services intermediated by a platform become commoditized. This is particularly relevant when the number of competitors is very high and they all provide a similar service. This is the case of taxi services included in Whim.

Service providers will have difficulties differentiating the service they provide from the service provided by competitors. Platforms try to standardize services as much as possible so as to simplify their intermediation services. They actively discourage differentiation by service providers. Platforms might even impose obligations on service providers to standardize their services along specifications required to join the platform. Usual strategies to differentiate the service, such as brand recognition, are discouraged.

Competition between the different transport service providers included in the platform becomes more intense. As services are standardized, it is easier to compare service conditions. In most cases, price is the main or even the only proposition to compete. This is increasingly the case in BlaBlaCar, as most passengers select drivers based on the price they have requested.

Some platforms even standardize the pricing of the underlying services. This is the case of Uber: all services in a particular category (SUVs, limos, etc.) are provided at the same price. The platform imposes a single price, or at least the maximum price to be charged. In this way, platform operators can fully automatize the matching of users and service providers, a condition necessary for the provision of services in real time, known as on-demand services.

In this way, the underlying services intermediated by the platforms become commoditized. Services can no longer be differentiated. Service providers lose control over pricing. They just passively wait for the platform to assign them passengers according to the algorithms. Transport service providers end up working for the algorithm.

The third effect is that platforms increasingly organize the service, instead of the traditional service providers such as railway, shipping and air companies. Platforms use their technology to identify and predict the individualized demand of passengers and shippers and they are in a position to call for a specific service to meet the request. This is already happening with Uber's ride hailing services and is starting to happen with larger coach services and even with rail services (Flixbus in Europe is an example). A MaaS platform might eventually become the organizer of the different transport modes in a city, deciding when demand is met with small vehicles as demand is low at night: at peak hours demand will again be met with larger vehicles.

One might wonder whether "platformization" will ultimately put at risk certain business models where the coordination of transport is central. We have seen that news pieces are unbundled and platforms are assuming the role of pulling together the most interesting pieces produced by third parties individually for each reader, making the role of the newspapers and TV stations as aggregators redundant, as editors decide how to bundle the different pieces to create a complete product. Networked media simply produce unbundled news pieces to be distributed by platforms. Transport platforms might eventually coordinate unbundled transport pieces – that is, cars, vessels, and aircrafts in wet lease (aircrafts including the flight crew and fuel) – making shipping companies and airlines redundant. As described above, these companies have created network effects by organizing standardized vehicles, routes, schedules, and so on. As platforms take on the role of coordinating third-party assets, and as they become able to create larger network effects thanks to algorithms, the existing system organizers might become redundant. Transport infrastructures and assets will always be necessary, but a company coordinating them, other that the platform, might not be necessary. Platforms might end up coordinating fully unbundled transport units.

Still, there is a limit to the role of platforms as organizers of transport systems. Some systems already have entities coordinating traffic for technical and safety reasons. This is the case of aviation. As well as airports, aircrafts, and airlines, there are entities controlling air traffic, mainly for safety reasons: The Federal Aviation Authority (FAA) in the USA, Eurocontrol, and the national Air Navigation Services Providers (ANSPs) in the EU and so on. Regulation protects these entities as technical coordinators of air traffic. Digital platforms can evolve into market coordinators, but only a political decision can substitute these technical coordinators. A similar role is played by infrastructure managers in the technical coordination of railway traffic. No similar role exists in the maritime industry or in the road industry, which are more decentralized.

16.2 Market Power

Once the service is commoditized, platforms have the power to set the conditions for service provision, particularly the pricing conditions. Over time, many platforms have reduced the revenue per service paid to the transport service providers. This happened with bike riders working with Deliveroo in London in 2016[1] and has happened a number of times with Uber.[2]

Prices can be reduced as the result of increases in efficiency due to network effects. As the cost of the provision of the service is reduced due to its more efficient provision through the platform (reduction in waiting times, reduction in empty-runs and so on), retributions for each service can be reduced. Platforms have aggressively passed on such cost reductions to passengers as a way of improving the competitiveness of the services against other competitors (taxicabs, for instance), to increase ridership, and to trigger further network effects for the benefit of the entire ecosystem: the passengers, the platform, and even the service providers, which could end up making more money by charging less for each service, but increase the number of services.

When platforms are growing, they have an incentive to aggressively share such benefits with the users in order to grow the customer base. It is even common to sell below price in order to grow. At such initial stages of the platforms, they must provide attractive terms to service providers to ensure that demand and supply are balanced. Service providers might confirm that despite reductions in fees per service, the total revenue actually increases due to efficiencies in the provision of the service.

As platforms become more stable they still have a strong incentive to share the benefits more fairly across the ecosystem. Platforms in multi-sided markets would need to keep all sides satisfied. If payments to service providers are reduced too much, service providers will start leaving the platform, the balance between supply and demand will break, users will have longer waiting times, meaning they will contract fewer services and the entire ecosystem, and the platform itself, will lose. Such an effect has been described as the "second invisible hand" in the platform economy.[3]

However, winner-takes-all dynamics might exclude competition and the fair distribution of the benefits created by network effects. Platforms achieve a level of market power that allows them to behave independently of the rest of players in the ecosystem, imposing on them terms and conditions in such a way that the platform itself monopolizes the benefits derived from the network effects. While no transport platform has yet arrived at this point, this seems to be the case with more mature platforms such as Facebook and Google.

Will mobility platforms eventually reach the market power that Facebook or Google already enjoy? Certainly network effects create concentrated markets with an oligopoly structure. Under particular circumstances, network effects lead to monopolies. It is too early to determine whether this will also be the case of mobility platforms, but some indications can be provided.

On thin routes where the number of transactions is not very high and network effects can only be triggered when a high proportion of transactions takes place in the platform, monopolization is probable. This seems to be the case of BlaBlaCar in Europe. On an average route, the number of daily transactions is limited. Passengers demand a minimum number of options to ensure they meet their travel plans. Full network effects are only possible if a platform accumulates all the existing supply and demand. Smaller platforms can compete in very dense routes between large cities, but at a national scale, the largest platform will eventually monopolize the market. Under these circumstances, platforms can become natural monopolies.

In thicker routes, where the number of transactions is very high, platforms might exhaust network effects as they reach a percentage of transactions that can be replicated by other platforms. In San Francisco, for example, Uber's scale is such that it can ensure that a driver can reach any passenger in less than one minute. If a second (even a third) competitor can reach the scale to replicate such level of service, there is room for competition. This seems to be the case, as Lyft has a substantial market share in a number of US metropolitan areas. Scale and network effects are always relevant, but they become exhausted at a certain point, so they can be replicated by more than one platform.

Thus, market power, and abuse of market power, will be among the fundamental debates around mobility platforms as they mature.

16.3 Vertical Integration

The market power of the platforms becomes an issue particularly when they integrate vertically and they not only intermediate but also compete in the provision of the underlying service.

Vertical integration triggered the regulation of aviation global distribution systems back in the 1980s. The Google Shopping case and the increasing calls to intervene against Google's expansion into the online travel agency industry through self-preference in searches are examples of conflicts triggered by vertical integration.

Such conflicts could become common in transport under different scenarios. One such scenario is when a platform that is active in a specific transport segment expands to become the platform that organizes mobility as a whole. This could be the case, for instance, with Uber expanding to include not only ride-hailing services, but also all other services. As already mentioned, Uber is already including other transport modes into its platform, starting with bike-sharing. Uber is also liaising with mass-transit authorities to better coordinate services, including the possibility of using the platform to acquire the tickets to use mass-transit services.

We think mobility platforms will grow organically, expanding from a successful transport mode to neighboring and complementary transport services, rather than just establishing themselves as the leading platform out of the blue; "top-down," so to speak. However, these organically evolving platforms will still face conflicts of interest: as they are more firmly established in a particular transport segment and take larger commissions for the intermediation of the corresponding services, they might be tempted to favor this segment over other transport services they intermediate.

Another scenario is that of a transport service provider scaling up to establish itself as the platform of reference, including not only its service, but services provided by other transport service providers. This could be the case with public authorities that operate large transport systems, often under legal monopoly rights, such as subway systems, suburban railways or bus services. If such a service is the backbone of urban transportation in a large metropolitan area, it will be in a good position to impose itself as the platform coordinating transport at a metropolitan scale. Other services, such as taxis, car rental, and shared mobility solutions, could be included in this platform in order to complement the backbone services.

Public authorities not only operate transport services, but also have the power to regulate transport services provided by third parties, particularly taxis and some shared mobility services (shared bike services are often owned and operated under concession by public authorities). As a result, they are in a particularly strong position to impose the conditions for the participation in the platform and even to impose an obligation on these third parties to participate in the platform.

In any of these scenarios, or others that might arise, vertical integration creates conflicts of interest and the need to increase transparency in the operation of the platform so as to avoid discrimination.

16.4 The Digitalization Dilemma

If transport service providers become commoditized and might lose their traditional position as coordinators of their systems, why should they digitalize and work with platforms at all?

First of all, companies across all industries are digitalizing; that is, they are introducing sensors to extract data in order to have a better understanding of the functioning of their systems and therefore improve them (cost reduction, automatization, AI, etc.). No one wants to miss out on the so-called Fourth Industrial Revolution,[4] including transport companies.[5] Digitalization brings evident opportunities to increase efficiency in the provision of each transport service taken separately; simply think of the automatization of the system.

Once a company starts to generate data, there is a risk that such data will become accessible to third parties, platforms in particular, even without the explicit consent of the transport companies. This was the case of music back in the early days of the internet. Music was digitalized and, subsequently, the owners of the rights over the music were unable to control the exchange of the music files through the platforms.

While transport services cannot be virtualized and provided over the internet like music files, data about the service can be accessible to the public and become the object of trading in a platform. As an example, the location of vehicles (buses, trains, etc.) is a reference that can be used to feed a platform with the possibility to coordinate multi-modal trips. This could be the beginning of a process of commoditization. It is no surprise, therefore, that transport companies around the world are becoming extremely protective of their data.

A key step on the process of becoming platformed is the ability of the platform to commercialize the service provided by a transport company, particularly in a bundle with other services. Data in itself might not be sufficient for a platform to integrate a transport service if the platform is not allowed to sell the ticket for the ride.

Still, transport companies have been collaborating with online intermediaries to improve their distribution networks. This is the case of railways in some countries (such as the UK) and definitely of air companies working with global distribution systems. The pressure that such intermediaries place on prices is well-known in the hotel industry. Online travel agencies might actually evolve into full platforms intermediating different transport services and becoming the organizer of the network of networks. They could do it on their own or collaborate with another player.

Collaborating with a platform can be the result of a more conscious strategy. For example, transport providers might identify the opportunity to trade their spare capacity through a platform in order to increase revenue without putting the bulk of their capacity for trade in a platform. This was the original strategy of gyms working with ClassPass. It seems to be the case of taxi companies and car rental companies working with Whim by MaaS Global.

Furthermore, fully collaborating with a platform, even taking the risk of being commoditized, can be the best strategy for a transport company. All competitors might refrain to work with a platform. However, if a service provider starts to work with a platform, it could monopolize the efficiencies generated by the network effects generated by the platform, provide services at a lower cost, expand capacity, and endup expelling from the market the competitors that initially refused to work with the platform.

This is what is happening with taxicabs all around the world. Taxicabs could have chosen to work with ride-hailing platforms such as Uber. They had the vehicles and the drivers. However, they preferred not to do so and to keep their traditional off-line business model. As barriers to entry were not relevant (other than regulatory barriers), other vehicle owners started to work with the platforms (private hire vehicles, limos, and private car owners). They ended up substituting taxicabs in a high number of services. Many taxicab companies went bankrupt and are now out of the market. It did not have to end up this way; taxicab companies could have faced up to the digitalization dilemma in a different way.

Should a transport service provider embrace change and be the leader in working with the platforms, or conversely, oppose change, thereby delay the development of the platforms and hope that platforms will remain irrelevant in their industry? This is the digitalization dilemma.

The stronger the competition (or the lower the barriers to entry), the more chances platforms have to find supply. This is the lesson from the taxicab industry. Sooner or later a competitor will be ready to embrace change and benefit from the efficiencies delivered by platforms.

The situation might be different in transport modes that are highly concentrated, with limited players and high barriers to entry. Let us not forget that some transport modes are still monopolized. These monopolistic players might evaluate whether their refusal to work with platforms might effectively avoid the emergence and growth of a platform in their business area. However, such an approach might backfire.

Even if an entire transport mode refuses to join a multi-sided platform, competing transport modes might do so and attract passengers. This might be the case of bus companies competing with railway companies with exclusive rights.

Finally, the obligation to engage with a platform can be legally imposed upon transport providers by the public authorities. Such an obligation can derive in certain jurisdictions from the application of competition law and arguments based on the essential facilities doctrine or the abusive refusal to deal, particularly when a service provider is dominant or even in case of joint dominance. There are increasing voices asking for transport data to be considered an essential facility, particularly when generated by monopolistic or at least dominant operators, which is often the case of railway undertakings. In 2019 the German Competition Authority opened a case against the German railway undertaking Deutsche Bahn for refusal to supply transport data.

The European Union is actively promoting data sharing in transport. The obligation has been imposed upon transport operators by the Delegated Regulation 2017/1926, forcing them to share a whole list of data, in standardized format, so that it can be accessible in so-called National Access Points. The first set of data were supposed to be made available on December 1, 2019, and further data would have to be made available in 2020, 2021, and finally 2023. Such data includes not only static information about timetables and rates, but also real-time information about operations.

This data is available to any third party: a start-up, a large platform (Google, Uber, etc.), and to competitors. The data can be reused. There is no limitation to the kind of reuse that can be made, but a license agreement will rule the terms and conditions. The guiding principle is that reuse will have to respect principles such as neutrality and non-discrimination, particularly when ranking travel options. An obligation not to mislead the end-user has been defined. These principles build on the regulation of Computerized Reservation Systems in aviation (Regulation 80/2009) and are, more broadly, based on the same principles defined for platform to business regulation adopted by the European Union (Regulation 2019/1150), applying not only to transport but also to search engines, social networks, and others.

Furthermore, the debate about data sharing regulation is often used to impose other regulatory obligations on transport operators. It is important to disentangle the data sharing debate by differentiating the various topics.

On one hand, the data sharing debate should be separated from the ticketing debate. Single ticketing was another traditional policy in Northern Europe to foster the use of mass-transit. Data sharing obligations are often connected to compulsory distribution agreements, so that platforms can not only provide information about travel alternatives, itineraries, rates, etc., but also sell tickets to use transport services provided by third parties. This has been the case in Finland and it is therefore no coincidence that Finland is where MaaS was developed. The Finish Transport Law imposes upon traditional transport service providers (railways, bus services, etc.) the obligation to allow third party distributors to commercialize their services, offering them all the available discounts. The legislation of the European Union foresees potential obligations on integrated and through-ticketing to be imposed on railway undertakings (Article 13a Directive 2012/34).

These obligations mirror the "net neutrality" obligations already imposed upon telecommunications carriers in favor of OTT platforms. In the web of the MaaS Alliance such an obligation has already been requested, under the reference to the need for openness and roaming. It could be argued that the net neutrality analogy is more appropriate than the roaming analogy, as roaming is a horizontal agreement among network operators, while MaaS requires vertical agreements between the platform and underlying network operators (transport providers with their own infrastructure, vehicles, etc.).

Digital platforms can certainly facilitate the contracting of transport services, but they can also disrupt the industry. It is not clear at this stage whether regulatory intervention is necessary or even desirable in order to facilitate the emergence of transport aggregators, or whether the market should evolve in a more organic way, through voluntary contractual agreements between transport service providers and potential aggregators.

Business-to-business data exchanges, particularly between infrastructure managers and transport service providers, are necessary for a more resilient transport system. Transport actors can drastically improve their efficiency by better coordinating though the exchange of information. Reluctance among players to digitalize and share data with other industry players is often a reaction against players with market power so as to protect their position. Obstacles to data sharing are often a reflection in silico of the refusal to better coordinate with other actors, for instance in the provision of access to infrastructure services.

Business-to-customer data exchanges, often connected to the distribution and ticketing of services, are usually perceived as more delicate, as they can modify the status quo for the benefit of new digital actors to the detriment of traditional players heavily investing in the provision of transport services. Data sharing for the full display of information for passengers and shippers, so they can better decide on their travel plans, seems a suitable objective and a balanced obligation to be imposed upon traditional player, particularly if they benefit from public funding, either in the form of exceptional State aid or EU funds or in the traditional form of compensations for public service obligations.

On the contrary, disguising a potential obligation to be imposed on transport service providers to sell their services through digital platforms as data sharing could unsettle the equilibrium between traditional and digital actors. The terms for platforms and aggregators to become distributors of transport services should be commercially negotiated. Compulsory commercialization of transport service through digital platforms should only be imposed upon traditional players under exceptional circumstances and only when the regulation of platforms is mature enough to avoid abuses by winner-take-all superintermediaries. Air transport provides the right model. There is widespread data sharing, but also a long lasting regulation of the activities of the intermediaries.

16.5 General Interest

As a final reflection, it is important to remember that transport is considered an activity of general interest with a decisive impact on the well-being of citizens and the competitiveness of cities and countries. Such general interest has been one of the traditional reasons for public intervention in transport markets all around the world. Platforms and the transformations brought by platforms should and will not change this.

Platforms might bring substantial benefits in terms of efficiency, cost reduction, and improvement of door-to-door services. Users can be expected to benefit from such improvements in efficiency. However, such efficiencies will not exhaust the challenges posed by the general interest. Affordability and universality of transport services, for example, will continue to be issues.

New challenges will arise, some of which can already be identified. Consider the following simple example. Leonia is a borough just across the Hudson River from Manhattan. It is mostly a residential area, but more than 300,000 vehicles travel though it to connect with the George Washington Bridge to get in and out of Manhattan via Interstate Highway 95. In January 2018, local authorities decided to ban non-residents from using residential streets during rush hour.[6] Traditionally quiet residential streets were being used as an alternative to the congested highway, particularly by Uber and Lyft vehicles directed to these streets by navigation devices. Platforms look for the fastest rides with the lowest costs. Their algorithms do not factor in general interest criteria, like respect for residential areas.

The increased efficiencies brought by ride-hailing platforms are actually increasing congestion in the densest urban areas of New York City, Boston, and San Francisco. Low prices attract more passengers and more passengers attract more vehicles. Algorithms do not consider the side effects of their ride-hailing platforms.

In a similar way, BlaBlaCar is detracting passengers from the environmentally more friendly railways, thus increasing the use of private cars for long-distance travel in France. The efficiency in the use of private cars brought by platforms has again increased congestion, damage to the environment, and risks for the safety of travelers, against long-standing public policies.

As platforms reinforce their role as system coordinators of the network of networks, more challenges will emerge. As platforms direct flows of traffic to one transport mode or another, they can reinforce or defy public policies that favor public transportation against private vehicles, environmentally friendly transport options against more polluting options, etc.

It has also been identified that the pressure of the platforms on transport service providers might affect the financing of transport infrastructure (including vehicles, but also roads, railway, etc.). Furthermore, platforms increase competition and reduce revenue for service providers. Commissions charged by platforms to service providers detract funds from the industry. Value erosion and revenue reduction are threats to the funding of the deployment and maintenance of transport infrastructure.[7]

It is still an open question how public authorities can enforce their policies under this new situation being created by platforms. Platforms are substituting service providers and regulators as coordinators and as managers of the transport system. New instruments have to be identified to ensure that platforms bring efficiencies while promoting the general interest. The general interest has to be included in the algorithm, and such inclusion must be supervised and/or regulated.

Notes

1 Osborne, H. (2016, Aug. 15). Deliveroo workers strike again over new pay structure, *The Guardian*, retrieved from www.theguardian.com/business/2016/aug/15/deliveroo-workers-strike-again-over-new-pay-structure

2 Krisher, T. (2017, Mar. 3). Uber drivers are growing angrier over price cuts, *Business Insider*, retrieved from www.businessinsider.com/uber-drivers-are-growing-angrier-over-price-cuts-2017–3

3 Goldman, E. (2010). The Regulation of Reputational Information. In *The Next Digital Decade. Essays on the Future of the Internet*, TechFreedom, Washington, p. 295.

4 Schwab, K. (2016). *The Fourth Industrial Revolution*. Penguin, New York

5 Montero, J. & Finger, M. (2021). *The Modern Guide to the Digitalization of Infrastructures*. Edward Elgar, Cheltenham.

6 Foderaro, L. (2018, Jan. 22). New Jersey town aims to keep app-guided outsiders off its streets, *The New York Times*, retrieved from www.nytimes.com/2018/01/22/nyregion/leonia-gps-navigation-apps.html

7 Finger, M., Bert, N., Kupfer, D., Montero, J. J., & Wolek, M. (2017). *Infrastructure funding challenges in the sharing economy*, Research for the TRAN Committee of the European Parliament, retrieved from www.europarl.europa.eu/RegData/etudes/STUD/2017/601970/IPOL_STU(2017)601970_EN.pdf

Part IV

Platforms in the Energy Industries

This Part will show how digitalization is and will increasingly penetrate the energy sector, and how it will transform it in the processes. However, the energy sector, at least its electricity portion, which is about to become the core vector of the transport of energy, has some fundamental differences from other network industries. These characteristics result from physics: because electricity cannot be stored on a massive scale, supply and demand must constantly be balanced. This requires a real-time system coordinator to ensure that the system works by keeping it balanced. This is a central coordinating function and it is questionable whether a virtual, digital platform, can take over this function, at least not in the immediate future. This is unlike transport, which is highly fragmented and where digital platforms can perform system coordination while delegating the operational parts to fragmented operators. Nevertheless, digitalization is an unstoppable process that will increasingly penetrate the electricity sector and the other energy sectors. In this Part, we will show how this is currently taking place and where this evolution might lead to.

Chapter 17

Network Effects in the Energy Industries

Network effects have played a fundamental role in the evolution of the electricity industry. Just as in posts, telecoms, railways, maritime transport, and aviation, the history of the electricity industry can be explained as a quest to fully exploit network effects by physically connecting a growing pool of assets in the most complementary way, for the benefit of all the users of the system.

Electricity was originally a distributed network, with each user producing electricity on its premises. Thomas Edison built the first integrated, centralized electricity system, producing electricity in one location and distributing it to a large number of users through an electricity grid. Complementarities would be exploited, as the grid would service electricity during the day to factories and offices, and during the night to private homes (bulbs). This proved to be the winning proposition in the industry. Over the decades, complementarities and the subsequent network effects would be exhausted by growing universal systems connecting electricity consumers of all types with all potential electricity generation modes, to ensure a balanced system.

Eventually, deregulation would transform the industry, introducing competition at least in the generation of electricity. The role of a system coordinator ensuring the balance of the system would always be necessary, contrary to other network industries.

17.1 Network Effects in the Early Electricity Industry

In early 1882, the financier J. P. Morgan had electric lighting installed at his private residence at 219 Madison Avenue in New York. He was one of the first individuals to benefit from such an innovation, thanks to the fact he was one of the angel investors in the company founded by the inventor Thomas Alva Edison. However, on the first night that the lights were switched on, the wiring at Morgan's library sparked a fire that destroyed his desk. He also had to deal with his neighbors' complaints about the noise, vibration, and smoke generated by the dynamo installed in a cellar below the stables at the rear of Morgan's property. However, such events did not discourage his support for electric lighting; in fact, he threw a party in his residence for 400 guests as a demonstration of the product and stated: "I hope that the Edison Company appreciates the value of my house as an experimental station."[1]

Morgan rarely invested in startups. He built his fortune by "morganizing" railways, telephony, and shipping; that is, consolidating existing competitors, reducing competition, and building scale and network effects. However, electricity, and electric lighting in particular, was an exception.[2] Morgan came to know about the experiments under development by Edison, the already famous "Wizard of Menlo Park." It is coincidental that Edison's laboratory was located in Menlo Park, New Jersey, just a few miles away from New York City, and that the headquarters of Facebook and venture capital firms Sequoia and Andreessen Horowitz are in Menlo Park, California.

Edison had invented the electric lightbulb in October 1879. After testing more than 3,000 possibilities, Edison identified the right carbon material to produce the wire filament to be heated by electricity to a high temperature to glow to produce the perfect light for households.

However, Edison did not only invent the lightbulb; he envisaged a whole electricity network to feed lightbulbs.[3] He envisioned a whole system formed with dynamos to produce the electricity in a central station, the copper wires to transport the electricity from the central station to the houses and offices, the meters to measure electricity consumption and bill the customer, and the equipment to regulate the voltage of the system. Edison's idea was to connect to the network not only electric lightbulbs, but all electric appliances. It may seem now like an obvious solution, but at that time electric motor devices such as elevators and electric trams often worked with batteries and were not connected to a network.

Electricity would not be produced at the customer's premises, as it was in Morgan's stables, but in a central station that supported hundreds of customers. The first central station started operations in Pearl Street, in downtown New York. It comprised four large coal-powered steam boilers, six stream engines, and six thirty-ton dynamos. The central station was connected with 18 miles of copper wire to a maximum of 7, 200 lamps half a mile around the stations. The whole network had a cost of $500,000[4] (around $12 million in today's money). On the afternoon of 4 September 1882, the system was switched on from the offices of J.P. Morgan at Wall Street.

Edison's proposal was a centralized network, contrary to the distributed model of private systems like the one installed in Morgan's residence. A centralized network could generate electricity more efficiently by using larger plants, and would save customers the hassle of managing their own generating system, as they would simply rely on the provision of electricity by a specialized company. At the same time, however, a centralized system would need a coordinator of the whole system, which could also be the company managing all of it and constitute the opportunity to grow a multimillion business out of electric lighting.

Edison's electricity network had a limited commercial success when it was launched. The number of customers was lower than expected, as was the revenue generated by the system. The second central station in New York, expected to be operative within months, only opened six years later. Even worse, competitors started to emerge, sparking the so-called "war of the currents."

The main challenge that Edison faced was the high cost of transporting electricity over copper wires. Edison's network was using low-voltage direct current (DC), running at 110 volts. Electricity transmission at low voltage was perfect for feeding lightbulbs, but proved to be very expensive to transport. Such an expense increased the cost of the centralized system, diminishing its competitiveness against private decentralized systems. Consequently, private systems were initially more popular than centralized systems. Edison installed more than 300 private systems while his central stations were growing at a very slow pace. Distributed generation was beating the centralized system.

A solution was developed outside Edison's system thanks to the work of engineers such as Nikola Tesla, a former employee of Edison. The solution was to increase the voltage for the transportation of electricity. Since the capacity of a wire is proportional to the square of the current traveling on it, each doubling of the voltage allows the same-sized cable to transmit the same amount of power four times as far, at the same cost. However, in order to safely use electricity in private premises, it was necessary to transform the current down to lower voltage once it reached the customer's premises. The solution was named alternating current (AC).

Edison was reluctant to switch to alternating current, for good reasons such as the fact that transformers were not a mature technology and AC motors were not available, so the network mostly lay idle during the day. The main reason for not switching was that meters for AC were not available, so the service could not be billed to customers. Edison started a crude campaign exaggerating the dangers of high-voltage electricity.

New competitors entered the market making use of AC technology. Up to 15 companies entered the market providing equipment and know-how to local licensees, often participating in the capital of such local utilities. From a regulatory perspective, the centralized model required local permits to install the wires distributing electricity from the power plant to the customer premises. Such permits were usually granted under exclusive rights, which led to the creation of local monopolies. The permits defined the terms of operation, such as the fee to be paid to the city and, often, the tariffs to be charged to customers.

As the technology matured, the new companies outgrew Edison's venture. Companies such as Westinghouse and Thomson-Houston were particularly efficient in low- and medium-density areas, as they could transport the electricity over longer distances at a low cost to reach the scattered population. Edison's company was stronger in dense urban areas where efficiency in transportation was not so relevant.

Over time, AC networks proved to be more efficient than DC ones. AC not only reduced transmission costs, but it also empowered companies to generate the electricity further away, benefiting from the most efficient electricity-generating opportunities. For instance, AC made it possible to use Niagara Falls to produce electricity for consumption in New York City.

17.2 Growing Network Effects

A second generation of networks[5] emerged in the 1890s as the market became mature for "morganization." Financiers, including Morgan of course, promoted mergers to concentrate Edison's company with other smaller firms and finally also with Thomson-Houston, creating General Electric in 1893.[6] General Electric concentrated two-thirds of the US market, with the rest under the control of Westinghouse. They both favored AC over Edison's preferred DC. Edison lost the "battle of the currents," but was awarded $3.5 million for his patents and his shares.

Local utilities, licensed by General Electric and Westinghouse, built universal systems providing electricity to residential users for lighting and small motors (for electric sewers and the like), but also to large industrial settings and to urban transport companies. Alternative current was transformed to the standardized voltages required by these different customers. Electrification was on its way, transforming industry and leisure. The parallelisms between electrification and digitalization have already been underlined.[7]

Having a diversified pool of users made the networks more efficient, as the load factor could be increased. Residential customers had their peak usage at night when they turned on their lightbulbs, while industrial customers had their peak usage during the day when the factories were active. Distributing the load during the day (and the different seasons in the year) increased the efficiency in the use of the fixed assets necessary to generate and distribute the electricity. Costs were reduced, increasing the competitiveness of the utilities' networks against the isolated systems used by factories and transportation systems. Over time, such isolated systems became the exception, and utilities' networks became the norm.

A key factor for the success of the utilities' networks was the creation of the "load dispatcher," the department in the utility that would coordinate the system by analyzing demand and ensuring the supply of electricity by the different generators in the system. Utilities would coordinate different generators, including ever-larger turbines and smaller back-up generators. The load dispatcher had to coordinate the operation of the electricity generation as to make sure that the dynamic demands of the heterogeneous customers would always be met. The load dispatcher was in charge of identifying the complementarities that would spark the network effects. The networks run by different utilities gradually became interconnected, which allowed them to exchange electricity and improve their load factors.

Large utility companies, managing increasingly large territories, became the center of the system. They benefited from economies of scale as they could build the largest generating plants, such as large hydroelectric plants. They also benefited from economies of scope, as the electricity was sold to more and more customers as electrification grew across industry and transportation. They also benefited from network effects; the more that users grew in number and diversity, the better the load factor could be distributed and the larger the complementarity, reducing the price of the service for all users. The power of network effects was also demonstrated in the electricity industry.

From a regulatory perspective, the control of larger utilities moved from local authorities, though installation permits, to state public utility commissions. Such commissions, well versed in the regulation of infrastructure networks due to their experience with water, railways, telecommunications, etc., applied the same logic to electricity networks.

17.3 Peaking Network Effects

The third generation of networks started around the 1920s, as utilities grew to control whole regions across the US, usually through intricate holding structures.[8] The drivers were to improve the load factor and to improve the generation mix. On one hand, technology enabled massive generation plants, both hydroelectric plants and carbon plants. Such plants created large economies of scale, but also required large pools of consumers that were beyond the reach of small utilities. On the other hand, it was necessary to balance the various generation plants, as hydroelectric plants might not be available during droughts, to ensure a constant supply at any time. The right complementarity in assets had to be identified to fully exhaust network effects.

Such large systems always required a system coordinator, an entity determining which plants would provide electricity at any given time, ensuring supply in peak periods, and the availability of back-up plants when other plants had to stop production for any reason. This role was played by an obscure alliance of interests around the holdings. Typically, in most countries, and in Europe across the countries, the industry coordinated itself (self-regulation). The role of the coordinator of the system was always at the center of the debate.

Electricity operators were nationalized in most countries, and some were merged into a single national monopoly that was then appointed to run the whole system. France is a paradigmatic case. The state-owned company Electricité de France (EDF) designed the generation mix, including large production plants such as nuclear power plants, built and operated the plants, was in charge of the long-distance transmission of electricity, and then the distribution to consumers. The public monopoly was in charge of coordinating the system.

In short, before deregulation, electricity was provided by regional and national vertically integrated monopolies that all benefited from network effects and economies of scale. Their limitation in size was not dictated by economics, but by politics, namely state boundaries. Most of these monopolies were publicly owned. They coordinated the balancing of supply and demand inside the vertically integrated company and then among themselves, factoring in big reserves.

17.4 Deregulation

Deregulation was a radical change for and a radical transformation of the electricity industry, one that would actually pave the way for digitalization. But more about that later. There are two versions of such liberalization: an American and a European one. The European version is more radical in the sense that it led to a more fragmented system, a more important role for coordination and subsequently for the regulator and finally bigger opportunities for digitalization. However, as we will see later, even digitalization cannot totally substitute for the role of coordinating the flows – supply and demand – of electricity.

The US started earlier, with its relatively soft version of deregulation of the electricity industry by allowing so-called independent power producers to feed electricity into the grid; or, in other words, by forcing grid operators to buy their electricity at a regulated price. This did not really change the role of the grid operator, except for that it has more input sources into the grid that need to be coordinated. It did not change anything on the distribution side either, where local distributors are still regulated monopolies. These are regulated by the so-called public utilities or public services commissions at the State level. And these utilities regulators (which not only regulated electricity) mainly, if not exclusively, focused on rates for end users. These rates should be devised in a way that they are socially acceptable (that is, protect the consumers), yet at the same allow for a fair return for the generators and the transmission system operator as well as the distributor(s). This is some sort of half-way liberalization, where regulation focuses only on the distributors and final consumers.

This model is being promoted by the US government, the World Bank, and related organizations worldwide, especially in the emerging countries, where they have identified investment opportunities for independent power producers. This model allows for relatively low-risk investments, as feed-in tariffs (into the grid) are typically guaranteed for a certain time and are typically low since they are calculated as net avoided costs. Competition among independent power producers is not really an issue, as there is typically no power market. This model also allows for low-risk investments into electricity-distributing companies, as these are not unbundled and customers typically remain captive. However, some political risks remain with the distributors and the regulators. Additional fragmentation is minimal, namely on the side of the power producers, but overall the systems stay relatively stable and the vertically integrated companies can handle the coordination of supply and demand.

By comparison, the European approach is much more radical and leads to a much higher degree of fragmentation. The European idea is one of a fully fledged power market, where customers can freely choose among different suppliers. This approach implies radical vertical unbundling; that is, a clear separation between the grid (transmission and distribution) on one hand and generation and retail on the other. It is the consumers (not just the retailers) who choose their suppliers, which can be power producers directly or intermediaries (retailers). This European approach also leads to much more elaborate trading platforms and mechanisms, which also have to be regulated.

The result is a highly fragmented system with power producers, transmission system operators (TSOs), distribution system operators (DSOs), retailers, intermediaries, and traders, including trading platforms. Power producers are, by definition, "independent"; that is, unbundled from the transmission grid operator, whether they are national (TSO) or local (DSO). As such, they no longer have to make power purchase agreements with the grid operator and can now sell their electricity directly to consumers, to retailers, to other intermediaries, or simply on a trading platform.

The TSOs no longer have to buy the electricity; they simply have to connect the power producers to their grid and transport their electricity at a regulated tariff. TSOs transport; they do not produce, sell electricity, or trade electricity. In this way, they now can focus on their core business, which is operating the high-voltage grid, balancing the system, and developing the grid. They are also the ones that constantly coordinate (balance) supply and demand. In all these activities they are regulated by an independent regulatory authority, which is another new actor resulting from liberalization.

DSOs should, in principle, also be unbundled from their retail activities, meaning they should only focus on the operations and the development of the distribution grid and make their money from the regulated grid usage tariffs. In reality, however, many if not most DSOs still service customers, typically households, with electricity, which they buy on the market or through bilateral agreements with producers. Overall, the electricity system is, after liberalization, highly fragmented. Retailers, intermediaries, and traders now incur pure market risks, as electricity has become a commodity. But they have to coordinate somewhat to make the system work. At the core of these coordination efforts is the TSO.

There are several consequences from this European approach to liberalization that is slowly but surely extending to all other parts of the world, including the US. The first and most immediate consequence is the fragmentation of the electricity system: there is an increasing amount of actors–generators, TSO, DSOs, retailers, and traders. Since electricity is a system that can only function as a system, all of these actors must be coordinated and the only actors capable of some sort of coordination are TSOs and regulators. However, because the regulator cannot act operationally, it typically delegates the operational coordination of balancing supply and demand to the TSO. De facto, the TSO becomes the system's coordinator.

The second consequence of deregulation is the return of politics: electricity becomes a commodity and the system is increasingly "regulated" by the market, and of course supervised by an independent regulator. But this is not exactly how it works. Electricity prices are often political prices, investments need to be made beyond a reasonable financial investment horizon which is too short term for large investments. Industry is risk-averse and the regulatory frameworks are not sufficient to reassure investors. This leads to the first security of supply issue, namely the question of long-term investments into grids and generation. Furthermore, a lot of electricity is still produced by various types of fossil fuels, namely coal and gas, which are scarce and display volatile price evolutions. This is the second security of supply issue. Also, nuclear energy is no longer an option for many countries and for those that continue with nuclear energy, this primary energy source also has supply or accident risks that only governments can bear. In other words, sooner or later politics returns to liberalized electricity, in a first step through more and more regulations, and in a second step through ownership and political control of operators. All this furthermore complexifies the system and moves it away from its optimal efficiency objectives and potential. In short, we are facing imperfect deregulation, where some half-baked liberalization measures are combined with regulated assets and trade. All this against the backdrop of an increasingly fragmented landscape of actors.

Notes

1 Stross, R. (2007). *The Wizard of Menlo Park*. Three Rivers Press, New York, p. 31.
2 Strouse, J. (2000). *Morgan. American Financier*. HarperPrennial, New York, p. 181.
3 Hughes, T. P. (1979). The electrification of America: The system builders. *Technology and Culture*, 20(1), 124–161.
4 Freeberg, E. (2014). *The Age of Edison. Electric Light and the Invention of Modern America*. Penguin Books, New York, p. 71.
5 Hughes, T. P. (1983). *Networks of Power. Electrification in Western Society, 1880–1930*. The Johns Hopkins University Press, Baltimore, p. 122.
6 Strouse, J. (2000). *Morgan. American Financier*. HarperPrennial, New York. p. 312.
7 Carr, N. G. (2008). *The Big Switch: Rewiring the World, from Edison to Google*. W. W. Norton, New York.
8 Hughes, T. P. (1983). *Networks of Power. Electrification in Western Society, 1880–1930*. The Johns Hopkins University Press, Baltimore, p. 366.

Chapter 18

Distributed Systems and the Need for Coordination

The electricity industry is rapidly evolving, mainly because of two parallel trends. On one hand, technology has created new forms of electricity generation, mostly solar and wind generation, which tend to be smaller in scale and better suited for installation at the customer premises, at least in a decentralized way. On the other hand, new policies oppose nuclear- and carbon-based electricity generation, which diminishes the importance of traditional centralized generation.

Electricity is increasingly generated at or close to the customers' premises, reversing the century-long trend towards centralization. Electricity is being transformed into a distributed network with an ever-larger number of electricity generation entities, even "prosumers"; that is, consumers who generate their own electricity in their premises and sell excess production.

The key question is whether this new industry structure requires a coordinator. We seem to be far from a fully distributed industry of independent off-grid consumers. Wind and solar electricity production is unreliable and batteries to store electricity are expensive, so the grid is still the most efficient system to provide electricity, at least as a back-up.

The role of the coordinator not only seems necessary, but also increasingly complex. System coordinators have to identify all electricity producers and prosumers, keep track of their unreliable production patterns, and match them to unreliable demand, as prosumers will only demand electricity when they cannot produce it themselves. This extremely fragmented ecosystem, with highly dynamic complementarities among a large pool of prosumers and consumers, seems to be the perfect opportunity for a digital platform to coordinate the market.

However, there is a natural limitation: the grid can only function if the electricity tension is kept stable, so a coordinator must balance the entries and exits of electricity in the grid. It is not only necessary to balance supply and demand from an economic perspective, but, more importantly, to have a totally balanced system for purely technical (that is, grid stability) reasons. Balancing the grid in real time is, so far at least, the main obstacle preventing full platformization of the industry.

18.1 Tesla's Ambitions

Elon Musk, who was born in South Africa and graduated in economics and physics at the University of Pennsylvania, was the main shareholder in PayPal when it was sold to eBay in 2002. He was a member of the group of influential individuals that would later exploit network effects in companies such as Facebook, LinkedIn, and Yelp.[1]

Musk invested the money he made from selling PayPal into three different companies. He invested $70 million, and later became the CEO, in an electric car manufacturer called Tesla Motors, named after the engineer who had invented the alternating current. He invested $10 million in a solar energy company, SolarCity, which would be later integrated into Tesla Motors. Finally, he founded and invested $100 million in SpaceX, a manufacturer of space launch vehicles with the ambition of building a "spacefaring civilization."

For a start, it seems difficult to identify the role of network effects in Musk's strategy. The three ventures were not focused on the creation of network effects at the data layer, but were instead focused on highly sophisticated manufacturing (cars, solar roofs, and space rockets).

A more detailed analysis demonstrates that Musk is at the vanguard of the platformization of the electricity industry, with the most ambitious strategy to coordinate electricity generation and storage and to couple the electricity industry with the transport industry. Tesla's competitive advantage in the car manufacturing industry was electrification, and batteries in particular. In fact, Musk was personally interested in batteries, since he had worked as an intern at an energy storage startup in California. As Musk declared in 2006, "the overarching purpose of Tesla Motors (and the reason I am funding the company) is to help expedite the move from a mine-and-burn hydrocarbon economy towards a solar electric economy. [...] in short, the master plan is: Build sports car. Use that money to build an affordable car. Use that money to build an even more affordable car. While doing above, also provide zero emission electric power generation options."[2]

Tesla is the leading producer of electric vehicles in the world. The first model, commercialized in 2008, was a high-performance electric sports car called the Tesla Roadster, only a few thousand of which were sold. As planned, Tesla has been launching increasingly affordable models, starting with a four-door sedan in 2012 (Model S), a mid-size SUV in 2015 (Model X), and then a $35,000 option in 2019 (Model 3). Despite delays, Tesla has delivered and moved from an elite small producer of electric vehicles to the mass market.[3]

An important handicap of electric vehicles is their limited autonomy. Electric vehicles have to be recharged, just as traditional carbon-based vehicles have to be fueled. However, being a new technology, no network of electric charging stations was deployed on the territory. It was necessary to develop the technology to accelerate charging times, and then deploy the technology across the territory. Network effects are evident. The more electric vehicles there are, the larger the incentive to install and operate recharging stations. The more stations, the better for the users. Tesla has improved the technology over the years: batteries are more efficient and larger distances can be driven without recharging. Rechargers are also becoming more efficient, reducing recharging times from hours to minutes. The network of rechargers is growing fast.

In order to overcome this chicken-and-egg challenge, Tesla ran its own charging stations. Should these stations be open for competitors, or should they be a closed system, erecting a competitive wall and forcing smaller competitors to build their own charging network? Should public authorities standardize charging technology and force interoperability, so that all chargers can be universally used by all cars, as is the case with tradition fuel stations?[4] These are similar to the questions that any company active in a network industry must ask.

The full plan was unveiled by Musk in 2016: "Create stunning solar roofs with seamlessly integrated battery storage. Expand the electric vehicle product line to address all major segments. Develop a self-driving capability that is 10X safer than manual via massive fleet learning. Enable your car to make money for you when you aren't using it."[5] This is when network effects became evident.

Musk is trying to become the intermediary in the most fragmented energy-generation system: individual citizens producing solar energy thanks to solar systems installed at their homes and offices' roofs. He is trying to build a complex ecosystem formed by potentially millions of small electricity producers, who are at the same time consumers of the electricity they generate, at the least in their electric cars, and who can trade it with other peers. This peer-to-peer model in electricity is known as the distributed electricity system and the peers are called prosumers.[6] Below we will go through the different pieces of the ecosystem and the role of Tesla as an electricity platform.

An important piece of the ecosystem is the solar roofs that need to be installed at everyone's home. Improvements in technology have increased the efficiency of solar technology as the costs have gone down and the amount of electricity generated has gone up. Tesla's solar panel's arm, which was created back in 2006, has developed the experience to provide "One ordering experience, one installation, one service contact, one phone app."

However, solar panels have fundamental and obvious downsides: they do not generate electricity at night, or even during the day if the weather conditions are not favorable. Also, electricity that cannot be consumed on the spot has to be sold or exchanged with other consumers. Coordinating the exchange of electricity generated at the edge of the system offers an obvious opportunity for a digital platform. Traditional players, which used to be in full control of the load factor by the coordination of large electricity generation plants with the consuming needs, must adapt to a new fragmented market with millions of small and utterly unreliable generation plants. This is the perfect scenario for a digital matchmaker with sophisticated algorithms.

The alternative is to store the electricity generated during favorable times, to be consumed when needed. Electricity storage becomes the cornerstone of the ecosystem. This is the fundamental competitive advantage that Musk built over the past two decades. Electric vehicles are electricity storage devices with four wheels. It is not only that electric vehicles consume electricity, with all the environmental advantages this entails. Furthermore, electric vehicles have the ability to store the electricity produced by the home or office solar roofs, in order to be consumed by the driver as required. In this way, the fundamental challenge of solar energy, the mismatch between generation and consumption times and consumption volumes, can be reduced. Even further, the wheels carrying the battery make it mobile, so the electricity can be moved and used at a different place. Electricity in the battery can be then sold to a third party at a different location. This creates still new opportunities for a matchmaker.

Since Tesla's electric vehicles are often away from home and from solar panels during working hours, when the sun is high, another piece of the ecosystem are the batteries sold by Tesla to be permanently installed at home. Tesla is increasingly commercializing solar roofs with home batteries to store electricity, even when the Tesla car is not at home. The pieces in the ecosystem increase in number, as the electricity can be stored both in home batteries and in the cars.

Vehicles are always a relevant piece in the ecosystem. Tesla is proposing to increase the complexity of the ecosystem by producing vehicles "to cover the major forms of terrestrial transport." As described in its 2016 plan, Tesla aims to produce heavy-duty trucks and vehicles for high-passenger-density urban transport. In this way, the ecosystem increases in complexity, but also in potential complementary relationships fueling network effects. The more vehicles there are, the more potential there is to exchange the electricity.

Finally, Tesla is working on the driverless car. One of Tesla's distinctive features is the Autopilot, technology that relies on algorithmic network effects to improve itself. The more distance is run, the better the technology gets. In his 2016 plan, Musk disclosed that 6 billion miles driven with Autopilot would be required for regulatory approval of an autonomous car.[7] The 3 billion mile threshold was reached in 2020.

Driverless cars are particularly fit for car sharing. The owner can use the car when needed, and the car can then drive itself to the location of a peer to provide a transport service. Thus, the electric vehicle becomes the cornerstone of a complex ecosystem coupling electricity and mobility. Complementary relations can be built across millions of electricity prosumers and sharing mobility users. Tesla would be at the center of it all, having created the ecosystem, managing the network effects, coordinating the newly created market. Tesla has the ambition of becoming an electricity platform, disrupting traditional electricity utilities, mobility companies, and dominating both markets by coupling them.[8]

This strategy is certainly a long shot. Car production has been slower than expected, solar roofs are taking time to catch on, and driverless cars are proving more challenging than expected.[9] In any case, this is a strategy that relies on network effects. Tesla's ambition is not only to produce and commercialize batteries, electric vehicles, and solar roofs, but then to connect all the pieces of the ecosystem, create direct, indirect and algorithmic network effects, and internalize them. Just like Edison over a century ago, Musk is not producing independent assets, but a whole system of complementary interoperable assets. What is not clear at this stage is the role of space rockets in Musk's strategy. Maybe Musk has the ambition to connect not only the world, but the whole universe, to build ever larger network effects.

Controlling the decentralized assets by building them is a good strategy for achieving platformization in an industry where the access to production and consumption data remains in the hands of the distribution system operator (DSO).

18.2 Decentralized Generation: New Technologies and New Policies

Tesla is just one example of a company that has tried to leverage the decentralization and growing fragmentation of the electricity industry. Such decentralization, especially decentralized generation, has been the defining feature of the electricity industry over the past 20 years or so. This decentralization comes on top of liberalization and is somewhat incorporated into the liberalization process. While decentralization is not replacing liberalization, it further complexifies the overall picture.

The first big changes emerged in the field of electricity generation in the form of decentralized wind and solar power generation. There has been a lot of technological innovation in these areas, as we have just seen in the case of Tesla, all leading to increased efficiency in the power production process, miniaturization, and lower production costs. Parallel, yet less spectacular innovations can also be found in gas turbines and in the generation of electricity from waste as well as from wood, along with innovations in geo-thermal electricity generation, and small hydro power plants.

The key question is whether decentralized generation, intermediated by platforms, has the power to substitute traditional electricity utilities. At this stage, it seems that distributed generation will not substitute traditional utilities managing the grid connecting electri-city consumers. Alternative distributed generation (wind, solar, etc.) is unreliable and is not in a position to guarantee the provision of electricity at all times. Tradeoffs can be made. A remote location off the grid can just settle with the limitations and be ready to go without electricity on occasions.[10] This is already happening in territories with non-reliable grids such as large parts of Africa. Massive investments in capacity and storage may guarantee independence for off-grid systems.

On the contrary, households deploying distributed generation assets are usually connected to the utility's grid as a back-up. While there may be some generation for self-consumption leading to a total disconnection from the grid, the reality is that all of these decentralized electricity generation units remain connected to the grid and usually not only satisfy the local consumption but also feed the electricity into the grid.

What is really new about distributed generation is its impact on the overall electricity system: it has first severe implications on grid investment, as the distribution network needs to be reinforced to be able to absorb the decentrally generated electricity. Secondly, and most importantly, decentralized generation creates a whole new set of challenges for the management of the overall electricity system, especially for balancing (that is, the coord-ination between supply and demand), with ensuing higher balancing costs. Paradoxically, decentral electricity generation increases the coordination needs and therefore also the coordination costs of the whole electricity system. Concretely, it leads to a bigger role for the system coordinator, the TSO: more balancing energy is needed and more balancing operations are required, as the system becomes more volatile. The TSO has a bigger role to play. Could it be replaced by a digital platform? Not today, for sure.

The evolution towards decentralized production is not just the result of technological innovation as well as of financial support for such innovation. It is also the result of concrete policies, namely policies favoring electricity generation thanks to renewables. These policies fit broadly into a climate mitigation agenda in the form of decarbon-izing electricity generation. Thus, in addition to financially supporting the develop-ment of decentralized electricity generation technologies, substantial financial support is also given to subsidizing the electricity production by way of renewables, including promoting electric vehicles that are supposed to run on renewables. This typically takes the form of so-called feed-in-tariffs, or other financing mechanisms to help make decentralized generation by way of renewables competitive with fossil- or nuclear-based electricity generation. Such policies typically help develop more decentralized electri-city production and lead to the above-outlined problem of increased system coordin-ation needs and costs.

Another relevant policy area pertains to electromobility; that is, the promotion of elec-tric vehicles. This promotion can take various forms, such as tax breaks or other incentives for buying electric vehicles, policies to make public transport electric, and many others.

Besides leading to the need for more electricity production, hopefully in a decarbonized way, this also leads to further decentralization, as charging stations need to be built up in a decentralized manner, further adding to the challenge of balancing the overall system. As seen above, grids need to be reinforced and more and more decentralized production and consumption units need to be coordinated.

18.3 Further Technological Innovations at the Interface of Generation and Transmission

Innovation – and financial support of innovation – is not limited to decentralized generation, mostly thanks to renewable energy sources. Because of unbundling, technological innovations emerge precisely at the interface between the grid and generation.

The first example of such an "interface technology" is the so-called "smart meter," which is a step further from automatic meter reading, as it communicates the consumption of electricity ideally in real time and thus allows more precise billing, in that prices can now be adjusted to times of consumption (such as peak-load pricing). This, in turn, makes it possible to create incentives for electricity consumption at particular times. But there is more to the smart meter, as the same information can also be given to the (distribution) grid operator, which obtains a much more precise knowledge about the state of his network. This in turn allows the grid operator to better balance the system. The smart meter is clearly useful both for generators and grid operators and, by doing so, establishes an information link between the two that was actually interrupted because of unbundling.

However, there is even more to the smart meter, as it also allows for two-way communication: smart meters that are attached to electricity consuming devices can also be used to stop or disconnect these very devices, in which case the smart meter can be useful for the consumer. Indeed, if electricity becomes too expensive, the smart meter can be programmed to stop a fridge or electric car charger from running, for example. Similarly, it can be useful for the grid operator (if the consumer agrees), who can use the smart meter to stop the connected device from running in order to better balance the electricity system. This is what is called "demand-side management" or DSM. Rather than adjusting production to the evolving consumption needs, smart meters can now also adjust consumption to the balancing needs of the grid or even to the available electricity. In other words, the smart meter can hold and transmit information that is highly useful for the coordination of an increasingly fragmented system.

Exactly the same considerations apply to batteries, another "interface technology" and area are where substantial technological innovation and subsequent cost-reductions are being made, as the case of Tesla demonstrates. Again, batteries are located at the interface between generation and grid management. They deploy their fullest effects when combined with smart meters. For example, batteries can store electricity when it is cheap and release it when it is expensive. Small-scale operations can run on batteries when the costs of electricity are too high or when demand must be curtailed. In that sense, they are electricity storage and generation devices at the same time, making them particularly useful for producers and for consumers, or both combined in the case of "prosumers." Batteries can also be used to balance the electricity grid; that is, store electricity when there is a surplus in the system (so-called negative balancing energy) and release it when the system needs balancing energy (positive balancing energy). As such, batteries add to the resilience of the system, and may even add to its cost-efficiency if they are cheaper than withholding electricity generation or other forms of electricity storage (such as hydrogen).

Overall, these "interface-technologies" such as smart meters or batteries add to the system's increasingly decentralized nature, a process that has already been enhanced thanks to decentralized (renewable) generation technologies and corresponding policies (and subsidies). As a result, the entire electricity system becomes increasingly complex, fragmented, and in need of coordination.

18.4 What Is the Smart Grid?

Decentralized generation, along with other technological innovations such as smart meters and batteries, all increase the fragmentation of the electricity system and therefore the need for and the cost of coordination. This is where the so-called "smart grid" comes into play. The smart grid is an answer to the coordination needs of the grid operator as created by unbundling and decentralized generation. The crucial question is whether the smart grid will be able to coordinate itself and thus substitute itself to the system operator; in other words, whether a digital platform connecting buyers and suppliers, including all the necessary information about the status of the grid will, in the long run, be able to substitute itself to the system coordinator.

To be clear, the "smart grid" is not about the grid; rather, it is about information management in an increasingly complex and fragmented electricity system. As such, it relies on the above presented smart meters; that is, on devices capable of measuring and interrupting electricity flows throughout the grid, mostly at the distribution level. More appropriately, we should talk about a "decentralized smart grid." In short, the smart grid is essentially a data-processing mechanism capable of integrating the information available throughout the grid, whether at its generation, transmission, or consumption levels. The information so collected is then essentially used for managing the demand of electricity at the decentralized level.

For example, industrial and domestic air conditioners, refrigerators, and heaters can be made to adapt their activation cycles to avoid activation when the grid is suffering a peak condition. There are now many examples of smart grid applications, whereby a certain number of homes, and in particular the appliances within these homes, using smart meters, are being connected via the internet and managed by a coordinating digital platform to make efficient use of both the consumed and (decentrally) generated electricity. However, all of these smart grid operations remain experimental and, as of today, cannot be scaled to larger regional and national levels.

Managing the data as collected through all these smart meters has numerous advantages. For example, smart meters can, to a certain extent, automatize such management by detecting errors automatically, interfering into the grid by stopping certain consumption or generation units, or even by helping repair certain things or routing electricity flows elsewhere. All this increases the resilience and reliability of the grid. It also helps reduce its vulnerability. In that sense, the smart grid – that is, a digital platform coordinating supply and demand as well as the status and the operations of the grid – can to some extent substitute or platform the decentralized coordinator, which, in this case, is the DSO. As explained above, it is the DSO that delivers electricity to the final household and must accept the electricity that is being generated at the decentralized point, such as from the solar panels on a rooftop. It is also the DSO that is directly affected if households self-consume ("prosume") and thus buy less.

Furthermore, the smart grid can better handle bi-directional electricity flows; in other words, it can manage decentralized generation much better. As such, it is actually a corollary, if not a condition for decentralized electricity generation. A smart grid can better allocate these flows and make the grid more flexible and more adapted to decentralized generation.

Probably the most important contribution of the smart grid lies in the efficiency gains for the grid operator, but also for the consumers who become also producers, the so-called prosumers. Better management of electricity flows leads to lower needs for investment and ultimately also to lower grid tariffs. This is like MaaS in the case of transport: smart primarily means "more efficient." Of course, these efficiency gains are offset by the cost of the smart meters and the management of the data, including the energy costs of managing and storing all these data.

Better management of electricity flows also leads to better load balancing and lower load balancing costs, at least at the decentralized level. Thanks to all this information, consumption patterns can be much better predicted using algorithms. Peak demand can be reduced and load curves can be much better balanced; this is also called peak-curtailing or peak-levelling. This again is both beneficial for the decentralized grid operators (DSOs) and for the consumers, but probably least to generators and retailers who are in the business of selling as many kilowatt hours as possible. Even they can take advantage of the smart grid, as they can better plan their generation capacities according to consumer demand, grid availability, and grid costs.

There may even be new business opportunities, both for generators and grid operators inasmuch as they can offer, thanks to all their data analysis, consulting services to customers to optimize their consumption profiles and demand response mechanisms. Similar consulting services can also be offered to government and policy makers. Corresponding information platforms can be a first step towards more sophisticated energy platforms.

18.5 What Is New? Platformization in Electricity Distribution

As in all the other network industries, digitalization leads to the platformization of the traditional infrastructure suppliers, albeit so far only at the decentralized level. In that sense, digital platforms have the potential to become the new interfaces with the prosumers. As in the case of mobility, digital energy platforms will now control the interface with the consumers (buying and selling) and appropriate the added value, whereas the historical DSOs will be reduced to becoming the transporters of the electricity. DSOs might well become commodified.

Currently, it is unlikely that digital energy platforms will coordinate entire national electricity systems. Given the current state of technology, digital energy platforms are most likely to be limited to coordinating decentralized electricity producing or consuming assets, including interface technologies such as batteries. And all this thanks to smart meters attached to all these assets. Figure 18.1 illustrates this evolution.

This will also lead to new interfaces: DSOs will be platformed, a new interface will be created between the digital electricity platform and the TSO, as can be seen in Figure 18.1. The platform will dynamically calculate the electricity flows: it will determine how much

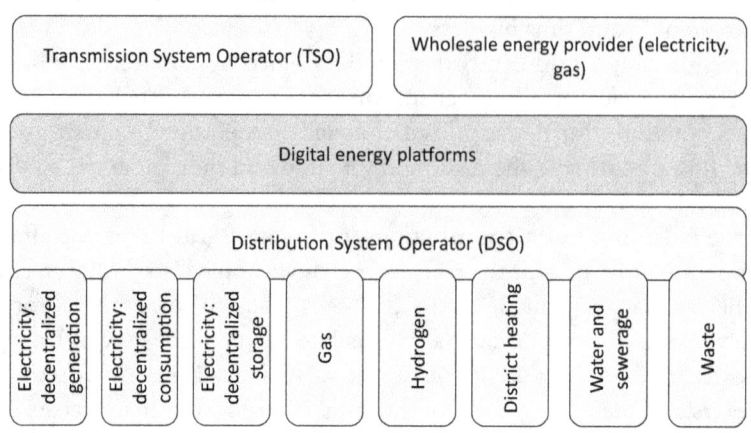

Figure 18.1

electricity is needed from the high-voltage grid at any given moment, considering the demand and the supply within its platform. Of course, this will be highly dynamic, as supply and demand evolve all the time, but there will be more to the activities of the platform: the platform will not only communicate to the TSO the electricity it will need from the high-voltage grid. It will also communicate to the TSO the positive and negative balancing energy it has available and coordinate with the TSO as to the TSO's needs in terms of balancing energy. Supplying positive balancing energy means that the platform can supply to the TSO the electricity the TSO needs to balance the grid because it may be short of electricity, for example by using electricity from the batteries of its customers. Supplying negative balancing energy means that the platform can turn off, thanks to its smart meters, the electricity-consuming devices of its customers according to the needs of the TSO. It will become the privileged intermediary between its customers and the TSO, as it has more to offer to the TSO than the DSO did. The DSO did communicate with the TSO as to its electricity needs (that is, the needs of its customers), but the DSO was unable to access the devices of its customers. At best, it was capable of accurately predicting and communicating to the TSO how much electricity it would need at any given time. The power of platforms will not stop at even the most complex and most demanding techno-logical systems, such as electricity distribution.

Notes

1   Soni, J. (2021). *The Founders: The Story of PayPal and the Entrepreneurs Who Shaped Silicon Valley*. Simon & Schuster, New York.
2   Musk, E. (2006). *The Secret Tesla Motors Master Plan (Just between You and Me)*, retrieved from www.tesla.com/blog/secret-tesla-motors-master-plan-just-between-you-and-me
3   Vance, A. (2015). *Elon Musk. How the Billionaire CEO of Spacex and Tesla is Shaping Our Future*. Virgin Books, London.
4   For a description of policies in Europe, see: Wappelhorst, S., Hall, D., Nicholas, M., & Lutsey, N. (2020). *Analyzing Policies to Grow the Electricity Vehicle Market in European Cities*, White Paper of the International Council on Clean Transportation, retrieved from https://theicct.org/sites/default/files/publications/EV_city_policies_white_paper_fv_20200224.pdf

5 Musk, E. (2016). *Master Plan, Part Deux*, retrieved from www.tesla.com/es_ES/blog/master-plan-part-deux

6 Baak, J. & Sioshansi, F. (2019). Integrated Energy Services, Load Aggregation, and Intelligent Storage. *Consumer, Prosumer, Prosumager. How Service Innovations Will Disrupt the Utility Business Model*, edited by F. Sioshansi, pp. 53–73, Academic Press and Elsevier, London.

7 For a critical appraisal, see: Kalra, N. & Paddock, S. (2016). Driving to safety: How many miles of driving would it take to demonstrate autonomous vehicle reliability? *Transportation Research Part A: Policy and Practice*, 94: 182–193.

8 Fuentes, R., Hunt, L. C., Lopez-Ruiz, H., & Manzano, B. (2020). The "iPhone effect": The impact of dual technological disruptions on electrification. *Competition and Regulation in Network Industries*, 21(2), 110–123.

9 For a critical appraisal of Tesla see Nidermeyer, E. (2019). *Ludicrous. The Unvarnished Story of Tesla Motors*. BenBella Books, Dallas (TX).

10 Couture, T, Cader, C., & Blechinger, P. (2019). Off-Grid Prosumers: Electrifying the Next Billion with PAYGO Solar. In F. Sioshansi (ed.), *Consumer, Prosumer, Prosumager. How Service Innovations Will Disrupt the Utility Business Model*, pp. 311–329, Academic Press and Elsevier, London.

Chapter 19

The Future of Energy

The power of digitalization, and thus the reach of the digital platforms, will not stop at electricity. In the same way as described above, other energy vectors, such as gas, hydrogen, district heating, water supply, wastewater management, waste treatment, and facility management, can and eventually will be integrated into the digital energy platforms in the medium and long term: thanks to smart meters placed in heating systems, in water distribution and wastewater collection tubes, in waste collection bins, and even in district heating tubes, digital energy platforms will be able to develop fully integrated energy services at a decentralized level, starting with a housing complex, but then extending to a district and maybe even an entire city. Like all the other platforms, there will be a natural tendency towards winner-takes-all, which might ultimately leave a single digital energy services provider in charge of a given geographical area.

19.1 From Electricity to Energy

Integrating the different energy vectors into a single digital platform will allow the platform to create an even bigger ecosystem and take advantage of even bigger inefficiencies by coordinating so-far uncoordinated sources of energy consumption and generation, perhaps even extending, as we have seen in the case of Tesla, to mobility.

This idea is not new and goes back to the energy crisis of the 1970s, when so-called integrated energy services companies (ESCOs) wanted to sell comfort, rather than kilowatts. ESCOs would take over entire buildings and offer a reduced price for the electricity to the users, while saving even more electricity. Initially, ESCOs took care of lighting, but over time they also moved into heating and management of buildings or even district heating. The focus was thus rather on larger customers or entire areas such as business parks, airports, and hospitals, which had larger contracts and larger potential energy savings.

This business model became particularly promising in the context of liberalization of the 1990s, as ESCOs focused on the services and especially broader energy services. This was a way, or rather a defensive strategy, of keeping customers from switching suppliers, since they were threatened by liberalization and thus in danger of losing electricity customers to competitors. It was not surprising, then, that ESCOs were often subsidiaries of local distributors, as for them selling integrated energy services became a means of preventing competitors from entering the market. However, digitalization did not yet play an important role. In other words, the business model of energy as a service had preceded digitalization and especially digital platforms. Also, ESCOs were still a quite defensive business model, basically a reaction against the possibility of losing customers due to liberalization. Not owning the assets, they needed to control them by way of heavy-handed contractual intervention, lacking the remote access to the assets that digitalization of the infrastructures would later offer.

Let us examine where this type of business model could lead to thanks to digitalization.

19.2 What Role for Digital Platforms in Energy?

We have seen above that digital platforms have the potential to platformize DSOs. The question is how far these new digital energy platforms can actually go? Do they have the potential, as in the case of Amazon, Airbnb, or Uber, to become global platforms, and thus eliminate most of their competitors? Or will they remain, as seems to be the case of the MaaS platforms, confined to the local or more precisely urban level?

As in all other network industries, digitalization enters because it makes operations, maintenance, and planning more efficient. This evolution is already well under way in the electricity and in most other energy sectors, such as waste management, district heating, gas supply, buildings and facility management, and mobility.

Will this lead to substitution or to platformization? We have argued that the power of digital platforms lies in the coordination of previously badly coordinated infrastructure assets, and there are many if one considers the broader energy system, especially at the distributed level: batteries in cars, bicycles and scooters, solar panels on rooftops, heat pumps, electric or gas heating in buildings, etc. Coordinating these scattered assets more efficiently can bring benefits to all actors involved: a car battery can be charged when the sun shines or when a heat pump is not working, heating and cooling can be much better coordinated with local weather patterns, etc. There are clearly many potential indirect and algorithmic network effects that can be leveraged by a digital platform coordinating a large number of fragmented energy assets.

As badly coordinated as they are, most of these infrastructure assets are already connected to the electricity grid, most likely at the distribution level. They are thus somewhat coordinated by the local distribution system operator (DSO), but this coordination is typically limited to tailoring electricity supply to somewhat unpredictable electricity demand. If mandated by law, the DSO will also accept that excess electricity produced by a household or a factory is being fed back into the grid. However, DSOs typically do not leverage the potential indirect and even less so algorithmic network effects, as they have limited control over most of these assets. The intelligence of the DSO is generally limited to the meter, which is usually not even smart, measuring electricity outflows from the distribution grid to the households and perhaps inflows in case the household produces excess electricity. The DSO has no access and even less control of the operations of the different devices inside a household or a factory, such as the heat pump, the electric vehicle charger, the buildings cooling system, etc. Thus, the potential to leverage network effects is huge, yet mostly untapped.

All this would be much easier in the case of purely off-grid operations: here, a platform could substitute a decentralized energy system and fully exploit the various network effects. Such off-grid operations do exist, but are limited to remote and rural areas, such as in Africa, where electricity is produced from decentralized solar power sources, somewhat stored, but mostly directly used for cooking and lighting, with limited potential to generate network effects.

A more promising approach is much larger greenfield operations, typically called "smart cities." Still in Africa, Konza Technopolis is a large technology hub planned by the Government of Kenya to be built 64 km South of Nairobi on the way to the port city of Mombasa. Similar projects are being planned and others have already been built and in operation in Africa, but mostly in Asia (Malaysia, Korea, China, India, Thailand, etc.). The obvious advantage of such projects is that digital devices (sensors, meters, etc.) can be built into every single piece of infrastructure from the very beginning; not just buildings, but also roads, water pipes, heating systems, and of course all the appliances that come with the building. A digital platform can subsequently coordinate all these assets for maximum efficiency. This is basically what a smart city and a smart city platform is all about. In short, a smart city will indisputably be a more efficient city.

It is not digital platforms that are building smart cities; real estate developers are too. Airbnb does not own property and Uber does not own any cars. While real estate developers are interested in more efficient cities, as it reduces their costs of building, maintaining, and operating them, they have no interest in being platformed either. Consequently, they want to remain in charge of the digitalization portion of their smart city, which they do – by definition – less efficiently than a platform would. Operating a digital platform is not their core competence, so it is unsurprising that labeling their real estate as a smart city is often simply a means to better market their buildings. In any case, the costs of an inefficient energy system can be passed on to the renters of their real estate.

Inversely, digital platforms have no interest in running smart cities either. They could be interested in running the platform of a smart city, if given unlimited access to the data produced by all these sensors, meters, cameras, etc. Of course, they do not want to own any of the assets and bear the associated infrastructure risks. However, the potential for reaping network effects becomes more limited in an already smartly planned and built city than in an originally highly fragmented system, as legacy cities are. Therefore, smart cities are probably not the first choice for a digital platform wanting to enter the energy industry, except perhaps for experimental, demonstration, and marketing purposes.

One important thing can be learnt from the various smart city experiments when it comes to the coordinating role of platforms: being set up as a greenfield operation, a smart city is in control of its electricity and energy systems. There is typically a single point of connection between the Smart city and the local or sometimes national electricity grid. Everything inside the city is managed by the smart city itself: regulated tariffs, grid charges, national consumer protection rules, etc. do not apply within the smart city, which handles all these issues bilaterally, via contract, with its "customers"; that is, the real estate owners (individual houses, buildings, shops, etc.). The smart city management, Konza Technopolis in our case, contracts with a wholesaler for the electricity it still needs, after all its internal efficiency measures and prosuming. It also coordinates with the local grid company, but most typically the national one, the TSO, for technical matters (grid stability, balancing, etc.), just like a DSO does with its national TSO. This is only in the case of greenfield operations, which are precisely what digital platforms are not interested in. In greenfield cities, the possibility to create efficiency gains thanks to coordinating fragmented and basically uncoordinated assets is much lower than in the case of legacy cities. Inversely, regulatory obstacles to reaping the full benefits of network effects are also smaller, if not inexistent, in greenfield cities.

19.3 How Far Will Platforms Be Able to Go?

Electricity at the local level is a complex ecosystem, constituted by an array of electricity consuming, of electricity generating and of electricity storing assets. This ecosystem becomes even more complex if other energy vectors, such as gas, district heating, or water are included. Coordinating such complex systems and intermediating with the corresponding energy suppliers (whether electricity, gas, district heating, water, or in the future even hydrogen suppliers) will already be a major challenge for a digital platform, but the digital platform is not alone and cannot do exactly as it pleases. Indeed, even if a (smart city) digital platform will one day coordinate entire smart cities, or preferably less smart cities, the platform will have to interface with a TSO, which will remain the ultimate coordinator of the (national) electricity system. It is highly unlikely that the TSO can itself be intermediated or platformed, at least not at the current stage of technological development. Indeed, the electricity grid needs to be coordinated by the same actor that is capable of balancing the system in real-time; so far, only the TSO can do this.

It is also unlikely, in our mind, that the TSO will evolve into a digital platform, at least not into the type of platform we have been discussing in this book. While TSOs are already highly digitalized in themselves, which is a prerequisite for them to manage the grid and especially to ensure grid stability, their core business remains the grid and its operations. TSOs already interact closely with electricity trading platforms, which are digital ventures in themselves, like any other financial market platform. In some cases, they manage or even own such trading platforms in order to better vertically integrate; that is, to be able to handle "implicit trading." Implicit trading is the most efficient form of electricity trading, as the commodity (electricity) and the grid capacity are traded together, ensuring maximum efficiency for both the grid and the electricity market. All this takes place at the national level, or at best at the European level, whereas the potential for platformization in electricity and in energy is, as we have seen, basically at the local, urban, and in any case the decentralized level.

In other words, platformization in electricity and in energy more generally will start from the bottom up by integrating at a digital level formerly fragmented assets, namely homes and SMEs. It will not be a top-down venture starting with an ambitious TSO. Tiko Energy Solutions AG is a virtual energy management company that claims its solution "is better than Tesla's." It was created a few years ago by the Swiss telecom operator Swisscom to aggregate flexible electricity consuming assets, such as home appliances, electric heating, battery chargers, etc. Subscribing homeowners, building managers, or operators of entire area networks (such as hospitals, airports, or shopping malls) connect all these flexible assets to the (Swisscom) telecom network, which then allows owners not only to monitor their energy consumption, but also to receive all kind of services, such as green electricity. Moreover, Tiko is capable of creating a so-called virtual power plant (VPP) by pooling all these flexible assets. Once pooled, Tiko can sell balancing energy, typically negative balancing energy to the TSO, by remotely turning off these assets when the TSO needs to diminish the load on the high-voltage grid. Thus, Tiko is already in the platform business: to the homeowners, building managers, or operators of area networks Tiko delivers efficiency gains and less costs, whereas to the TSO it delivers balancing energy. The more houses, buildings, and area networks are connected to the Tiko platform, the better the offer it can make to the TSO, and the better it can sell its virtual power plant to the TSO, the cheaper it can serve its customers. Thus, Tiko already platformizes the DSO, which is basically reduced to transporting electricity as well as to making sure that the distribution grid remains stable and functions efficiently.

The model is extendable, both in terms of customers and energy vectors, as it can easily be extended to gas and water. The model is also extendable beyond a telecom operator and could easily be taken over by a pure digital platform: it is sufficient to install devices capable of controlling electricity, gas, heat, and water consuming assets, manage supply and demand, and optimize this complex system thanks to sophisticated algorithms. Thus, the platformization of decentralized electricity- and energy-related assets is simply a matter of time. Tiko is showing the way in this regard, as are many other similar energy platforms.

19.4 What Does This Mean for Regulation?

Among all the energy vectors, electricity is one of the most regulated industries, both at the high-voltage transmission and at the distribution levels. As in the other network industries, existing regulation of the electricity sector can slow down the emergence of digital platforms, as the absence of digital regulation can facilitate the emergence of such platforms. We will look first at existing regulation.

The current method of calculating grid charges is a challenge for the emergence of digital energy platforms. Today, a consumer pays a fee for electricity to be transported to his or her premises. This fee represents roughly 50 percent of the final costs, the other 50 percent being the energy consumed. Electricity is transported at different levels of tension, level 1 being the highest tension level covering the costs of the transmission grid as operated by the TSO and tension level 7 being the lowest level, covering the costs of the distribution grid as operated by the DSO. In between, there are two more levels of tension, plus three levels of transforming the electricity from one level to another. Thus, the grid tariff a final user pays is actually composed of seven components. This makes perfect sense when electricity was generated centrally by a huge power plant and distributed through a capillary system down to the household.

However, with decentralized generation, electricity is now also being fed into the grid at the lowest level; for example, by one's rooftop solar power panel or by your neighbor's windmill. Such locally produced electricity will typically also be consumed locally, sometimes directly by the prosumer, in which case no grid tariffs are due, but most often within a local community. This is the case within so-called "local energy communes," as favored by newly emerging electricity sharing platforms, also called microgrids. Exchanging electricity at the local level currently still incurs grid costs that cover all seven levels, thus discouraging local consumption, energy communes, and digital platforms facilitating both.

Such tariff structures can easily be changed by political and regulatory decision. The problem does not lie here. However, if prosumers, energy communes and decentralized electricity generators no longer pay the grid tariffs beyond level 7, the costs of the existing grid not only remain the same, but must now be borne by those consumers that are not privileged enough to be able to take advantage of these new evolutions in the electricity sector. Also, the entire electricity infrastructure will still be needed to a certain extent – as a back-up, so to speak – enabling the local systems to function. Would it be fair to benefit from the entire electricity system, like a form of insurance, while not paying for it?

This is a similar problem to the one we have already encountered in telecommunications and especially in transport, namely that platforms will make use of the infrastructure – fiber networks, roads, trains, electricity grids – without adequately contributing to their financial sustainability. While regulation regarding grid tariffs is currently still tilted towards the legacy infrastructure, thus somewhat slowing down the emergence of digital energy platforms, the regulatory alternative of doing away with grid tariffs for the back-up electricity infrastructure (levels 1 to 6) will jeopardize the financial viability of the electricity systems, which are a general interest infrastructure.

This is just one of the challenges created by decentralized generation and emerging digital energy platforms taking advantage of this evolution. As noted earlier in this book, digital platforms will have to share their benefits fairly with the entire ecosystem in order to also contribute to the long-term viability of this general interest electricity infrastructure; that is, to the general interest. Regulating platforms to do this will require transparency of their algorithms and fairness, among other things – a truly enormous regulatory challenge, as we will see in the last part of the book. Many other regulatory challenges, which we have already encountered in the other network industries, could be mentioned as well, such as data privacy (who owns the smart meter data?), liability, security (hacking of the smart meters and all other devices), market power of the digital energy platform (winner-takes-all), non-discrimination among the different energy sources (but also among the different customers of the platform), etc.

These are typical challenges of digital platforms in infrastructures, and they will be discussed in more detail and more systematically in the final part of the book. In this, respect, digital energy platforms will not be very different from digital mobility or even digital telecommunications platforms. However, there is a still a fundamental difference between electricity and the other traditional network industries, inasmuch as, until today, it has not been possible to substitute the coordination of a national electricity system with a digital platform. A system coordinator, the TSO, will keep its central function, which is to assure the electricity system's stability and security. In electricity lingo this is called "system adequacy." Consequently, the interface between the digital energy platform(s) and the TSO will become a regulatory challenge.

As long as the platform is only optimizing the electricity consumption of its community, things will not change much: prosumers will consume less, albeit in a more volatile manner. This will mostly be a challenge for the DSO, which will have to make greater efforts to forecast electricity flows. Still, the DSO will continue to serve as the interface with the TSO, as has always been the case. Things are already different in the case of Tiko, as we have seen above: the digital platform actively manages the electricity consumption and presumption of its community and start to function as a virtual power plant (VPP). In this function, the platform enters into a direct, contractual, and heavily regulated relationship with the TSO. In regulatory terms, the VPP can still be assimilated to a contractor of balancing energy, and will be regulated as such. However, applicable regulation has been made for real, that is physical, and not virtual power plants. Regulation will have to be adapted accordingly.

In the not-too-distant future, digital energy platforms will move beyond being virtual power plants; they can easily evolve into large consumers that will sometimes consume a lot and at other times consume less. They will contract the electricity for their users directly on the market, probably also in flexible ways. This will create the need for a much closer interaction between the platform and the TSO as the TSO will want to plan corresponding electricity flows over its grid together with the platform. This will lead to an almost vertical, or at least very symbiotic, relationship between the TSO and the platform, which will almost certainly raise regulatory scrutiny. If the platform becomes highly powerful (that is, if it aggregates substantial amounts of electricity), it will come to have power over the TSO, leading to market distortion and net neutrality issues on the side of the TSO.

Part V

Regulating Platforms as the New Network Industries

Our analysis of the communications, transport, and energy platforms has shown that network effects are central for understanding the success of digital platforms. It has also shown that digital platforms are building network effects on top of the traditional network industries, as traditional players are being platformed. Finally, it has shown that massive value is being created by digital platforms.

However, it has become clear that the existing legal and regulatory framework is not ready to face the challenges that the digital platforms pose. Existing rules were defined for competition between firms integrating assets and employees in large organizations in order to avoid transaction costs in the market. When the integration is not done inside a firm, but by a platforms integrating third-party assets and employees, the existing legal framework does not know how to classify this new form of industrial organization. The mere application of the existing rules might be an obstacle for the creation of value by the platforms. In the specific case of the network industries, the existing regulation can benefit platforms, as traditional players are subject to strict regulations regarding market power, and platforms can benefit from competition in an unleveled playing field.

We have also shown that digital markets have a tendency towards concentration, winner-takes-all, as network effects require scale and only a few platforms can exhaust the economies of scale. In some cases, markets are tipping and one platform has become dominant. Platforms in a dominant position can abuse their power: access is denied, excessive prices are charged, there is a lack of transparency, discrimination, self-preferencing, and so on. The existing rules on competition and economic regulation do not provide solutions for the existing conflicts.

There are increasing calls for the regulation of the digital platforms. It can be taken for granted that some of them are the voices of the traditional players disrupted by the digital platforms. Recall how newspapers sued free newspapers for their use of the word "newspaper" and even for littering. However, it is already clear that there are valid reasons to increase the regulatory attention on digital platforms. Different regulatory authorities, particularly in Europe, have issued reports identifying market failures in the advertising market, in the app store market, and so on. Antitrust cases against the leading platforms proliferate, particularly in the European Union, shedding light on market dynamics and on abusive business practices.

Our analysis is structured into four different levels. First, digital platforms will be analyzed as providers of intermediation services. All jurisdictions have legislation, case law, and codes of conduct regulating intermediation services, as they have always been prone to conflict. Second, we will discuss how the scale at which digital platforms intermediate makes them superintermediaries, requiring extra protection for the intermediated parties. Third, we will analyze the special situation of digital platforms with market power. Finally, the need for regulation to consider the entire ecosystem around digital platforms will be underlined.

Overall, we propose that the regulation of the traditional network industries provides the best guide for regulating digital platforms, which we believe can and should be considered as the "new network industries."

Chapter 20

Platforms as the New Network Industries

Given the growing calls for digital platforms to be regulated, we think that it is important to understand the business model and the specificities to be taken into account when regulating them, and in particular, the central role of network effects. Consequently, we believe that the century-long experience of the regulation of the network industries provides the most valuable guide for the regulation of digital platforms. We argue that digital platforms should be considered the new network industries.

20.1 Digital Platforms and Network Effects

A new type of industrial and social organizational model is emerging rapidly in the form of digital platforms in multi-sided markets. These digital platforms allow groups with millions and even billions of users to interact via the internet and match their respective needs by way of sophisticated algorithms. Network effects play a central role.[1]

Firstly, platforms can create larger direct network effects. Traditional network industries created direct network effects by investing heavily in infrastructure to be shared by users. Network effects were driven by the supply side. They were usually limited in geographic scope to a country (or even to a segment of a country). Platforms, on the contrary, build network effects by pooling together demand and then connecting it to supply. Investment is necessary to aggregate demand, but not at the level of investments required to create and maintain infrastructure. Aggregation takes place at the data layer and the low transaction costs in the virtual world make it possible to aggregate demand faster and more widely than did traditional players in the network industries. This is why platforms do not usually operate at a local or even at a national level, but typically have a regional and even a global reach.

Digital platforms vastly outnumber traditional players in number of customers. WhatsApp has more than 2 billion users worldwide, more than the largest telecom carrier in the world, China Mobile, with 925 million customers, and certainly more than the largest carriers based in Europe (Vodafone has more than 600 million users) and the US (Verizon has around 150 million users).

Secondly, platforms create indirect network effects. Platforms can connect not only large numbers of homogenous individuals, but two different groups (two-sided markets) and more than two groups (multi-sided markets). A good example is transport platforms connecting passengers and different transport providers (taxis, scooters, mass-transit services, etc.). Again, this is possible because of the reduction of transaction costs. Technology makes it possible to connect large and heterogeneous crowds.

There are two reasons why indirect network effects are so powerful. Firstly, it is self-evident that the more groups can interact in a platform, the larger the scale and the larger the network effects. Secondly, and not so obviously, the possibility of having very different groups interacting creates new complementarities, a coordination that did not previously exist and might create new value on top of the value already created by the mere aggregation of a large number of users.

Platforms and indirect network effects in multi-sided markets allow the coordination of previously fragmented complex systems. The creation of value by coordinating different groups of users is the most profound transformation created by digital platforms in the network industries. Network industries often form complex ecosystems with a large number of players.

In complex systems like the network industries, only the coordination of the different players can ensure the most efficient outcome. It has been pointed out how digitalization allows infrastructure managers a more efficient management of the load factor through the use of predictive algorithms and the dynamic management of capacity in the infrastructure. However, there are limits to the results that can be obtained individually by an infrastructure manager that is disconnected from the rest of the players. The dynamic coordination of a complex ecosystem with multiple players requires the coordination of the different players. This is the opportunity for a platform in a multi-sided market.

Thirdly, platforms create algorithmic network effects, as ever-larger pools of data improve algorithms thanks to machine learning technology. For example, a search engine improves as more users search on it; translation software improves as users make more and more automatic translations. The more data a platform has, the better it can play its role as coordinator of a complex system.

Algorithmic network effects are fundamental for the operation of platforms. The scale and complexity of the interactions in sophisticated multi-sided markets require the most efficient management to the platform. Negative network effects such as congestion and poor coordination results can destroy a platform. Only the best predictive algorithms, fed with large amounts of data, are in the position to efficiently manage the massive amount of interactions among massive pools of users of a much different nature.

This is clearly the case of platforms in traditional network industries. A platform that dynamically matches passengers and different transport service providers must take a massive amount of data into account: the individual choices of passengers, the availability of supply of each transport mode, the state of traffic, etc. Similarly, an electricity platform must dynamically take into consideration the evolution of the different suppliers of electricity (will the sun shine in two hours as to generate enough electricity in the panels installed in the roofs of a neighborhood, or will it be necessary to rely on a large natural gas power plant far away from the city?) and the erratic demand of such different users as businesses, homes, electric cars, etc.

Algorithmic network effects are relevant, as the more data a platform has, the better positioned it will be to dynamically coordinate complex systems thanks to predictive algorithms. Uber explained this concept as follows: "Our massive, efficient, and intelligent network consists of tens of millions of drivers, consumers, restaurants, shippers, carriers, and dockless e-bikes and e-scooters, as well as underlying data, technology, and shared infrastructure. Our network becomes smarter with every trip."[2] Note how Uber refers to itself as a network, a network that becomes smarter with every trip: algorithmic network effects are key.

These three network effects reinforce each other. The more users, the better the algorithm; the better the algorithm, the more users and the more categories of users can efficiently interact through the platform.

20.2 The Business of Creating Value through Network Effects

Digital platforms are creating colossal value as they empower new interactions resulting in massive positive network effects. The role of platforms as value creators has to be understood and honored.

Positive network effects create value. Everyone understands that network effects in traditional network industries create value. The shared use of a common infrastructure creates value for all the users. A universal telephone network creates a lot of value. A dense transport network creates value. The electricity grids have created a lot of value. The network effects created by infrastructure networks a century ago were so powerful that the whole production system was transformed, countries developed both economically and socially.

The value created by digital platforms might not be as obvious, but it is as real as the value created by the traditional network industries. As digital links connect more and more users, the value of these virtual networks increases for all the users. Communications are cheaper and more effective thanks to communications platforms, door to door transport is made possible at a lower cost thanks to transport platforms, greener electricity can be locally produced exchanged thanks to energy platforms, and so on. "Network effects are not simply abstractions."[3]

A concrete example of the value created by platforms can be described for ride-hailing platforms such as Uber, Lyft, and Didi. Platforms make isolated cars work as a network. They identify the location of drivers and riders and they efficiently match drivers and riders, reducing empty runs. The higher the number of drivers and riders, the shorter the drive to pick up the rider, so the shorter the waiting time for the rider, but also costs are reduced for the driver. Cost reductions have been aggressively passed to riders. Lower prices generate new demand, igniting a virtuous cycle. This is not related to license costs, labor conditions, or regulated tariffs. Network effects are even more relevant if riders are ready to share a vehicle for all or part of a ride. The platform identifies the complementarities in demand and coordinates different requests for transport in the same area or direction. Large pools of riders make it possible to match different travel requests to be served by a single driver. Rides take a little longer, but the cost reduction is evident, as the cost is distributed among more than one rider. In terms of prices for the rider, it can add a further reduction of up to 50 percent, and again, lower prices generate new demand. A further twist was introduced in late 2017, when Uber launched Express POOL. If the rider is ready to compromise with the pick-up and drop-off points, by walking to more efficient locations in order to streamline routes, the price can be 75 percent lower than the regular Uber service. This is the power of network effects.

In other words, efficiency is generated by the better coordination of previously fragmented and isolated assets. Thanks to platforms, new complementarities are identified and a more efficient coordination is possible, which creates value. As the value generated by network effects is related to the volume of users, the massive scale of the platforms is creating colossal value. As potentially all humanity, all organizations, and all devices and things can be interconnected, it is easy to underestimate the value that can be created by platforms. Substantial research has focused on the calculation of the value derived from network effects. In Silicon Valley, venture capital has fully understood the value of these network effects.

It is important to understand that building network effects around a platform is a business in itself, a business that requires a good idea, large investment, and the best execution. Network effects do not appear magically; they are the result of the entrepreneurial adventure, which entails a lot of risk.

Firstly, a good idea is necessary. It is necessary to identify an opportunity to create value by establishing a multi-sided market that will benefit from new complementarities. It is necessary to identify an opportunity to introduce efficiency in the provision of goods and services.

Secondly, capital is necessary to build the multi-sided market. It is not always understood how much investment is required to create network effects. A mere idea will not attract users to the platform. As network effects require a minimum volume of users; this is the "chicken-and-egg" challenge for all new platforms. As the most obvious multi-sided markets have already been created, the new ones pose even lager challenges. Investment is necessary to attract users. Investment is necessary to advertise the platform, to pay for the creation of content, and to take the cost of discounts and subsidized services to attract users.

It is obvious that building an infrastructure network requires the mobilization of large amounts of capital. To recall, capital markets (banks, stock exchanges, corporations, etc.) were created to aggregate capital to invest in a network industry: railroads. Platforms might not require investments as large as traditional infrastructure networks, and they certainly do not need to stretch the reach of the current capital markets. Nevertheless, a large investment is necessary to build a multi-sided market of global reach. The main challenge is not the availability of capital, but the availability of capital smart enough to understand the business model and the profit opportunities derived from investment in the creation of something as intangible, as virtual, maybe as volatile, as network effects.

Thirdly, having the right idea and capital does not guarantee success. Execution is just as important. The construction of multi-sided markets requires a constant attention to small details, to take into consideration the needs and demands of the different sides of the market, in order to keep a balance between the different interests. The challenge is not only to create value, but to identify the right distribution of such value across the ecosystem, to ensure the right incentives for all the participants to join and to remain in the platform. Furthermore, such interests evolve over time, in such a way that the balance in the distribution of value across the market is a dynamic balance that requires permanent adjustment by the platform.

The creation of network effects by platforms in multi-sided markets is a business in itself, a complicated and risky business that has the potential to create, but also to destroy colossal value. Public authorities need to understand this new business model and understand that the creation of network effects is a business that requires heavy investment and entails significant risk. Public authorities must understand the enormous value that can be created by platforms. Society at large will benefit from the efficiency these platforms create.

Network effects lead to concentration in the market. As scale is necessary to build any kind of network effect, concentration is the unavoidable consequence in network industries. The number of competitors will always be low. Markets in platforms, just like those in infrastructures, will always be imperfect.

After an initial period of competition among literally thousands of competitors in telecommunications and electricity (hundreds in railways) in the US, as market players identified the relevance of scale and network effects, consolidation took place and the number of competitors was severely reduced. At a later stage, a further round of consolidation led to monopolies in many network industries, either private in the US or state-owned in Europe and the rest of the world.

However, deregulation has shown that monopolies are not the unavoidable consequence of the existence of network effects. Deregulation has challenged the assumption that all infrastructures are natural monopolies and will have to remain so. Some infrastructures seem to be natural monopolies, as network effects do not get exhausted and an extra user always reduces the average cost of the use of the network. This seems to be the case with railway infrastructures and electricity transmission grids, but the experience in telecommunications shows that infrastructures that have traditionally been considered natural monopolies might not be such. Network effects are always relevant, and scale is necessary to ensure the efficiencies derived from them, but once a certain threshold is reached, the efficiencies derived from network effects get exhausted. There is no substantial cost reduction in adding new customers. More than one company can deploy parallel infrastructures and compete in equal terms. These markets will certainly always present a reduced number of players, forming an oligopoly, maybe even a tight oligopoly, but not necessarily a natural monopoly.

Digital platforms present such relevant network effects that they deserve to be included in the category of network industries. Therefore, it is reasonable to question whether such network effects lead to market concentration and even to natural monopolies. Automatisms in answering these question should be avoided. Just as the answer is not the same for all traditional network industries, it seems that not all platforms will lead to the same market structure. The specificities of each market have to be considered. A factual analysis of each market is necessary, taking into consideration that these are highly dynamic markets and, in many cases, are in their infancy. Therefore, it might be too early to determine whether they are natural monopolies or not.

It seems clear that concentration is common in all digital platform markets. The number of relevant players is always small. This is the case of social networks, dating webs, communications platforms, food delivery apps, ride-hailing apps, etc. Network effects require scale and only a few players can reach the necessary scale in each market. Mobile telephony is the perfect illustration: only a small number of players ensures that all the players generate sufficient direct network effects. The number of start-ups might be large at the initial phase of a new market, as a lot of companies try to grow larger and build the network effects. Aggressive competition during this early period helps the market to mature and a few successful players, usually early movers with deep pockets, emerge as the market matures. Oligopoly seems to be the most common market structure in the new network industries.

In some markets, a single platform builds a leadership position. It might not become the only player, but a significant difference in market share and market power is created from the rest of competitors. This has been the case, in particular, with the leading non-transactional platforms: Google and Facebook. In most countries, Google has a market share above 90 percent in the online search market. The multiplatform strategy followed by Google has granted it strong positions with YouTube, Gmail, Google Maps, etc. Facebook also enjoys a strong position in the social networks market, reinforced by acquisitions such as Instagram and WhatsApp.

The market power of the most mature platforms, those that were created more than 15 years ago in the pure content industries, has raised valid questions about the winner-takes-all nature of the platform economy. The example of these gigantic companies has pushed founders and investors to adopt aggressive strategies to grow as quickly as possible, certainly faster than the competitors, growing the largest network effects, and thus conquering an unrivaled position. The winner-takes-all. This is not a new strategy. AT&T's Theodore Vail had already identified the importance of outgrowing competitors in the network industries more than a hundred years ago. Maersk had done the same in the maritime industry, as had J. P. Morgan in the railway and electricity industries.

However, just as in the traditional network industries, a monopoly might not always be the unavoidable market structure in the new network industries. Network effects might be exhausted at a threshold that allows replication by more than one player in each market. This might be the case with ride-hailing platforms. The case of San Francisco, where both Uber and Lyft were born, might be of interest. After ten years of fierce competition, no monopoly has emerged in the provision of ride-hailing services in San Francisco. Uber is the market leader, but Lyft claims that its share of the ride-sharing market grew to 39 percent in 2018, up from 22 percent in 2016.[4]

It is not easy to identify whether an aggressive investment strategy to grow scale and network effects is the way to reach a solid monopolistic position, or simply a foolish way to burn capital. Investors might just be subsidizing services in a never-ending effort to outgrow competition, as smaller competitors might have the ability to benefit from the same network effects at a lower cost. San Francisco became a bargain hunter's paradise during the past decade, benefiting from the possibility to get goods and services below cost, and even for free, thanks to the generosity of venture capitalists. Who needs to buy for instance a mattress or even pay to have it moved to a new home when an aggressive app delivers a new one to you for free as it is building scale and network effects?[5]

20.3 Digital Platforms Are the New Network Industries

It is important to realize that digital network industries rely on network effects just like the traditional network industries did. Direct, indirect, and algorithmic network effects are the reason behind the efficiency and the competitive advantage of digital platforms. Just as traditional network industries transformed the economy and the entire society a century ago, digital network industries are transforming the economy and society today.

Like traditional network industries, the digital network industries rely on standardization and scale. Standardization is a requirement to build scale. Telephone, railways, and electricity infrastructures had to be standardized in order to be integrated into a large-scale network. In the same way, standardization is necessary in order to build large digital networks. For instance, transportation services mediated by digital platforms have to be standardized so that they can be presented to passengers as recognizable services. Even the price must be standardized in order to streamline transactions. Even if digitalization allows the personalization of the services made available to the demand side, such personalization is always a pale mirror image of the rich and often gaudy diversity in reality, based on a profiling strategy that standardizes preferences according to presumed patterns.

Scale is the defining attribute of network industries. Network effects grow as scale grows. Traditional network industries create scale on the supply side by integrating supply into a single monopolistic infrastructure. Demand had to adapt to the single supplier, not always without opposition, as raising prices were feared. Digital network industries build scale as they integrate demand and then connect it with supply. Scale is no longer found in a large supplier, but in a large pool of users of the platform, both on the demand side and on the supply side. It is the size of the platform that creates the network effects, not the size of the suppliers of the underlying service.

These new network effects bring dramatic increases in efficiency. The transformations brought by such efficiencies are as powerful as the transformations brought by economies of scale thanks to industrialization and the corresponding emergence of the traditional network industries during the turn of the twentieth century. Back then, industrialization created tensions and gave rise to social movements as farmers and artisans were displaced. Millions of workers and organizations are again being substituted and displaced today (for example, written and audiovisual content producers, taxi drivers) and millions of individuals and organizations start working for obscure algorithms. Social movements might again be on the rise.

Now, like then, the new network effects are leading to a new round of concentration and market power, this time no longer only at the national or regional level, but also at the global level. Growing concentration and winner-takes-all effects in media, telecommunications, and transport platforms pose a fundamental challenge not only in economic, but in broader social terms, both with obvious political implications.

The concentration of economic and political power seems to be a source of concern today as it was a century ago when the traditional network industries emerged. There are growing concerns about increasing concentration and the role of antitrust to control it.[6] With the traditional networks it took more than 30 years to understand the ultimate reasons behind the concentration of power, both the legitimate (scale and network effects) and the illegitimate ones (systematic strategies to monopolize markets) and to define the legal remedies that allow society to fully benefit from efficiency while ensuring the common good. It should not take as long to adapt the existing regulatory framework to the new network industries, once it is understood that they are not that different from the old network industries.

This situation is becoming urgent given that the concentration of economic and political power today is creating similar problems to those faced a century ago: "Corruption dominates the ballot-box, the Legislatures, the Congress [...]. The people are demoralized; [...] The newspapers are largely subsidized or muzzled, public opinion silenced, business prostrated, homes covered with mortgages, labor impoverished, [...], imported pauperized labor beats down their wages, [...]. The fruits of the toil of millions are badly stolen to build up colossal fortunes for a few, unprecedented in the history of mankind; and the possessors of these, in turn, despise the Republic and endanger liberty." This is not the reactionary and xenophobic declaration of a twenty-first-century anti-globalist. This the text of an 1892 Manifesto by I. Donnelly, leader of the Populist Party.

Network effects are at the center of the digital revolution, which is why we propose to label platforms as the new network industries. We believe that digital platforms – for example, in the areas of mobility (such as mobility platforms) or energy (such as energy services platforms) – constitute a new type of infrastructures, albeit a digital form of infrastructure.

20.4 Differences between the Old and the New Network Industries

However, one should not conclude that the traditional and the new network industries end up being identical in all their dynamics. Just as there are significant differences among the different traditional network industries, there will be differences in the new network industries as well.

The main specificity of the new network industries is the relevance of indirect network effects. Value is created by coordinating multiple groups (at least two) rather than one as is common in traditional network industries. This distinguishes these companies not only from other network industries, but from most of the firms in the industrial economy.

Digital platforms do not need to build and own the assets they are coordinating. As they reduce transaction costs, they can rely on the market to make the assets available for other users. They do not need to integrate the assets into the firm under a hierarchical structure. They can build the complementarities empowering network effects without owning the assets.

This is actually the only way to build the complex ecosystems around digital platforms. Traditional network industries were mostly a single infrastructure, a single service market, exploiting direct network effects. Because the new network industries build complementarities between multiple types of assets and services, indirect network effects are relevant. Value is created by coordinating a large and varied constellation of players around the platform. Any analysis of a platform, and also the regulatory analysis, must take into consideration the entire ecosystem around them.

Another fundamental difference is that the complementarities empowering network effects in the platform markets are built in a data layer, while in the traditional network industries they were built on the infrastructure layer (single infrastructure or interconnected infrastructures) or on the services layer (scheduled coordinated services such as in air and maritime transport).

As a direct consequence, network effects can be more easily built in the data layer. Digital platforms require a smaller investment to create network effects than the infrastructure-based network industries. Traditional network industries, railways in particular, pushed the frontier of the capital markets and were behind the creation and growth of such relevant capitalistic institutions as corporations, banks, stock exchanges, etc. The industrial age infrastructures required a concentration of capital that was previously unknown. Digital network effects also require massive capital to grow to a global presence. As explained, growing platforms is a business in itself, a capital-intensive business, but the amount of capital required to build digital platforms is far from the amount needed to build physical infrastructures.

Another direct consequence is dynamism. Physical infrastructures are static and, once deployed, usually have a long life-cycle that can easily run into decades. Technological developments and progress certainly plays a significant role in the infrastructures and many improvements have taken place over the last century. But improvements have been incremental, not revolutionary; they have been slowly implemented. Digital network industries rely on software that can be easily modified. Actually, platforms are constantly testing new features; responses by users are constantly monitored and measured. New features that are deployed and tested with limited parts of the customer base are expanded to the entire community if successful, from small changes in color in the graphical interface to new incentives for the different sides in the multi-sided market. In any case, it is important to recognize that dynamism is more relevant in the early stages of a new digital market, and that once a business model proves successful, platforms become more stable. This has been the case with search engines, social networks, etc. The short life-spans of the early internet frontrunners (Netscape, Lycos, MySpace, and so many forgotten internet stars) have given way to the solid position of companies such as Google, Facebook, and Amazon, which have been leaders in their segments for ten or 15 years.

The geographic scope differs between traditional and new network industries. Infrastructure-based networks have always tended to be limited within the borders of a country. Large countries such as the US or China have large networks. Some operators have expanded across jurisdictions, with irregular success, but no global players have emerged in telecommunications, energy, and most of the transport industries (maritime is the exception). Platforms, on the contrary, have easily grown global. This is the case with multi-platforms around Google, Facebook, Amazon, and Apple. The only limitation seems to be political, as some governments in large countries have blocked the expansion of US-based platforms to support their own local platforms, particularly in China and Russia (with great success in China, the most populated country in the world).

Finally, regulation is another significant difference. Infrastructure-based networks are very heavily regulated by the authorities of the territories where they are deployed. Much of the regulation has been designed over the decades in order to maximize network effects: monopoly rights, access and interconnection obligations, etc. By contrast, digital networks are subject to very light regulation. There are various explanations for this phenomenon. First, deployment usually does not require the previous consent of the authorities in the form of licenses, concessions for the use of public land, etc. Second, market failures can only be identified once new services mature and are better understood by the regulators, so it is understandable that only at this stage, once platforms start to mature and the downsizes of their operations become apparent, does regulation start to be discussed. Third, the large scale and multinational presence of the leading platforms are a challenge for national authorities. Only very large countries or regional institutions such as the European Union have the power and reach to really subject the new giants to their regulations.

20.5 Implications for Regulation

Needless to say, at this stage, these powerful network effects in the new network industries require regulation. On one hand, regulation must acknowledge the new model of industrial organization. Digital platforms rely on the market to pool together the complementary assets and services for the creation of network effects. They intermediate between third parties to pool them together.

On the other hand, the network industries' paradigm has proven useful to understand and regulate the different industries that share the fact that network effects are relevant. Telecommunications, transport, and energy present common traits due to the relevance of network effects, the tendency towards concentration (with economic and political consequences), and the relevance of these services for society. Of course, the different sectors present specificities derived from the technologies they use and the economic conditions that are specific to them. However, the network industries paradigm has proven useful for defining the regulatory conditions for the exploitation of these networks. This does not mean that the same regulatory conditions have been applied to all these industries. However, common themes have been identified and similar solutions have been identified for the different industries.

In the same way, it is our understanding that the network industries' paradigm is the most useful framework for understanding and regulating digital platforms. The market failures that are emerging in the new network industries are too familiar to the network industries' expert and, in particular, to the academics and practitioners following the telecommunications' regulation debate during the past decades: how to reconcile vigorous competition with the efficiency derived from network effects, scale, and concentration, at the same time ensuring social objectives such as universal access to services under fair and equal terms or other public policy objectives.[7]

This does not mean that the regulatory framework developed over decades for a competitive telecommunications industry, for example, should be automatically extended to digital platforms. Traditional and new network industries are different, just as the traditional network industries were different among themselves and the new platforms are also different among themselves.

In short, we think that the expertise built over a century of competition and regulation in the traditional network industries constitutes a solid basis when it comes to facing the challenges of the new network industries. Expertise in the traditional network industries provides the most solid knowledge of network effects, whether they are direct or, in the case of digital platforms, indirect, as well as the new algorithmic network effects. Experts in the network industries know how these industries have been regulated in the past and can therefore assess how they are affected by digital platforms. Consequently, they can contribute to the discussion of how the interfaces between the traditional network industries and the digital platforms should be regulated. Finally, they are sensitive to public service considerations and the need to introduce such considerations into the regulatory equation of the new network industries, and this goes further than the regulation of network effects and market power.

Notes

1   The relevance of network effects has been widely identified as the key feature in digital platforms, from academics Shy, O. (2011). A Short Survey of Network Economics. *Review of Industrial Organization*, 38, 119–149, to public authorities such as the German antitrust authority Bundeskartellamt (2016). *The Market Power of Platforms and Networks*, working paper, available at www.bundeskartellamt.de/SharedDocs/Publikation/EN/Berichte/Think-Tank-Bericht-Zusammenfassung.pdf?__blob=publicationFile&v=4

2   Uber Form S-1 for the IPO, April 11, 2019.
3   Cusumano, M., Gawer, A., & Yoffie, D. B. (2019). *Business Platforms, Strategy in the Age of Digital Competition, Innovation, and Power.* Harper Business, New York, p. 59.
4   Carson, B. (2019). Lyft's revenue doubled in 2018 as it gains on Uber in U.S., but losses still growing, *Forbes*, March 1, 2019, available at www.forbes.com/sites/bizcarson/2019/03/01/its-official-lyft-files-to-go-public/
5   Suich Bass, A. (2019, Mar. 4). How to hack the freeconomy. the techies who figured out how not to pay for an Uber, or a mattress, ever again, *The Economist.*
6   Baker, J. B. (2019). *The Antitrust Paradigm. Restoring a Competitive Economy.* Harvard University Press, Cambridge, MA; Wu, T. (2018). *The Curse of Bigness. Antitrust in the New Gilded Age.* Columbia Global Reports, New York.
7   Bostoen, F. (2019). *Regulating Online Platforms: Lessons from 100 years of Telecommunications Regulation.* Harvard Law School, Boston.

Chapter 21

Regulating Platforms as Intermediaries

One of the fundamental debates around digital platforms is the nature of their services. Are communications platforms providing telecom and media services, or are they merely intermediating such services? Are transport platforms providing transport services, or are they intermediating transport services provided by third parties?

The answer to these questions carries the most relevant consequences: if they are considered providers of the underlying services, they would have full liability for them, just like editors, transport service providers, and so on.

We believe that a sound understanding of this model of industrial organization naturally leads to the conclusion that platforms are intermediaries facilitating the interaction between third parties. Platforms are not media editors, they are not telecom carriers, nor are they transport companies and so on.

However, platforms as intermediaries are not exempted from regulation. Intermediation has always been an activity prone to conflicts of interest and abuses, so there are many precedents for regulating intermediation services.

We believe that digital platforms should be subject to legislation that imposes upon them the basic obligations traditionally imposed upon intermediaries: namely, the obligation to protect and promote the interest of their users.

21.1 Platforms Provide Intermediation Services

Digital platforms provide intermediation services. The regulation of digital platforms must acknowledge the new model of industrial organization. Platforms create network effects by facilitating the interaction of third parties in multi-sided markets. Platforms are not replicating the traditional industrial model of building large, hierarchical, vertically integrated organizations accumulating assets to provide goods and services to consumers. On the contrary, platforms create value by providing the structure that facilitates the most efficient interaction, the optimal coordination among third parties.

The most evident form of intermediation is that of matchmakers. Matchmakers rely on the aggregation of supply and demand and then intermediate so that supply and demand meet in the most efficient way. The larger the pool of suppliers and consumers, the more powerful the network effects. The only way to grow the massive scale to exhaust the network effects is to rely on third-party suppliers. Airbnb would not be able to obtain the investment it would need to build the assets that have been made available through the platform. Uber would not be able to acquire the number of vehicles across the world it would need to provide the transport services made available through the platform. The ambition of the platforms is not to be the providers of goods and services, but to mobilize the goods and services provided by third parties to coordinate them for the customers in the most efficient way.

Intermediation can take other forms. Search engines such as Google are also facilitating the interaction among third parties. The algorithms developed by the platform facilitate the interaction between the individuals looking for content and the content producers. Interaction with advertisers is then facilitated to finance the platform. Non-transactional platforms facilitate the interaction of advertisers and eyeballs. The audience interacts with the ads produced by advertisers, even if such interaction does not take the form of a contractualized service provision between them. From a legal perspective, platforms are not providing an intermediation service, but they are certainly facilitating interactions among third parties. In this sense, innovation platforms such as operating systems are also intermediating between software developers and users.

Intermediation must be considered as a service in itself. There are many examples of individuals and companies providing intermediation services: real estate agents, insurance agents, stock exchanges, marriage brokers, etc. The disputes and conflicts of interests derived from the provision of intermediation services are well-known. There is legislation, case law, and even codes of conduct governing the provision of intermediation services. The conflicts of interest and potential abuses of platforms as intermediaries are not greatly different from those faced by traditional intermediaries. The regulation of online platforms should be built upon these precedents.

Ancillary services are often provided together with the intermediation service. Platforms often provide information about the service providers, in the form of ratings, comments by users, etc. It is also common for platforms to manage payments on behalf of the service providers. These services are complementary to the intermediation service and they do not modify the fact that intermediation is the main service provided by digital platforms.

In some cases, digital platforms not only intermediate services provided by third parties, but also provide their own services. As already described, this is the case with Amazon, which started as a retailer and was later opened up to other retailers in the Amazon Marketplace. Apple and Google intermediate their own apps in their app stores. Uber started as a platform, but then evolved to include bikes and scooters provided by Uber itself. This evolution does not undermine the nature of the new model of industrial organization.

Once the business model of this form of organization is fully understood, it becomes evident that platforms should not be regulated as the providers of goods and services. They are not providing the underlying goods and services, but they are intermediating in the provision of the underlying goods and services. This is a business and a business model in and of itself.

A basic mistake when regulating platforms is to automatically extend the regulation of the underlying good or service to the platform intermediating it. Platforms are not providing the good or service; they are intermediating it. It is true that platforms have developed contracting procedures that work with minimal friction, in a way that might even make users think they are contracting the provision of the service with the platform itself. Regulators should not fall into such a trap.

Liability is one of the main consequences of the attribution of the provision of the service to the platform. Platforms become fully liable for the services provided when they are considered not an intermediary, but the service provider. The cost of operation is increased as a consequence.

The application of labor law is another consequence of considering the platforms as the provider of the good or service. In most jurisdictions, if the platform is considered the provider of the service, the individuals executing the service are not independent providers, but employees of the platform. This could be the case of Uber drivers, bikers intermediated by delivery platforms, etc. Again, extending labor law to the relationship between platforms and individual service providers increases the cost of running the platform.

Increasing the costs of running platforms by attributing full liability for the provision of the underlying service and characterizing the individuals using the platform to provide services as employees reduces the speed at which platforms create network effects. Network effects can be created quite quickly by mobilizing third-party resources. The higher the regulatory costs imposed on platforms, the slower the speed at which network effects are created.

The creation of network effects can even be made impossible if platforms are declared to be the service providers and high costs are imposed on them. The potential for a platform to grow can be limited to the point of making it unsustainable if, for instance, it is forced to contract as employees all the individuals who have been using the platform to intermediate their services.

If platforms in multi-sided markets are understood as a new model of industrial organization, and if the value derived from the creation of the network effects is honored, it is almost an automatic consequence that the regulatory framework should acknowledge the intermediary nature of the activity of the platforms, and not impose unnecessary regulatory burdens to the creation and growth of network effects. Users and society as a whole will benefit from the value created by platforms.

Regulating platforms as intermediaries does not exempt platforms from regulation. Intermediation services are subject to abundant regulation in all jurisdictions. There is legislation, case-law, and codes of conduct setting rules to ensure fairness and transparency in the provision of intermediation services, as well as specific redress mechanisms. This is necessary because intermediation is an activity prone to conflicts of interest and abuse.

US advertisers were shocked in 2016 when it was disclosed that media agencies they had hired to secure best-performing advertising slots in media were receiving large rebates from media companies.[1] When an agency contracted ad space on behalf of an advertiser, the media company would grant a rebate to the agency of 15–20 percent of the total expenditure. Rebates should have been disclosed by agencies to advertisers, if not directly passed on to them. Were agencies acting in the best interest of their customers? Was the undisclosed rebate giving the agencies an incentive to act in the interest of the media company? Was the agency putting its interest ahead of the interest of the customer?

Mistrust in intermediaries is universal, and often justified. Taxicabs traditionally complained about taxi dispatchers manipulating the assignment of services in favor of certain taxi drivers. Real estate agents, insurance brokers, and many other intermediaries are mistrusted to the point where conflicts of interest and abuse become the subject of bad jokes.

Therefore, intermediation activities are often regulated. Most jurisdictions have adopted specific legislation to regulate specific intermediation activities. This is the case of small-scale intermediation, which includes real estate agents, insurance brokers, travel agents, freight forwarders, etc. This is also the case of larger-scale operations, such as payment cards and stock exchanges. There is abundant case-law on conflicts with intermediaries, and codes of conduct self-regulating the activity of intermediaries are also common.

It is surprising that such a vast amount of precedents has not informed the regulation of digital platforms. Platforms have always maintained that they are mere intermediaries, not the providers of the underlying services they intermediate. Nonetheless, as intermediaries, they have at least the same conflicts of interest as those identified for traditional intermediaries. When intermediation is undertaken at a large scale, automated by algorithms, conflicts of interest do not disappear; they grow in scale.

Regulation of intermediation services provided by digital platforms can be inspired by the existing rules on intermediation. For example, an interesting lesson can be drawn from the rules defined for one of the most popular intermediation services, that of real estate agents. The Code of Ethics and Standards of Practice of the US National Association of Realtors[2] can be a useful guide to identify the most common conflicts around the intermediation activity, and so are the rules imposed to ensure fairness and transparency in the provision of intermediation services by digital platforms.

21.2 Obligation to Protect and Promote the Interests of Their Clients

The first duty imposed upon real estate agent by the Code of Ethics is "to protect and promote the interests of their clients." A similar obligation is imposed upon intermediaries in most jurisdictions, although it can take different forms. This obligation is sometimes defined in even stricter terms, as the obligation to put the client's interest above those of everyone else, including the agent's.

For a start, fairness requires intermediaries to be specifically obliged to provide the intermediation service. Real estate agents "shall not offer for sale/lease or advertise property without authority." An agent cannot intermediate without the consent of the intermediated party. As platforms need scale, they have the incentive to include in their ecosystem goods from a large number of suppliers in a market, even if they do not give their consent. The link wars were described in the first part of this book: Google included newspapers' headlines and snippets in the Google News service without the consent of newspapers and without any compensation. Platforms should not be allowed to act as interlopers, intermediating against the will of the intermediated parties.

The fairness requirement goes further. Real estate agents have a duty to "submit offers and counter-offers objectively and as quickly as possible." This is the main consequence of the obligation to promote the interests of their clients. Agents facilitate transactions by passing the offers of sellers and buyers to each other. Speed is certainly not a problem for digital platforms. As transactions are automated with the use of algorithms, the intermediation service is provided on-demand and in real-time if so requested by the customer.

Automation, by contrast, might pose a challenge to the objective display of the underlying services. Supply has to be standardized in order to be automatically sorted, ranked, and displayed to interested users. Platforms should not distort the display of the underlying service. It should not be manipulated either by making it more attractive to the potential user or by diminishing its attractiveness to favor other options.

The following example illustrates the kind of conflicts that platforms face as intermediaries. When Netflix was launched, it sent movies in DVD format to customers by post. Inventory was an issue, as customers tended to order the same recent hits. The solution was to nudge customers towards older movies. One of Netflix's founders explained it in as follows: "If we could show customers what they wanted to watch, they'd be happier with the service. And if we could also show them what we wanted them to watch? Win-win. Put it simply: even if we were ordering twenty times more new releases than any Blockbuster (an enormously expensive gambit), we wouldn't be able to satisfy all demand, all the time. And new releases were expensive. To keep our customer happy and our costs reasonable, we needed to direct users to less in-demand movies that we knew they'd like – and probably like even better than new releases."[3] Are customers well served when they are directed to older movies rather than blockbusters?

The parties in a transaction usually have opposing interests: one party wants to sell at a high price and the other wants to buy at a low price. When intermediaries work for one party, it is easier to identify, protect, and promote the interest of such party. When intermediaries work as "dual agents" – that is, for both parties in a transaction – conflicts of interest are structural.

Most digital platforms usually work for both parties in a transaction. While it might be that only one party is paying for the service, this does not mean that the other party is not a customer benefiting from a service. In fact, most platforms also disclose terms and conditions on their websites for the parties benefiting from the service at no charge. This is the case of Google and its search engine, Facebook, Uber, etc. Platforms have to serve the interests of all parties.

The position of the platforms is particularly prone to conflicts of interest, as they not only act as dual agents, but do so for millions and even billions of individuals at the same time. There are potential conflicts of interest between platforms and intermediated parties and between parties on each side of the transaction: competing service providers, and final users competing to benefit from a service.

Intermediation services must be provided in an objective form in order to serve the customers' interest, in all the sides of the multi-sided market. To bias intermediation would go against the interest of the intermediated parties if there is no objective reason to justify the manipulation of the ranking, matching, and so on.

As transactions are automated by algorithms and undertaken on a massive scale, it is not simple to identify potential biases in the provision of the intermediation service. This basic issue is often ignored in the public debate. The use of mathematical algorithms is sometimes presented as a guarantee of objectivity in the management of the transaction. However, at this stage it is clear that algorithms are not neutral and objective tools, but mere instruments for the execution of company policies. Algorithms can be built with a bias in favor of one of the intermediated parties.[4] This is particularly the case when the platform vertically integrates.[5] It has been argued that algorithms might be built on an implicit bias with unintended discriminatory consequences.[6] The regulation of algorithms will be analyzed in Chapter 24.

Intermediation bias is the key issue in the regulation of intermediaries. Not all traditional intermediaries are subject to the same standard. For instance, US financial regulation draws a distinction between intermediaries under fiduciary obligations (financial advisors) and intermediaries under suitability obligations (brokers). Fiduciary obligations impose a duty on a company to defend the best interests of its customer over its own interest. Suitability obligations merely require the intermediary to have a reasonable basis to believe that the proposed product is suitable for a specific customer, as having knowledge about the investment profile of the customer.

Not all digital platforms should be subjected to the same intermediation standard. Even if platforms are intermediaries, not all platforms provide the same intermediation service. Search services are different from social media platforms and from matchmakers. Let us analyze some of the proposals.

The principle of "platform neutrality" was proposed by the French Digital Council (Conseil National du Numérique), an advisory body of the French Government set up in 2011. In its first report in 2013, the Council identified the central role of platforms in the digital transformation. It initiated the first organized and systematic analysis of online platforms before introducing regulation. In 2013 a ten-member working group held four workshops and a series of meetings were organized with economic and legal experts to develop a proposal for the regulation of online platforms. The result was an opinion published in 2014 under the title "Platform Neutrality. Building an open and sustainable digital environment."[7] The position of the Council was further refined in a report published in 2015.[8]

"Platform neutrality" has been inspired by the "net neutrality" obligation imposed upon telecommunication carriers when they provide access to internet services. Platform services would be provided to all, with no discrimination, at reasonable conditions, as a common carrier service.

The first digital platform regulation we have identified, that of computer distribution systems in aviation, imposes on platforms the obligation to "load and process data [...] with equal care and timeliness" and "use a neutral display,"[9] as described in Chapter 12.

Nevertheless, intermediation services can hardly be provided under strict equal terms. For example, the Code of Ethics for real estate agents includes no such requirement. Matching a specific supplier with a specific buyer of the service entails an unavoidable ranking of the best matching option, excluding other matches. This is why the "platform neutrality" proposal evolved into fairness and transparency in platforms to business relations.

The difficulty of imposing strict neutrality has created confusion. It has been common to look for alternatives in other legal frameworks, such as competition law, other regimes in economic law, and regimes protecting civil rights.[10] Competition law has obvious limitations for defining the standard of objectivity in the provision of intermediation services, as it only applies to platforms in a dominant position. It also has a greatly reduced scope, which is the protection of the competitive process. Specific figures such as that of "economic dependence" originate in German law,[11] and provide an alternative legal basis, but are limited to those countries, limited in scope, and not focused on the nature of the intermediation service itself. The same can be said about unfair competition practices that are declared illegal in some European countries.

Privacy is also relevant to ensure fairness. Real estate agents have a duty to "preserve confidential information." Information about a specific transaction can benefit a third party, and sharing that information can damage the client's position. Intermediaries acquire a privileged position in the market, as intermediation in a high number of transactions gives them better information about market conditions.

The situation is no different when transactions are facilitated by digital platforms. Platforms must preserve the confidentiality of the information provided by the client. Clients should at least be aware of the protection granted (or not granted) to their data. Intermediaries are not protecting the best interest of their clients if they share confidential information or, worse, if they actually sell it to third parties.

The issue is particularly relevant because data obtained from clients is often personal data under the protection of privacy legislation. The General Data Protection Regulation (GDPR) of the European Union[12] is becoming the global standard for personal data protection. The principle of "privacy by design" imposes upon platforms the obligation to protect personal data as a key feature of the application.

The legal basis for defining the standard in the provision of the intermediation service should not be in some legal framework defined for some alien purpose, but in the specific set of rules defined for the provision of the specific intermediation service, inspired by the centuries-long tradition of regulating intermediation services.

We argue that the basic standard should be to provide the intermediation service for the benefit of the customers. Then, certain services could be subject to suitability obligations, being obliged to make recommendations that suit the best interests of their users. This should be the case, for instance, with matchmakers. Only in exceptional circumstances should platforms be placed under fiduciary obligations.

Digital intermediaries are required to provide their services for the benefit of their customers. A platform promoting behavioral addiction to the service does not seem to be promoting the customers' interests.[13] A platform surveilling customers and selling customers' data without informing them is not promoting the customers' interests.[14] A platform imposing arbitrary restrictions to developers to include an app in the app store is not protecting the customers' interests.

21.3 Transparency

Transparency is necessary to reduce conflicts of interest. The realtor's Code of Ethics imposes transparency obligations to reduce conflicts of interest. Real estate agents "may represent the seller/landlord and buyer/tenant in the same transaction only after full disclosure to and with informed consent of both parties."

Platforms should be obliged to disclose the existence of these conflicts and should inform customers about how they manage the conflicting interests of the different parties they mediate for. Do they merely use a first-come-first-served rule? Do they prioritize some provider or customer over another? What are the prioritization parameters? Are they mere objective parameters such as vicinity to the customer (for transport services), previous ranking by the customer, or similar? Are there other parameters, such as commissions paid to the platform, priority given to customers that make a more intense use of the platform (pay more), or priority to new customers to "hook" them?

Algorithms are constantly adapting their matching parameters. It has been disclosed that Google modifies its search algorithm more than 100 times a year.[15] Customers should be informed about changes in the algorithm to ensure they are not negatively affected.

Maximum transparency is necessary when platforms intermediate their own products in competition with the platform's customers. Real estate agents "shall not acquire an interest in or buy or present offers from themselves [...] without making their true position known to the owner [...]." Transparency is particularly necessary when the intermediary is not merely representing a third party, but it is actually acting as a principal, at least in some cases.

Conflicts of interest multiply in intensity when the intermediary is also acting as principal in the transaction. Digital platforms are increasingly intermediating their own goods and services, in competition with their customers. Amazon competes with other retailers in Amazon Marketplace. Apple has its own music app that competes with other music apps available in the Apple-managed app store. Uber has launched its own shared bicycle and scooter services, which are included in the platform in competition with riding services provided by third parties. In this situation, disclosure is important, but a strict definition of an obligation to intermediate for the benefit of the customers is of the utmost relevance.

Finally, real estate agents have some basic fundamental duties in the provision of the intermediation service: disclosing their identity when providing the service, providing agreements drafted "in clear and understandable language expressing the specific terms, conditions, obligations and commitments of the parties," informing customers about their compensation, and presenting a "true picture in their advertising." These are all obligations that can be perfectly extended to digital platforms.

21.4 Liability as Intermediaries

The liability of intermediaries is another permanent source of conflict. Real estate agents are "obliged to discover and disclose adverse factors reasonably apparent to someone with expertise." While intermediaries are not liable for the underlying good and services they intermediate, they have an obligation to discover and disclose "adverse factors" that should have been identified by an agent with the expertise expected in such a professional, and they are liable if they do not disclose such "adverse factors" to clients.

Platform liability is one of the most debated issues in platform regulation. Both in the US and in the EU, existing legislation and case-law has granted a somewhat privileged status of liability exemption. In the framework of the definition of liabilities for internet players in the early days of the internet, platforms were exempted from the liability for identifying illegal content and goods that they intermediated. Google and Facebook, but also eBay benefited from such a privilege. They were subject to the same liability regime as internet access providers, providers of internet content hosting services, etc.

Intermediation, particularly the matching of supply and demand, requires the active participation of the intermediary in the evaluation of the characteristics of the intermediated good or service. While automation implies that the platform does not make as a detailed an analysis of the intermediated good as a real estate agent, the platform cannot intermediate without a minimum examination of the intermediated good or service.

Consequently, platforms must be considered liable for some of the "adverse factors" in the goods and services they intermediate. They should at least be liable for the "adverse factors" they identify and they should identify these adverse factors based on their expertise. The online provision of the intermediation service should not exempt platforms from this basic liability.

21.5 Redress Mechanisms

Finally, the Code of Ethics and Standards of Practice of the National Association of Realtors provides redress mechanisms in case of conflict in the provision of the service. Clients can benefit from the services of mediators before they go to court to solve disputes.

Digital platforms have been very slow to respond to the demands of intermediated service providers about the reasons why their service has been excluded from the platform, demonetized (in platforms passing some advertising revenue to content producers), or simply ranked more poorly and therefore losing traffic (in search engines, for instance).

Algorithms and automation cannot be an excuse for not providing redress mechanisms for unsatisfied parties. Algorithms are not neutral. They are designed to meet the priorities of the designer. Machine learning might make it more complex to identify the specific reason for the ranking/matching of some content or company, but it should not be an excuse for not providing any explanation. Automation cannot be a reason for excluding any kind of human intervention to solve requests of explanation from customers who think they have been discriminated against or that their interests have been poorly served by a platform.

Redress mechanisms can take different forms. Initially, simple procedures can be designed by the platform for customers to communicate their requests and for the platform to respond in a timely fashion with an explanation about their behavior and a potential amendment of behavior if necessary. Conflicts can then be solved by independent third parties. Public intervention would be the intervention of last resource.

In conclusion, most of the existing rules governing traditional intermediation services can be extended to services provided by online platforms. However, it is surprising that these basic principles have not yet been extended to platforms in the form of legislation. It would have been useful to impose on platforms the most basic principles of serving the users' interests, objectivity and non-discrimination, as well as some of the instrumental obligations such as implementing redress mechanisms.

Notes

1   Auletta, K. (2018). *Frenemies. The Epic Disruption of the Ad Business (and Everything Else)*. Penguin Press, New York.
2   Code of Ethics and Standards of Practice of the National Association of Realtors, version of 1.1.2019, retrieved from www.nar.realtor/sites/default/files/documents/2019-COE.pdf
3   Randolph, M. (2019). *That Will Never Work. The Birth of Netflix and the Amazing Life of an Idea*. Endeavour, London, p. 214.
4   Intermediation bias was first identified in searches: Hagiu, A. & Jullien, B. (2011). Why do intermediaries divert search? *The RAND Journal of Economics*, 42(2), 337–362.
5   Feasey, R. & Krämer, J. (2019). *Implementing Effective Remedies for Anti-competitive Intermediation Bias on Vertically Integrated Platforms*. Report for CERRE, retrieved from www.cerre.eu/sites/cerre/files/cerre_intermediationbiasremedies_report.pdf
6   Pascuale, F. (2015). *The Black Box Society. The Secret Algorithms that Control Money and Information*. Harvard University Press, Cambridge. MA.
7   Opinion no. 2014–2 of the French Digital Council on platform neutrality, May 2014, available at https://ec.europa.eu/futurium/en/system/files/ged/platformneutrality_va.pdf
8   Conseil National du Numérique (2015). *Ambition numérique: Pour une politique française et européenne de la transition numérique*. Report, retrieved from www.entreprises.gouv.fr/files/files/directions_services/secteurs-professionnels/economie-numerique/CNNum--rapport-ambition-numerique.pdf

 9 Regulation (EC) No 80/2009 of the European Parliament and of the Council of 14 January 2009 on a Code of Conduct for computerised reservation systems and repealing Council Regulation (EEC) No. 2299/89, OJ L 35, 4.2.2009.

10 As an example see Krämer, J., Schnurr, D., & de Streel, A. (2017). *Internet Platforms and Non-Discrimination*, CERRE Project Report, retrieved from www.cerre.eu/sites/cerre/files/171205_CERRE_PlatformNonDiscrimination_FinalReport.pdf, and Expert Group for the Observatory on the Online Platform Economy (2020), *Work Stream on Differentiated Treatment*, Progress Report, available at https://platformobservatory.eu/app/uploads/2020/07/ProgressReport_Workstream_on_Differentiated_treatment_2020.pdf

11 German antitrust authorities initiated actions against Amazon for abusive practices as "the sellers are dependent on Amazon"; see press release Bundeskartellamt (2018). Bundeskartellamt initiates abuse proceeding against Amazon, 29 November 2018, retrieved from www.bundeskartellamt. de/SharedDocs/Meldung/EN/Pressemitteilungen/2018/29_11_2018_Verfahrenseinleitung_Amazon.html;jsessionid=28428957AB1B6DED02C51B292680654A.2_cid387?nn=3591568

12 Regulation (EU) 2016/679, General Data Protection Regulation, OJ L 119, 04.05.2016; cor. OJ L 127, 23.5.2018.

13 Alter. A. (2017). *Irresistible. The Rise of Addictive Technology and the Business of Keeping Us Hooked*. Penguin, New York.

14 Zuboff, S. (2019). *The Age of Surveillance Capitalism. The Fight for a Human Future at the New Frontier of Power*. Profile Books, London.

15 Feasey, R. & Krämer, J. (2019). *Implementing Effective Remedies for Anti-competitive Intermediation Bias on Vertically Integrated Platforms*. Report for CERRE, retrieved from https://cerre.eu/publications/implementing-effective-remedies-anti-competitive-intermediation-bias-vertically/

Chapter 22

Regulating Platforms as Superintermediaries

While digital platforms are intermediaries, they are in a special position due to the scale of their operations. They automate the interactions, aggregate a substantial amount of supply and demand in a market, and often even create the markets around them. They have more power over the intermediated parties, both business and consumers, than traditional intermediaries, which are usually small craftsmen representing a small number of customers. Therefore, regulation should be adapted to the special role of digital platforms as superintermediaries, as we call them.

On the demand side, platforms are taking a more active role in determining the products to be selected by consumers. Platforms increasingly have the role of standardizing the products and the contracting conditions, as well as nudging consumers to select a specific provider. The role of a platform goes beyond that of a traditional intermediary, so it is reasonable to impose greater liability on platforms than on traditional intermediaries. This does not mean that platforms have to be made fully liable for the products they intermediate. This is the approach the EU authorities have already taken in the audiovisual and telecom markets. Platforms active in these markets have greater liability than a mere intermediary, but less than an editor or a telecom carrier.

On the supply side, businesses are increasingly dependent upon digital platforms in their relationship with consumers. They depend upon the platform's algorithms for their offer to be ranked and displayed to consumers. The conflicts of interest that are so common with intermediaries are exacerbated by algorithmic intermediation. However, there has so far been little transparency when it comes to the workings of the algorithms. The European Union has started to adopt legislation to increase fairness and transparency in the relationship between businesses and digital platforms. Most of the obligations imposed by the new legislation are somehow inspired by the traditional obligations on intermediaries. They are adapted to algorithmic intermediation. Still, European regulation is a first step.

22.1 Digital Platforms as Superintermediaries

The nature of the conflicts between platforms and users might be similar to that of traditional intermediaries, but the scale of the activity of the platforms certainly makes a difference. Platforms operate at a scale previously unknown for intermediaries. They have the power to determine the conditions for the operation of the market and they often define the terms and conditions for the provision of the service, including the price. In many cases, they have created the markets in which they operate. Platforms do intermediate, but as superintermediaries. Superintermediaries should be subject not only to the traditional obligations applied to intermediaries (see above), but to some additional and new obligations derived from the central role they play in the digital multi-sided markets.

The first difference between the traditional and the new digital intermediaries is the scale itself. Traditional intermediaries mostly have a small scale. Matching supply and demand was often a craftsman activity, developed by individuals with a good knowledge of the market. Real estate agents would work on a local scale. This was the case of most brokers. Intermediation was an activity that was difficult to escalate; only in thin markets with little activity would the intermediary be in a position to concentrate a large percentage of activity and manipulate the market. In regular markets, fierce competition existed between high numbers of intermediaries. Market entry for intermediaries was often easy.

Digitalization has enabled automation and automation has empowered intermediaries to scale up. Digital technologies have empowered the construction of massive multi-sided markets with millions and even billions of users at a global scale. Operational limits to scale-up have been removed, which has meant that the intermediaries creating the largest network effects have succeeded.

The second difference is standardization. Intermediating on a large scale is only possible because the intermediation activity is automated. Algorithms are developed to match supply and demand and create new complementarities. Automation is only possible if a certain degree of standardization in the trading conditions is introduced. The more the platforms automate the matching, the more standardized the object of intermediation will have to be.

Network effects require standardization of the underlying assets. Standardization is a common feature in the creation of network effects. As examples, railway time, container size, and electricity current had to be standardized to make networks out of fragmented infrastructures. Platforms have to standardize the underlying content and services they mediate in order to be effective.

The case of Uber is particularly illustrative. In order to match supply and demand, Uber determines the standard conditions of the service. Various categories of service are created according to the quality of the vehicle and the service. Then the platform defines the maximum price to be charged for the ride. On-demand services cannot be left to negotiation by the parties. There is no time. Transaction costs are reduced by standardizing the service, including the price. This does not mean that Uber is transforming itself from an intermediary into a service provider. It is just that the intermediary is standardizing the service to reduce transaction costs for the benefit both of supply and demand. It becomes a superintermediary.

The standardization role of the platforms has been labeled as a "regulatory role."[1] It has been argued that platforms regulate the market, as they determine the conditions of the provision of the services they intermediate. We do not believe that platforms regulate; rather, they merely undertake the economic activity to create network effects by standardizing the terms of the transactions they intermediate. The term "regulation" here is confusing and should be avoided. It is the role of the real regulatory authorities to regulate – that is, to control the economic activity of the platforms, including standardization activity – to exclude market failures and abuses. Standardization enables the new multi-sided markets and, at the same time, shapes the market.

The third difference is that superintermediaries have the power to shape the market in which they operate, a market that they have often created around themselves. The role of platforms as system coordinators requires a regulatory recognition and specific rules adapted to these circumstances.

The first platforms simply displayed the intermediated goods and services, as would be done on a billboard or in the classified ads section in a newspaper. This was the case with Craigslist. These platforms play a very limited role in the organization of the market.

Overtime, platforms became more sophisticated in how they displayed supply to the demand side. They would provide a safe contracting environment, with tools such as reviews, ratings, secure payments, etc. However, these platforms still did not actively match supply and demand; examples include eBay and Airbnb.

The most popular platforms today actively match supply and demand. This is the case with YouTube, Facebook, or Spotify, which actively propose content to viewers. They do not merely let the user browse the available content and do not merely inform about the popularity of the content (the most viewed, etc.). They actively take the role of a curator, selecting content for the demand-side user. Sophisticated algorithms have the power to personalize selections based on previous choices made by the user, choices made by other users with a similar profile, or simply based on the interest of the platform to display a specific content. Platforms have the power to nudge users towards new content that the user would not have otherwise selected. Search engines are in the same position. Google personalizes search results, predicting content that it believes will be of interest for each specific individual making a search.

It has been argued that digital platforms "curate" the underlying services, "coordinate" the system, or even act as operators.[2] This goes beyond the traditional role of an intermediary such as a real estate agent, and even beyond the role of intermediaries constituting marketplaces, such as stock exchanges. Platforms are not providing the underlying products, but they go beyond the traditional role of an intermediary. They "superintermediate."

The special role of platforms is not specific to the largest platforms such as Google or Facebook. The special role of platforms does not derive from market power, but from the special nature of intermediation when it is automated and uses algorithms. Therefore, regulation of platforms as superintermediaries should not be limited to platforms with market power, but it should apply more generally to all digital platforms, at least those that reach a certain maturity and scale.

Regulation should be adapted to the new reality. The regulation of platforms should be adapted both in the relationship with consumers and in the relationship with the intermediated businesses.

22.2 Liability of Superintermediaries

Regulation should take into account the particular roles of platforms as superintermediaries with a preeminent role in the markets they intermediate. Such a role should have consequences regarding liability in the relationship with the user of the intermediated service and in terms of fairness in the relationship with the services provider.

The role of platforms as mere intermediaries becomes increasingly blurred as platforms lean forward to set the streamlined conditions of the intermediated transaction. They actively make a personalized selection of supply for each individual and even nudge users into specific content or services. Public authorities have more and more difficulties to determine the legal conditions applicable to such an activity.

The Court of Justice of the European Union considered eBay as a mere intermediary with no liability for the counterfeited goods sold in the platform.[3] In the same way, the Court considered Airbnb to be a mere intermediary of an underlying accommodation service, arguing that the facts do not establish that Airbnb "exercises such a decisive influence over the conditions for the provision of the accommodation services to which its intermediation service relates, particularly since Airbnb Ireland does not determine, directly or indirectly, the rental price charged."[4]

However, the Court of Justice could not admit Uber as a mere intermediary because a more active role in the definition of the transaction conditions was identified. The Court had a point. They identified Uber as a provider of intermediation services, but also identified that Uber was doing something more: "Uber exercises decisive influence over the conditions under which that service is provided by those drivers. [...] Uber determines at least the maximum fare by means of the eponymous application, that the company receives that amount from the client before paying part of it to the non-professional driver of the vehicle, and that it exercises a certain control over the quality of the vehicles, the drivers and their conduct, which can, in some circumstances, result in their exclusion."[5] Therefore, the Court refused to consider Uber as a mere digital company (providing "information society services" in EU lingo) and classified Uber's service as "a service in the field of transport."

The most sophisticated courts and public authorities have identified that platforms play a larger role than traditional intermediaries. Platforms play a larger role in the definition of the terms and conditions of the goods and services to be intermediated. Furthermore, platforms often create the market (the multi-sided market) around them. Real estate transactions existed before agents started intermediating, but many of the services intermediated by platforms did not previously exist or have been fundamentally transformed or even created by platforms.

This difference should have legal consequences. Since platforms are determining some of the conditions for the provision of the underlying service, they should have some liability for such services. For example, they could be obliged to provide insurance and take secondary liability in case the service provider is not in the position to meet its liabilities.

It is common for platforms to assume these kind of commitments on a voluntary basis. These commitments increase security in the transaction, reducing transaction costs and making the platform more attractive. Credit card companies are a good example here: they take responsibility if your card has been used fraudulently. This kind of commitment helps a platform to grow network effects. This is why platforms have been doing some content filtering (Facebook, YouTube), filter drivers with criminal records (Uber) and provide insurance (Airbnb).

The paradox is that the more platforms take voluntary liabilities, the more authorities consider that platforms are not intermediating but actually providing the underlying good or service. This is a mistake that works against consumers and should be avoided. Platforms should not be automatically considered as the providers of the intermediated services and be fully liable for the provision of such ancillary services. Such a position would ignore the reality of the new model of industrial organization and hinder the development of platforms and the benefits derived from them

A new balance has to be identified. Platforms should not be fully exempted from liability, but they should not be made fully liable for the intermediated goods and services. Legislation can impose some obligations and liabilities on platforms. Imposing obligations and liabilities that are already common in the market excludes the risk of over-regulation. At the same time legislation could ensure that these obligations and liabilities do not transform an intermediation service into a fully owned service with full liability.

The European Union is taking this path. Sector-specific regulation in media and telecommunications has extended obligations from editors and carriers to platforms. They are not considered to be the providers of the underlying service, so no full liability for the service is imposed, but some obligations are still directly imposed up platforms.

In media, the review of the Audiovisual Media Service Directive has introduced a new category of service: the "video-sharing platform service," defined as "providing programmes, user-generated videos, or both, to the general public, for which the video-sharing platform provider does not have editorial responsibility, in order to inform, entertain or educate, by means of electronic communications networks [...] and the organisation of which is determined by the video-sharing platform provider, including by automatic means or algorithms in particular by displaying, tagging and sequencing."[6]

Even if providers such as YouTube do not have editorial responsibility, they still "organize" the service, and are therefore obliged to take measures to protect minors, and more broadly, to protect the general public from content inciting to violence or hate against minorities, or in the case of general illegal content. Such measures specifically include age verification mechanisms, mechanisms for users to flag illegal content and redress mechanisms.

In telecommunications, the new European Electronic Communications Code[7] defines a specific category for Over The Top service providers (OTTs) such as Skype and WhatsApp, which are also known as "number-independent interpersonal communications services." These service providers are not subject to all the regulatory obligations imposed upon traditional telecom carriers, as it is recognized that they do not control the conveyance of the signal; however, some security obligations and some obligations on access by disabled users are imposed on them as well as to respect fundamental civil rights. They will not be able to discriminate based on nationality or place of residence. The providers of these services might be even obliged to interconnect their services with other service providers to ensure interoperability.

These might seem like timid steps that do not have much impact on platforms, but they certainly open the way for a more robust regulation of platforms in sector-specific regulation.

22.3 Fairness and Transparency for Business Users

The superintermediary role of platforms should have legal consequences in the relationship of the platform with the intermediated companies. This is the ambition of the European Union Regulation 2019/1150[8] for the promotion of fairness and transparency for business users of online intermediation services, adopted in 2019. The scope of the Regulation is to protect intermediated service providers from excesses by superintermediaries. Parallel obligations are imposed in Directive 2019/2161[9] on consumer protection.

Regulation 2019/1150 acknowledges that "intermediation of transactions through online intermediation services, fueled by strong data-driven indirect network effects, leads to an increased dependence of such business users, particularly micro, small and medium-sized enterprises (SMEs), on those services in order for them to reach consumers. Given that increasing dependence, the providers of those services often have superior bargaining power."

Online intermediation services are defined as "services that allow business users to offer goods or services to consumers, with a view to facilitating the initiating of direct transactions between those business users and consumers, irrespective of where those transactions are ultimately concluded." This category includes e-commerce marketplaces, online software applications services, such as app stores, and service marketplaces such as Airbnb. Search engines also fall under the scope of the Regulation, even if the obligations imposed on them are similar but not identical.

The point of departure of the Regulation is the acknowledgement that platforms provide intermediation services, not the underlying products. Digital platforms are considered online intermediaries. However, it has been considered necessary to adopt special rules for digital intermediaries, which we think is a recognition of the special power of platforms as superintermediaries.

The EU Regulation imposes obligations on platforms to ensure fairness in the provision of intermediation services. In their terms and conditions, platforms must define the main parameters determining the relative prominence given to intermediated goods and services, as well as the reasons for the relative importance of such parameters compared to others. If one of the parameters is the remuneration to the platform to improve the ranking of the services of the paying intermediated party, the platform is obliged to disclose it and explain how remuneration affects the algorithm. However, the Regulation does not require platforms to disclose algorithms.

The Regulation correctly identifies that algorithms are not neutral. Algorithms rank the intermediated product, whether it is a video, a ride, or an accommodation. Ranking largely determines the result of the transaction, as consumers tend to select the product proposed by the platform and are often just matched automatically, with no option to choose the service provider. Transparency must be introduced in the ranking of options, to ensure fairness for service providers. However, the Regulation does not impose obligations on the ranking other than transparency.

The EU Regulation does not prohibit differentiated treatment by digital platforms, but they are obliged to disclose this in their terms and conditions. Platforms are obliged to describe differences in ranking, remuneration, access to data, as well as any condition in the provision of the good or service that might be connected or ancillary to the intermediation service. The Regulation is particularly concerned with the vertical integration of platforms and the potential discrimination in favor of the goods and services provided by the platform in competition with the goods and services provided by third parties intermediated by the platform.

A parallel provision has been adopted to protect consumers. Directive 2019/2161 obliges search engines to inform customers about the general parameters determining the rank of the search results as well as of their importance. In particular, they must inform consumers about payments made to achieve higher rankings. More generally, consumers must be informed when prices are personalized through algorithms and profiling.

The Regulation identifies the risk of suspension or termination of the intermediation service. A statement of reasons must be provided to the intermediated party, with reference to the specific facts and circumstances that led to the decision to suspend or terminate the service. The Regulation imposes a 15-day notice period before terms and conditions for suspension and termination are modified by a platform.

The EU Regulation imposes upon platforms the obligation to justify the commercial or legal consideration that ground the potential restriction imposed upon the intermediated parties to provide the intermediated goods and services at different conditions through other means. Most Favored Nation clauses, or the prohibition to provide better rates outside the platform have been common in travel and accommodation platforms, triggering divergent responses from antitrust authorities across the world.

Finally, the EU Regulation obliges platforms to establish redress mechanisms. Platforms are obliged to establish internal complaint-handling systems and the possibility for business users to engage mediators to attempt to reach an agreement to settle the dispute out of court.

The EU regulation is a first step in the regulation of platforms. It correctly identifies that platforms are intermediaries and not the providers of the underlying service, and therefore focuses on the conflicts of interest between platforms and the intermediated service providers. The EU Regulation also correctly identifies that digital platforms have a privileged position in the relationship with service providers, which is their position derived from being superintermediaries. The transparency that is introduced has to be welcomed. However, the Regulation does not really impose obligations to ensure fairness. Not even the most basic obligation to protect and promote the interest of the user is introduced, let alone further obligations such as suitability and fiduciary duties.

The regulation of platform to business relationships will expand in the coming years. One specific scenario is the relationship between platforms and individuals providing business services, whether it is content producers for YouTube or individuals driving for Uber. Since platforms have an unbalanced power relationship with individual providers making use of the platform to reach customers, it might be sensible to protect such individuals, imposing some regulatory obligations upon the platforms. The most exploitative instructions in the algorithms could be prohibited (such as unrealistic expectations in terms of timing of the services or penalties in case an individual rejects a service), minimum payments could be imposed upon platforms, and platforms could be forced to provide insurance to such individuals in case of accident or illness. Regulation could be inspired by the traditional objectives of labor law, but without automatically extending all the restrictions of labor law to a business relationship, which is different in nature.

22.4 Network Industries and Superintermediaries

The interplay between traditional network industries and digital platforms is determined by the role of digital platforms as superintermediaries, both from the demand perspective and from the supply perspective.

From the consumer perspective, digital platforms are not mere intermediaries facilitating the contracting of the service provided by a traditional player. As has been analyzed in telecoms, media, transport, and energy, platforms have the ability and the will to go beyond mere intermediation. They often monopolize the relationship with the customer, build a new service on top of the services provided by traditional players, and aim to end up coordinating the entire sector.

Digital platforms should be regulated like the superintermediaries they are. Their role as system integrators and coordinators should be the focus of the attention by the regulators. While they might not be producing the underlying content or transport service, they are the ones promoting it for consumers, nudging them to a specific product instead of an alternative. This should be the core of the regulation on digital platforms: the algorithmic ranking of underlying services.

The regulatory framework imposed upon digital platforms active in the traditional network industries should ensure a level playing field. No artificial incentives should exist to contract through platforms, rather than directly contracting with the underlying service providers. Existing regulatory obligations imposed upon traditional players, such as those regarding liability, should be re-evaluated to ascertain whether they are still proportionate. It should be evaluated whether regulatory restrictions are a burden for traditional players to compete with platforms. If this is the case, but obligations are still necessary, it should be considered whether similar restrictions should be imposed upon the digital platforms as well.

From the supply perspective, the coordination role should also be the focus of the regulatory intervention. Objectivity in algorithmic ranking – or, in other words, non-discrimination against a specific service or service provider – should be ensured. This is particularly the case when digital platforms are vertically integrated and they intermediate their own products in competition with the products provided by traditional players.

The balance in the vertical relationship between platforms and the underlying network industries is particularly relevant in traditional network industries, as they are all vested with a general interest. The traditional obligations on the network industry players to ensure the universality of the service, its affordability and sustainability might not be sufficient once the service is being coordinated by a digital platform. A similar obligation might have to be imposed upon the digital platform to ensure that the general interest objectives are reached. The final chapter in this book covers this issue in more detail.

Notes

1   Crémer, J., de Montjoye Y.-A., & Schweitzer, H. (2019). *Competition Policy for the Digital Era*, Report to the European Commission, p. 60, retrieved from https://op.europa.eu/es/publication-detail/-/publication/21dc175c-7b76-11e9-9f05-01aa75ed71a1

2   Knieps, G. (2019). Network economics of operator platforms. *Networks Industry Quarterly*, 21(3), p. 17.

3   Judgment of July 12, 2011, *L'Oréal v. eBay*, C-324/09, ECLI:EU:C:2011:474.

4   Judgment of December 19, 2019, *X/Airbnb*, C-390/18, ECLI:EU:C:2019:1112.

5   Judgment of December 20, 2017, *Elite Taxi/Uber*, C-434/15, ECLI:EU:C:2017:981.

6   Article 1 Directive 2010/13/EU of the European Parliament and of the Council of March 10, 2010 on the coordination of certain provisions laid down by law, regulation or administrative action in Member States concerning the provision of audiovisual media services (Audiovisual Media Services Directive), OJ L 95/1, 15.4.2010.

7   Directive (EU) 2018/1972 of the European Parliament and of the Council of December 11, 2018 establishing the European Electronic Communications Code (Recast), OJ L 321, 17.12.2018, pp. 36–214.

8 Regulation (EU) 2019/1150 of the European Parliament and of the Council of June 20, 2019 on promoting fairness and transparency for business users of online intermediation services OJ 2019 L 186/57.
9 Directive (EU) 2019/2161 of the European Parliament and of the Council of 27 November 2019 amending Council Directive 93/13/EEC and Directives 98/6/EC, 2005/29/EC and 2011/83/EU of the European Parliament and of the Council as regards the better enforcement and modernisation of Union consumer protection rules, OJ 2019 L 328/7, 18.12.2019.

Chapter 23

Regulating Platforms with Market Power

Network industries have a tendency towards concentration. As network effects require a critical mass, only a limited number of players can reach the minimum size to be competitive. That was the case of the traditional network industries and is the case today with the new network industries. In some cases, digital markets might even evolve to become quasi-monopolies.

Competition authorities have started to impose fines on the largest platforms for anticompetitive practices, particularly in Europe. However, there is growing consensus that competition law has not managed to impose an effective control on platforms with market power. Mergers leading to market power have been allowed. Once platforms have tipped into dominance, competition authorities struggle to stop strategies to reinforce market power, as well as to stop exploitative practices damaging platform users.

This is exactly the same dynamic that led to the regulation of the traditional network industries a century ago, despite the availability of antitrust law. When there are structural obstacles to competition, such as powerful network effects, antitrust rules are not in a position to guarantee competition in the market or to ensure a fair deal to the customers.

A number of reports around the world from reputable academics and public authorities have concluded that it is necessary to regulate digital platforms with market power. The European Union is leading the way in defining a regulatory framework for digital platforms with market power.

First, a toolkit of regulatory obligations is being proposed to promote competition in digital markets. Inspired by the regulation of liberalized network industries, again with telecoms taking the lead, obligations being considered include data portability to avoid lock-in effects (inspired by number portability), prohibition of exclusivity to foster multi-homing, interoperability to share network effects, as in telecoms, and data pools to share algorithmic network effects.

Second, structural measures, such as the horizontal divestiture of the largest digital platforms, or the vertical unbundling of vertically integrated platforms, have been proposed. Again, these are well known measures in the traditional network industries.

Finally, a toolkit of regulatory obligations is being proposed to exclude exploitative behavior by platforms with market power. The toolkit is heavily influenced by regulations regarding fair, reasonable and non-discriminatory access to telecom carriers with market power.

23.1 Winner-Takes-All?

Winner-takes-all is the most feared evolution of platforms in multi-sided markets. Network industries tend towards concentration and platforms in multi-sided markets are no exception. In some cases, network effects might lead to natural monopolies. Telecommunications, transport, and electricity provide abundant examples. Some of the most mature platforms, such as Google and Facebook, might be approaching a similar outcome. However, this is not the unavoidable evolution of all platforms in multi-sided markets.

As the internet has matured, the position of the largest platforms has become stable and some platforms have enjoyed large market shares and market power. When the internet went mainstream in the mid-1990s, new companies experimenting with new business models grew to leadership positions being surpassed in a matter of a few quarters by other companies with new technology and new business models. Netscape, Yahoo, and AOL are good examples. However, the days when empires were created and then vanished in a matter of months are long gone. Google has been in a leadership position in searches for more than 15 years and Facebook has been leading social networks for more than ten years. At this stage, it is clear that competition is not "a click away."

It is well-known among network experts that network effects lead to market concentration. As large volumes of users are necessary to create network effects, only a limited number of players can be active in the long term. Fierce competition between numerous platforms can take place in the market during the initial stages. For example, a high number of platforms are active in the home delivery market. They are all competing to pool the largest volumes of users to beat the competition by having the largest network effects. Users (both consumers and service providers) benefit from such competition in the form of discounts and prices that might even be below cost. Fierce competition accelerates the penetration of the service. This happened with the telephony service a century ago and it is happening with new platform services today. Competition for the market benefits consumers,[1] at least in the short term. The smallest platforms will not be able to remain in the market as they cannot compete on price due to the small network effects they create. Consolidation is also a common feature in these markets. In the medium term, the number of players comes down to the number of players that can reach the minimum network effects to be competitive.

The more difficult it is to replicate the pool of users of the leading platform, the larger the market power of such a leading platform will be. This is the case, for instance, when users have difficulty using more than one competing platform in parallel. In some cases, such as Uber and its competitors, users can participate in more than one platform at the same time. It is common for both drivers and riders to participate in more than one platform, so newcomers can grow by sharing users with the market leader. When this is not the case, for whatever reason, the position of the leading platform becomes more solid. This is the case with Facebook. As it is cumbersome to feed the platform with updated information, pictures videos, etc., users tend to use only one platform to display their content.

However, the fear of winner-takes-all platforms might have been exaggerated due to the extraordinary success of specific platforms, namely Google and Facebook. Both multi-platforms are massive in terms of number of users (in the billions), so they have created colossal network effects and they are difficult to replicate. They have extraordinary market shares in their most relevant activities (search, social network, mobile operating systems).

Most of the multi-sided markets they have created are non-transactional markets, as they are financed by advertising revenue rather than by payments from final users to service providers intermediated by the platform. The volume of the advertising revenue is impressive, particularly compared with traditional media revenue. Facebook makes more advertising dollars than all the newspapers in the world combined. Google is not far from surpassing the advertising revenue of all the world's television stations combined. It can be concluded that, in their core activities, Google and Facebook are leading examples of winner-takes-all.

However, it seems clear at this stage that other platforms have not tipped as Google and Facebook have. Uber and Airbnb, for instance, have large market shares, but face strong competition.

Winner-takes-all is more probable if network effects are not exhausted when a certain threshold is reached, but can keep growing until all users join the platform. For example, the value of Facebook continues to grow as more users join the social network, and it would reach its maximum value if all users would connect. But this is not always the case. For instance, the value of a ride-hailing service tops out for riders when it can be ensured that an available vehicle is no more than 1 minute away from the user's location. If this threshold is reached with only 30 percent of the vehicles in a city, then alternative platforms can be formed and they can reach the necessary threshold to compete with the market leader. Just like economies of scale, network effects can be exhausted at a certain threshold.

It is a common mistake to assume that winner-take-all and a monopoly is the unavoidable result of platforms in multi-sided markets mature. This might be the case in some markets, but it is not the unavoidable structure of all multi-sided markets. If network effects are exhausted at a low market share, a healthy number of platforms can compete. Building the necessary network effects takes time and money. Therefore, it cannot be automatically assumed that all platforms are always in a winner-takes-all game.

In any case, some platforms are growing to winner-takes-all positions. A platform with market power is in a position to completely change the game. As discipline from competition is reduced, such platforms have less of a counterbalance to the incentive to monopolize the value generated by the network effects they have created.

A simple way to identify whether a platform is capturing an exceptional proportion of the value generated by network effects is to identify the net profit margin of the platform. Net profit margin measures the amount of net profit a company obtains per dollar of revenue gained. The profit margin is equal to net profit (also known as net income) divided by total revenue, expressed as a percentage. Mature large platforms have a net profit margin that is substantially above the market average, and in some cases, are among the corporations with the largest net operating margin in the world. Facebook has reached a 40 percent net profit margin, which is exceptional. Other mature platforms such as eBay (27.22 percent) or Apple (close to 40 per cent) are also at exceptional levels of profitability.

Furthermore, dominant platforms can engage in strategies to exclude competitors and to expand their dominant position to neighboring markets. Some of these practices have been identified in this book. Exclusivity clauses imposed by digital platforms prevent multi-homing and growth of competitors. Self-preferencing behavior by digital platforms facilitates expansion into downstream markets.

This is not the first time in history that network effects have winner-takes-all champions of unprecedented size and reach. Telecommunications, railways, and electricity networks reached this point more than a century ago as competitors consolidated into monopolies. The regulation of these monopolies in network industries provides the most interesting insights into the regulation of network effects.

23.2 Competition Law and Regulation

As digital platforms have matured, some of them have built market power positions. Users are increasingly describing platforms as gatekeepers: customers can only be reached through the platform. Furthermore, users complain that platforms are abusing their market power both to impose exploitative conditions and to consolidate and expand their market power. While public authorities have started applying competition law to the largest platforms, there is increasing consensus that competition law is not sufficient to discipline platforms with market power.

It is not surprising that authorities, particularly in Europe, are using competition law to discipline the abuses of the largest platforms. This book has analyzed antitrust cases in the EU and the US against Google, Facebook, Apple, and Amazon. Even China initiated antitrust actions against Alibaba in 2020. Competition law is flexible; as it is a horizontal legislation applicable to all business activities, it can be applied to new business realities such as platforms in multi-sided markets.

Nevertheless, competition law faces three fundamental limitations in terms of governing platforms in multi-sided markets. Firstly, competition law faces the challenge of adapting itself to the new model of industrial organization platforms multi-sided markets represent.[2] Competition law evolved as a response to market (and power) concentration due to industrialization and economies of scale. The concept of a "relevant market" was developed to identify the sphere of activity to be analyzed by authorities and the concept of "dominance" in such relevant market was developed as the threshold to trigger the special obligations of companies with market power. The precedents defining anticompetitive (and therefore prohibited) behavior evolved for a model of industrial organization based on building scale by vertically integrating a company. Within this framework, it was relatively simple to identify market power in a given relevant market.

Multi-sided markets pose a relevant challenge to the established principles of competition law. It is complex to define the relevant market in a multi-sided market, as has been well represented in the *American Express* case in the US, which sparked a fierce debate around the definition of the relevant market in the multi-sided market of payment cards (keeping in mind the relevance of the positions for digital platforms).[3] The plaintiff and the *amici curiae* supporting the plaintiff – a long list of respected antitrust law and economics professors – claimed that it is sufficient to consider one side of a multi-sided market as the relevant market and identify whether there is market power. As they stated, "we define markets in antitrust analysis to determine whether market power exists in the relationship between two sets of economic counterparties."[4] The defendant argued that precedents have traditionally underlined the need to understand the business reality behind a case and that sound economic analysis should be employed. In multi-sided markets, the business reality was that the conduct that affects one side of the market will probably affect the other side (or sides) of the market. An analysis that ignores such effect ignores a highly relevant part of the picture.

In 2018, the Supreme Court decided in favor of the defendant: "the fact that two-sided platforms charge one side a price that is below or above cost reflects differences in the two sides' demand elasticity, not market power or anticompetitive pricing. [...] Thus courts must include both sides of the platform – merchants and cardholders – when defining the credit card market."[5]

There are strong arguments in support of defining the relevant market as including all sides in a multi-sided market, as it seems the only way to consider the complex reality of this model of industrial organization. However, the practical effect of this approach is that the application of antitrust law is weakened. As is usually the case, the larger the market, the greater the chances of having the market share of the potentially dominant firm diluted in it. Furthermore, and this is new, the difficulty of analyzing complex multi-sided markets might become an unsurmountable burden for the plaintiffs, up to the point that it might "devastate antitrust."[6]

Secondly, competition law has a narrow interest to protect. Over recent decades, such interest has been reduced to protecting the effective competition structure in order to ensure consumer welfare in economic terms. This narrow objective poses a challenge in the application of antitrust to digital platforms, as platforms have reduced prices to consumers, to the point of providing services for free in the case of the most popular non-transactional platforms that are financed with ads. The difficulty of applying competition law under such a narrow purpose to a digital behemoth such as Amazon has already been underlined.[7] Even if antitrust expands its scope, as is increasingly supported by a group of antirust professors – disparagingly referred to as "antitrust hipsters" – the scope of antitrust will always be limited to economic objectives.

Other values, such as pluralism in media or privacy are foreign to and cannot and should not be protected by competition law. European competition authorities are trying to creatively expand the scope of competition law, taking into consideration other interests such as privacy. A prime example is the decision of the German antitrust authority in February 2019[8] that Facebook abused its dominant position by conditioning the use of the social network to accepting an extensive permission to collect and process personal data. However, this approach risks denaturalizing competition law and introducing an undesirable legal uncertainty.

Ex-ante regulation is better suited to protect interests foreign to economic consumer welfare.[9] This is certainly the case of pluralism in media and privacy, but also of other economic values, such as the sustainability of digital ecosystems, as will be discussed in the next chapter. Ex-ante regulation can be specifically adopted to meet any political objective that merits protection according to the democratic process.

Third, competition law faces limitations in the prescription of behavior to companies with market power. Competition law is proscriptive in nature; it prohibits business practices and imposes fines in case of breach. Competition authorities can impose upon platforms a fine for an illegal exclusionary practice, and can even impose a fine for refusal to deal under the essential facilities doctrine. However, competition law was not designed to prescribe the conditions under which services have to be provided. Antitrust has limited instruments to define the conditions under which platforms must provide access to other platforms to ensure interoperability. It does not have the instruments nor the institutional structure to control markets in a continuous and concentrated form. This is a job for ex-ante regulation and ex-ante regulatory agencies.

The prescriptive definition of behavioral obligations, particularly in dynamic markets, requires the continuous supervision of market conditions and flexible mechanisms to adapt obligations to an evolving reality. This is the same reason behind the creation of regulatory agencies in the US a century ago, and is why competition law and competition authorities are not well placed to regulate platforms.

Reform of competition law to adapt it to digital markets is underway. A plethora of reports issued over the past years in the EU (Crémer Report),[10] the UK (Furman Report),[11] the US (Stigler Report),[12] Australia,[13] Japan,[14] and elsewhere[15] have identified the need to review existing antitrust law and practice to better control anticompetitive behavior by digital platforms. The following shortcomings of current antitrust rules have been identified: difficulty determining the relevant market for digital platforms, difficulty analyzing zero-pricing with traditional antitrust tools, difficulty identifying the role of data from an antitrust perspective, and weak merger control.

There is a growing consensus that competition law is not sufficient to regulate digital platforms with market power. This was the conclusion of the Furman Report in the UK, whose recommendations to implement an ex-ante regulatory framework and to establish a corresponding regulatory unit have been upheld by the government and supported even by the antitrust authority. The European Union is following the same path: a consultation was launched in 2020 in order to define the specific form of ex-ante platform regulation.[16] In Australia, the antitrust authority made a similar recommendation.

For all these reasons, competition law and competition authorities are leading in the understanding of the competitive dynamics around platforms and will block the most blatant abuses to be committed by platforms. However, competition law will not be in a position to define and enforce a coherent regulation on digital platforms. It is the job of ex-ante regulation to control structural market power, as is the case with the market power generated by network effects.

It is well known that certain industries display structural limitations to competition. This is the main lesson that can be derived from the 150 years of network industries. It was identified early on that competition law is in a position to protect market dynamics from anticompetitive practices, but it is not in a position to substitute such dynamics. When economies of scale and network effects lead to market concentration, even to almost-monopoly, the rules on competition are not in a position to reverse this trend and to create competition in the market. It is the role of ex-ante regulation, of specific state intervention imposing obligations on companies active in the market, to guarantee the general interest.

Originally, ex-ante regulation was adopted to substitute competition, as network industries were monopolized by private companies in the US (see Chapter 6 for telecoms). It was only in the 1980s that such regulation changed its approach and, instead of substituting competition with rate-of-return rules, pursued the aim of promoting competition even in the network industries. This is the approach that is under consideration for the digital platforms with market power.

Platform regulation could take the form of asymmetric regulation.[17] Some obligations could apply to all platforms. This would be the case with the regulation analyzed for superintermediaries, as is the case with the nature of the relationship with individual service providers, the liability regime and so on. The most cumbersome regulatory obligations would be reserved to a limited number of platforms with market power and would primarily aim to promote competition.

An asymmetric regulation with a more burdensome framework for platforms with market power requires the definition of parameters to consider a platform as having market power. We have already seen that the definition of the relevant market and the identification of dominance according to traditional competition law is highly complex and constitutes one of the difficulties of applying competition law to digital platforms.

One option is to define a very basic and simple threshold. The first EU Directives on telecoms imposed asymmetric access obligations to carriers with a market share of more than 25 percent in predefined product markets.[18] However, such a rigid approach had to be reviewed, as the predefined markets did not include, for instance, broadband internet access, a market that eventually evolved to be even more relevant than telephony. Directive 2002/19/EC defined an open-ended but more complex procedure to first define the product markets where ex-ante regulation is necessary and then define the carriers with significant market power, based on criteria similar to those of dominance in competition law.[19]

An alternative is to define a more sophisticated threshold, forcing authorities to demonstrate the proportionality of the regulation triggered by the declaration of a platform as having market power.[20] In the UK, the Furman Report has proposed a "strategic market status" test for "those in a position to exercise market power over a gateway or bottleneck in a digital market, where they control others' market access."[21] The French telecommunications regulator has proposed a "structuring digital platform" test for "online platform operators or operating system providers which, in particular because of their intermediation activity in accessing internet services and content, and because of their importance, are able to significantly limit the ability of users to engage in economic activity or communicate online,"[22] based on seven parameters such as the control of bottlenecks, number of users, access to data and so on. In the US, the academics in the Stigler Center at the University of Chicago have proposed the notion of "bottleneck power," which "describes a situation where consumers primarily single-home and rely upon a single service provider, which makes obtaining access to those consumers for the relevant activity by other service providers prohibitively costly."[23] European academics have proposed a three-criterion test: (1) non-contestable concentrated market structure, (2) a digital gatekeeper that is an unavoidable trading partner, and (3) ineffectiveness of competition rules.[24]

The EU telecommunications regulation provides the best model for this kind of regulatory approach. First, it defines the threshold and the criteria to define a carrier as having significant market power in a product market. The Commission regularly publishes a list of product markets in which it is presumed that there are significant market power carriers. Second, the regulation defines a procedure to define the specific obligations to be imposed upon such carriers. Out of a list of obligations defined in the EU Directives, national regulatory authorities must determine which obligations are necessary and adequate to solve the identified market failure, to ensure proportionality in the regulatory obligations. The proposal of the national authorities is then put up for review by the Commission. Then interaction of the Commission and the national regulators create checks and balances to avoid over-regulation.

23.3 Pro-competition Behavioral Obligations

Behavioral obligations can be imposed on platforms with market power to promote competition and make the markets contestable. Traditional network industries have experimented with different behavioral obligations to promote competition. In the early days of the network industries, monopolies were promoted by public authorities, and subsequently subjected to control. More recently, regulation has focused on the re-creation of competition in the different network industries, if necessary through the regulation of the existing bottlenecks, including the regulation of the network effects. This experience seems to be of the highest value to define the regulation of platforms in multi-sided markets. At least four obligations can be identified, many of which replicate regulatory measures already implemented in the traditional network industries, mostly in telecoms: data portability, prohibition of exclusivity to foster multi-homing, interoperability, and data openness.

First, data portability is an instrument to reduce lock-in effects.[25] Reducing switching costs is a traditional strategy for fostering competition in deregulated network industries. Lock-in effects can be identified and then eliminated through regulatory obligations. The best example is number portability in the telecommunications industry. It was identified that telephone users were reluctant to change carrier if they would have to change their telephone number. All carriers were then obliged to work together to ensure that customers would have the right to take their telephone number with them when switching from one carrier to another.

Data is the raw material in the digital industries and has been identified as one of the barriers to switch platforms, as well as a barrier to entry for newcomers. Data fed by customers is often the obvious barrier to switch platform. Users might have spent years feeding their social network profiles, actually becoming the repository of the content they produce (pictures videos, etc.). They will hardly be in the position to switch to an alternative social network if they cannot take such content with them. The same problem is faced by all kinds of application fed with user data, such as applications tracking exercise, sleep patterns, and so on.

Data portability, clearly inspired by number portably, has been proposed as a solution for this obvious lock-in effect. Platforms would be obliged to create a feasible way for customers to take their data with them to another platform. Data portability is also a consequence of more structural debates on data ownership, namely as to whether the owner of the data is the person generating it, or the platform enabling the management of the data.

Data portability is already included in the EU General Data Protection Regulation (GDPR): "The data subject shall have the right to receive the personal data concerning him or her, which he or she has provided to a controller, in a structured, commonly used and machine-readable format and have the right to transmit those data to another controller without hindrance from the controller to which the personal data have been provided."[26] Data portability under the GDPR applies only to data uploaded by users and not to data that is inferred about them, which is most likely the majority of the data that platforms possess about their users.

However, data portability requires sophisticated management of data, coordinated across platforms by setting common standards. It is necessary to structure data in a way that can be processed by the new platform. Regulatory intervention is necessary to make data portability a reality.

Second, the prohibition of exclusivity clauses and the promotion of multi-homing is a pro-competition intervention. Multi-homing is the parallel and simultaneous use of competing platforms by users, either service providers, final users, or both. Hosts often display their accommodation units in Airbnb and other competing platforms at the same time. Guests often look for accommodation on Airbnb and also in competing platforms. The term "multi-homing" has its origin in computer lingo, when a computer was connected to more than one network, usually to increase reliability.

Multi-homing reduces barriers to entry as newcomers can build scale by selling services to customer already engaging with another platform. Multi-homing enables newcomers to replicate the scale of incumbents, as there is no obstacle to sign as many customers as the largest platform. All market players can build similar network effects by pooling together the same customers.

Multi-homing imposes a burden on users (both service providers and consumers) as they must make concurrent use of more than one system. They have to work with two or more applications, learn to use them, take the potential costs, etc. This is sometimes a cost that can be perfectly assumed, as multi-homing is simple and does not trigger high costs. In the early days of telephony in the US, merchants had to subscribe to two or three carriers in order to be able to reach all telephone users in a town, multiplying the number of contracts. This is still the case today in some African countries.

The burden on users is sometimes so onerous that multi-homing is not feasible. It was already identified how gamers are reluctant to acquire more than one videogame console due to the high cost of the hardware. Gamers tend to avoid multi-homing due to the high cost it entails. Social network users are also reluctant to multi-home, as uploading information and replicating the upload of the same pictures, stories, etc. is perceived as too time-consuming and of little interest, as most users are already on the largest platforms. Therefore, alternative platforms focus on niche markets instead.

Sometimes multi-homing is not feasible because the dominant platform actively disincentivizes it. Dominant platforms can obstruct multi-homing by imposing exclusivity on customers, mostly to service providers. As long as it is feasible, service providers usually prefer to work with more than one platform, so that they do not depend on an exclusive intermediary to commercialize their services. However, facing the choice to exclusively work with the market leader (with the largest pool of users) or be excluded from the largest pool of users and contracting only with a smaller platform, service providers will rationally choose the largest player. This strategy accelerates market concentration, reaching the tipping point where network effects enjoyed by the largest platform cannot be replicated by competitors. There are more subtle ways to impose exclusivity. Platforms can grant rebates to customers, such as to guests in accommodation platforms, incentivizing to grow the volume of business made with the platform. Such rebates disincentivizes working in parallel with another platform.

Competition authorities have identified exclusivity clauses imposed by platforms as the most relevant threat to competition in multi-sided markets. As a consequence, the most common remedy in platform mergers has been to prohibit exclusivity clauses to facilitate multi-homing. This has been common in food delivery[27] and ride-hailing mergers.[28]

This obligation could be extended beyond mergers as a regulatory obligation imposed on platforms beyond certain thresholds in terms of revenue, number of users or market shares. Obligations could also be imposed to actively promote multi-homing.

Third, another pro-competition behavioral obligation to be imposed on network industries is interoperability, which is defined as the "capability to communicate, execute programs, or transfer data among various functional units in a manner that requires the user to have little or no knowledge of the unique characteristics of those units."[29] Interoperability allows the simultaneous use of more than one network, not in parallel – as in the case of multi-homing – but thanks to the forced cooperation of the competing platforms. A network manager is allowed access to the network built by a competitor so that a service can be built for the final customer including elements of both networks.

Interoperability is the key to protecting existing network effects while avoiding the market power that derives from them. Telephony provides the clearest example. Network effects are expanded, as users can reach a larger volume of individuals. As described in Chapter 6, AT&T's Theodore Vail wanted "one system" ensuring "universal service." He assumed that "one system" meant a single provider, a monopoly. However, one system can be built by interconnecting all the telephony companies and ensuring the interoperability of their services. All users of a telephone company could contact any other user of the telephony service, even if served by another company. It was only necessary to ensure that all telephony networks would be interconnected. Vail's AT&T consciously obstructed all request to interconnect telephony networks.

Deregulation of telephony services was based on the obligation to interconnect all telephony networks to ensure interoperability. Such an obligation would be imposed on the incumbent with the largest market share and also on the rest of players to ensure the full interoperability of all the existing networks. Access regulation is imposed on railway infrastructure managers, airports, electricity grids and so on. This form of network access regulation reinforces network effects, as the telephone system is universal in reach: anyone can call anyone else. However, the network effects are not monopolized by a single company; they are shared among all the market players as they are all interconnected and their services are interoperable. Of course, small carriers are happy to share their small network effects with the rest of the carriers. The largest carriers, who would have a competitive advantage due to their larger networks, have little incentive to surrender their competitive advantage. They only provide access to their networks to competitors because they are obliged to do so by the existing regulatory framework.

Both in the US and in Europe, regulated competition in the traditional network industries is focused on the same objective. Market power has to disappear or at least be disciplined by competition from other service providers. Network effects are recognized as a leading source of market power. The objective is not to destroy the network effects, but to impede them from becoming a competitive advantage. The value of such effects is recognized, honored, and protected. Regulatory policy is focused on keeping network effects, while avoiding them becoming a source of market power over other market players.

Regulating interoperability is technically complex. It was not easy to regulate it in telecoms or electricity, it appears to be very difficult in railways, ports, or airports. After more than 40 years of interconnection regulation in telecommunications, one might think that such regulation was easy to implement, but this was not the perception back when access regulation had to be developed after the breakup of the Bell System. For AT&T, even something as simple as a plastic device to be wrapped around the fixed phone to reduce noise and protect the confidentiality of the conversation was a potentially dangerous access to the network that required meticulously detailed technical specification.[30] Full regulation of the interoperability of telephony networks was a technical and economic challenge.

Platforms can be forced to share their network effects by allowing interoperability with competitors. From a technical point of view, platform interoperability requires the development of specific APIs (application programming interfaces). No physical access takes place, as is the case with access to infrastructure networks. It all happens at the data layer, in the form of software. APIs allow third parties to interface with proprietary software owned by a platform. The more complex the software of the platforms, the more complex the APIs, but there is no fundamental technical obstacle to open the platforms to third parties, to competing platforms.

Most platforms already have APIs to interoperate with third parties. Mobile operating systems allow third parties to develop apps. Facebook allows third parties to develop games and other applications to be integrated into its platform. Throughout this book we have given examples about how newcomers built network effects on top of pre-existing networks. WhatsApp extracts contacts from a smartphone's directory. It used to be common to identify contacts in a platform to build new networks, but incumbent platforms tend to block this kind of interoperability.[31]

Incumbent platforms are understandingly reluctant to develop APIs to interoperate with competitors in such a way that network effects are shared with newcomers. This can happen in different ways.

The most obvious example of platform interoperability is that of the services provided by communications platforms. Just as emails can be exchanged across platforms, instant messages could be exchanged between communications platforms such as WhatsApp, Google Hangouts, Apple's Messages, and WeChat. This is not the case at the moment. Facebook even used Federal Criminal Law in the US against a small startup called Power Ventures, which was trying to make different platforms interoperable.[32] All the leading communications platforms are closed systems, competing to outgrow other platforms, following Vail's strategy.

The first example of regulation imposing platform interoperability can again be identified in the European Union. As already described, the new European Electronic Communications Code foresees the possibility of imposing interoperability on "number-independent interpersonal communications services"; that is, platforms such as WhatsApp and Skype "which reach a significant level of coverage and user uptake."[33]

Interoperability can be expanded to more complex interactions intermediated by platforms. In social networks, content uploaded in a platform would be accessible from another social network.[34] Even further, a small accommodation platform could interoperate with Airbnb and provide its final users access to hosts in Airbnb. In this way, small providers would have access to Airbnb's large pool of hosts. The network effects created by Airbnb would be shared with smaller competitors.

From an operational point of view, access to online platforms poses the same challenges as those originally posed to infrastructure networks. Third-party access would force platforms to standardize their rules and procedures to make them available to third parties. Such standardization would introduce an element of rigidity in the management of the platforms, as changes have to be extended to third parties, giving them time to adapt their software and operations. This is more of a problem for platforms than for infrastructure networks, as platforms are more dynamic and evolve faster than infrastructures. However, platforms tend to become more stable as they mature. Small platforms can constantly change conditions and internal rules and procedures as they look for the most effective algorithms to match supply and demand. Larger platforms have massive customer bases and face more rigidity when looking to making substantial changes to their rules and operations. Therefore, and because there is no reason to regulate small platforms with no market power, interoperability regulation should be limited to large mature platforms.

In any case, the main operational challenge derives from the possibility of imbalances derived from feeds introduced by interconnected platforms. For example, small interconnected platforms might increase the number of potential guests in Airbnb's platform, unbalancing the equilibrium that Airbnb has been between guests and hosts. Again, this is not specific to platforms when compared to infrastructure networks. Telephony networks have always raised the risk of traffic overflow if they would be open to competitors. Specific mechanisms were introduced to avoid such risks, such as in the form of contractual commitment forecasts regarding the traffic to be passed from one network to the other, higher prices and penalties in case of not meeting the forecasts, financial guarantees, etc. Unbalancing incumbent platforms is as dangerous as congesting physical infrastructure with non-forecasted traffic. Even if – or maybe because – they are virtual, digital networks rely on a delicate equilibrium among all the parties in the ecosystem. Opening a digital network to third parties certainly poses a risk to the existing equilibrium, but experience in physical networks has shown that such a risk can be managed.

Network effects would not be diminished by mandating interoperability to platforms, but would actually be reinforced, as more users would join the platform. The experience of traditional network industries suggests that a whole regulatory framework, including institution building, is necessary in order to make access regulation effective. Competition law cannot guarantee the continuous control that is necessary in order to enforce interoperability rules.

Antitrust has instruments to mandate access. The essential facilities doctrine was built in US antitrust to ensure access to facilities that cannot be duplicated and are considered essential for competition by third parties in a downstream market. Actually, railroads were at the origin of the essential facilities doctrine (*Terminal Railroad Association* case in 1912), which has subsequently been applied in the electricity industry (*Otter Trail in Power Co. vs. United States* in 1973); telecommunications has provided the leading cases decided by the Supreme Court (*MCI vs. AT&T* in 1979 & *Verizon vs. Trinko* in 2004).

However, the essential facilities doctrine is exceptional in US antitrust. Rather than preventing monopolization through the proscription of exclusionary practices, the essential facilities doctrine comes as close as antitrust ever has to condemning "no-fault" monopolization, to judge the fairness of the monopolist's behavior, and to impose the traditional common carrier obligation to serve everyone on reasonable terms. Platforms are again raising the interest of antitrust lawyers in the essential facilities doctrine. There is growing academic literature on the topic.

Finally, data openness is a behavioral pro-competition measure proposed in the Furman Report.[35] As described above, algorithmic network effects rely on the availability of massive datasets to be used for machine-learning purposes. Access to such datasets is a competitive advantage, such as in the search market. Sharing datasets would reduce the barrier to entry for newcomers and allow them to challenge the position of incumbent platforms.

Being a highly intrusive obligation, and being risky in terms of personal data protection, the Furman Report advised certain checks and balances: data exchanges could be made in a controlled environment with pre-approved entities, under the supervision of the data protection authorities and be limited to opening up raw underlying data that is an input to the service, rather than processed information where companies have invested further in deriving insights and inferences from the original data.

23.4 Pro-competition Structural Obligations

Breaking-up platforms is a structural measure of controlling market power in digital markets. It is structural because it either puts an end to market power, if a horizontal divestiture is imposed, or, in the case of vertical unbundling, eliminates the risk of leveraging into connected downstream markets.

Divestiture can be identified as the most radical regulatory measure against winner-takes-all dynamics in network industries. Divestiture is the ultimate antitrust remedy: it forces a company to sell off some of its assets. Divestiture was imposed a century ago against the largest industrial trusts, such as Rockefeller's Standard Oil.[36] It was also imposed in the 1980s on the Bell System that ran telecommunications in the US, which kicked off deregulation,[37] and it has been proposed as a solution to market power in the platform economy.

In November 2014 the European Parliament voted in favor of a non-enforceable declaration to "unbundle search engines from commercial services." Even if Google was not specifically mentioned, the divestiture of Google was in the mind of the members of Parliament. It was only a declaration, with non-binding effects, and no further action was taken after the declaration. In March 2018, the European antitrust Czar, Margrethe Vestager, insisted that the possibility of breaking up Google was "open and on the agenda." This possibility has been increasingly discussed in the US.

There have been growing criticisms of the weak antitrust enforcement in the digital world,[38] particularly the approval of previous mergers that have reinforced the largest multi-platforms.[39] For example, reversing previously approved mergers could be applied to the acquisition of Instagram and WhatsApp by Facebook and to YouTube and DoubleClick by Google. Reversing these mergers is an increasingly common proposition,[40] even by candidates for the US Presidency such as Elizabeth Warren.[41]

The antitrust suit filed by the Federal Trade Commission against Facebook in December 2020 formally request the divestiture of Facebook's Instagram and WhatsApp. The FTC identified that the scope of both acquisitions was to obstacle the growth of potential competitors.

Divestiture might be a sound policy when a company reaches a position of extraordinary market power that becomes an obstacle in itself to competition, particularly when it affects markets related to the monopolized activity. Antitrust law prohibits such dominant positions being built by merging previously independent companies. Divestiture builds on the same reasoning, but it is more intrusive as it goes against an established venture, interfering in the exercise of the ownership and free enterprise rights.

Horizontal divestiture – that is, the break-up of a company to diminish its size – is rarely a good decision in the network industries. While divestiture is certainly an effective measure to reduce market power also in network industries, horizontal divestiture affects the source of market power, the scale that ignites the network effects that provide the ultimate competitive advantage in the network industries. However, by reducing or even by eliminating the network effects, value is destroyed not only for the divested company, but also for users and society as a whole. The divestiture of digital platforms is not recommended as a policy, as it is not recommended in general for the network industries. It removes the network effects and the value they create.

Furthermore, horizontal divestiture might be useless if a market naturally leads to concentration as it creates a competitive advantage, since the market leader with the largest network effects has lower costs or another competitive advantage. The market will again concentrate. The experience of the break-up of the Bell System shows that even though the monopoly was broken into to seven regional monopolies (the Baby Bells), they ended up concentrated again a couple of decades later in the form of two carriers.

There are some cases in which horizontal divestiture might have beneficial effects. For example, this could occur with multi-platforms active in different multi-sided markets. Full network effects might be exploited in each market without the need to mutually reinforce them with the position in another market. Furthermore, efficiency and the creation of economic value is not the only reason to intervene in the market. There are other values that justify public intervention against the concentration of power, both economic and other kinds. Pluralism has justified in the past public intervention against concentration in media. The same logic could apply to the concentration of advertising revenue in the largest non-transactional platforms.

Vertical separation is a more nuanced approach to fighting some of the negative effects of market monopolization. Digital platforms would be allowed to intermediate in the provision of good and services, but they would not be allowed to provide the underlying service in competition with third parties. Separation can be total, when activities would be undertaken by independent companies, or only functional, in the sense that different corporations are required to provide the intermediation and the underlying service, even if they belong to the same business group.

Vertical separation is a regulatory measure adopted in Europe in different network industries, particularly railways (separation of infrastructure management and transport service provision) and electricity (separation of electricity production, transmission, and distribution).[42]

Vertical unbundling does not always reduce network effects. They are fully protected, so separation does not reduce value for users and society as a whole. Vertical unbundling does not have the negative effects of horizontal divestiture. On the contrary, the scope of vertical unbundling is to structurally impede the extension of market power generated by network effects downstream to connected markets. For example, a company managing an app store would not be allowed to commercialize its own apps.

Vertical unbundling is the structural obligation to be adopted when behavior obligations do not work. Discrimination in favor of the downstream arm is excluded (ownership separation) or at least made more transparent (functional separation) as the unbundled arms have to keep separate accounts, managers, incentives, etc. As data is key for digital platforms, data separation has been proposed as a remedy. Specific units in a platform company would not be allowed to share data. This was the remedy imposed by the German antitrust authority on Facebook, forbidding it from pooling together data from Facebook and data from WhatsApp.[43]

Vertical unbundling has been called for in the Amazon market place.[44] For example, authorities in India enacted a regulation, effective in February 2019, prohibiting e-commerce marketplaces such as Amazon from selling their own inventory. Amazon is only allowed to operate marketplace platforms where other parties sell goods to retail consumers. In this way, the authorities expected to put an end to Amazon's supposedly predatory practices of duplicating successful products commercialized by third parties and then discriminating in favor of such products.

Vertical separation is not limited to matchmakers; it has also been proposed for non-transaction platforms. Google would have to separate the ad exchange platform it manages from the underlying ad display ventures, such as the search engine and YouTube.

In conclusion, horizontal divestiture is not recommended as a regulatory policy in network industries such as platforms in multi-sided markets, because this measure not only reduces market power but also the value created by network effects, reducing consumer welfare. On the contrary, vertical separation does not necessarily affect network effects and is therefore not a threat to the value generated by network effects. Vertical separation, known as unbundling in the traditional network industries, is the ultimate instrument against the expansion of market power stemming from the intermediation activity.

23.5 Fair, Reasonable and Non-discrimination Obligations

Market dynamics are sometimes not sufficient to discipline players with market power. This is the case of the traditional network industries and also of digital platforms. In these circumstances, users must be protected from exploitative practices.

A set of behavioral obligations can be imposed upon digital platforms with market power in order to rule out exploitative abuses; that is, unfair conditions imposed upon users, such as excessive pricing, tying, and so on. In such cases, platforms could be forced to trade with users under fair, reasonable, and non-discriminatory ("FRAND") conditions.

FRAND conditions, and the FRAND acronym, evolved around patent licenses, in particular for standard setting.[45] The European Commission[46] and the Court of Justice of the European Union have supported FRAND remedies in mergers and the refusal to contract under such terms as proof of abuse of a dominant position.[47] A FRAND framework has been proposed for the regulation of digital platforms.[48]

Of course, there is a long tradition of FRAND-like sector-specific regulation in the traditional network industries as regards network access conditions. What is specific to access regulation is the role of sector specific regulators to set the FRAND conditions (or in other terms objective, transparent, proportionate, and non-discriminatory access conditions).[49]

Non-discrimination seems to be the most immediate obligation to be imposed upon digital platforms with market power, particularly as they vertically integrate and start competing with other intermediated businesses. Objectivity can be imposed on all platforms considered to be superintermediaries, based on the obligation to protect the users' interests.

Competition law has also been used to impose non-discrimination in the Google Search (Shopping) Case and there are a number of open cases regarding discrimination against Apple, Google, and Amazon. However, an ex-ante regulatory non-discrimination obligation could support a more straightforward enforcement.

On one hand, a regulatory obligation could explicitly impose both an external and an internal non-discrimination obligation. While competition law has been traditionally understood to impose external non-discrimination (discrimination between third parties) as abusive, internal discrimination (discrimination between third parties and the service provider itself) has been considered to be prohibited only in exceptional circumstances, mostly in the case of essential facilities in vertically integrated industries. However, traditional network regulation has explicitly imposed internal and external non-discrimination obligations. A recent example is the European Electronic Communications Code: "Obligations of non-discrimination shall ensure, in particular, that the undertaking applies equivalent conditions in equivalent circumstances to other providers of equivalent services, and provides services and information to others under the same conditions and of the same quality as it provides for its own services, or those of its subsidiaries or partners."[50]

On the other hand, a regulatory obligation could define the specific differentiations that are considered non-discriminatory. Non-discrimination does not mean that services are provided to all users on exactly the same terms. Objective reasons can justify different treatment. However, the definition of the borderline between objective and non-objective differentiation is prone to conflict.

Fair and reasonable contracting terms in the form of regulatory obligations could also be imposed upon dominant digital platforms. Such obligations could apply to all the contractual conditions that bind platforms and users (in particular business users). For instance, such obligations could outlaw exclusivity clauses prohibiting multi-homing. They could also limit excessive prices, such as the allegedly excessive fees imposed by Apple and Google in their app stores.

A full, continuous, and concentrated control of FRAND network access conditions has been implemented in most countries in most infrastructures. Independent regulatory authorities have been established, with different powers to control access conditions, from solving access disputes when they arise, to the ex ante definition of access conditions in the form of regulated offers previously approved by the authorities, or even defined by the authorities themselves.

A FRAND regime for dominant digital platforms would not necessarily require a fully developed regulatory framework, as is the case for the traditional network industries. A lighter framework could be defined, in the form of an obligation to agree on industry codes of conduct in which the dominant platforms and the users would define the FRAND conditions. Mechanisms could be defined for regulatory intervention where agreements cannot be reached or where the codes of conduct are not respected by the dominant firm. In favor of such an approach is the information asymmetry existing at the moment, as authorities have a limited understanding of the functioning of the platforms.[51] On the contrary, personalized algorithmic ranking and matching might make it difficult for users to control the effective application of FRAND frameworks, requiring the intervention of public authorities to introduce transparency into the algorithms (see the next chapter).

Notes

1 Weyl, G. & White, A. (2014). Let the Right "One" Win: Policy Lessons from the New Economics of Platforms, *Competition Policy International*, 12(2), 29–51.

2 OECD (2018). *Rethinking Antitrust Tools for Multi-sided Platforms*, retrieved from www.oecd.org/daf/competition/Rethinking-antitrust-tools-for-multi-sided-platforms-2018.pdf

3 Follow the debate in Evans D. & Schmalensee, R. (2019). *Antitrust Analysis of Platform Markets. Why the Supreme Court Got It Right in American Expre*ss. Competition Policy International, Boston.

4 Ibid. p. 167.

5 *Ohio v. American Express Co.*, 138 S.Ct. 2247 (2018), Point II.A.

6 Wu, T. (2018, Jun 26). *The Supreme Court Devastates Antitrust Law, New York Times.*

7 Khan, L. M. (2016). Amazon's antitrust paradox, *Yale Law Journal*, 126, 710.

8 Decision of the Bundeskartellamt, *Facebook*, B6–22/16, of February 7, 2019.

9 Lynskey, O. (2017). *Regulating "Platform Power."* LSE Law, Society and Economy Working Papers 1/2017, retrieved from http://eprints.lse.ac.uk/73404/1/WPS2017–01_Lynskey.pdf

10 Crémer, J., de Montjoye Y.-A., & Schweitzer, H. (2019). *Competition Policy for the Digital Era*, Report to the European Commission, retrieved from https://op.europa.eu/es/publication-detail/-/publication/21dc175c-7b76-11e9-9f05-01aa75ed71a1

11 Furman, J. et al. (2019). *Unlocking Digital Competition*. Report of the Digital Competition Expert Panel, retrieved from https://assets.publishing.service.gov.uk/government/uploads/system/uploads/attachment_data/file/785547/unlocking_digital_competition_furman_review_web.pdf

12 *Stigler Committee on Digital Platforms*, Final Report, September 2019, p. 32, retrieved from https://research.chicagobooth.edu/stigler/media/news/committee-on-digital-platforms-final-report

13 Australian Competition & Consumer Commission (2019). *Digital Platforms Enquiry*, retrieved from www.accc.gov.au/system/files/Digital%20platforms%20inquiry%20-%20final%20report.pdf

14 Japan Fair Trade Commission (2019). *Report Regarding Trade Practices on Digital Platforms*, retrieved from www.jftc.go.jp/en/pressreleases/yearly-2019/October/191031Report.pdf

15 Competition Authorities Working Group on Digital Economy (2019). *BRICS in the Digital Economy. Competition Law in Practice*, retrieved from www.cade.gov.br/acesso-a-informacao/publicacoes-institucionais/brics_report.pdf

16 European Commission (2020). Digital Services Act package: Ex ante regulatory instrument for large online platforms with significant network effects acting as gate-keepers in the European Union's internal market, retrieved from https://ec.europa.eu/info/law/better-regulation/have-your-say/initiatives/12418-Digital-Services-Act-package-ex-ante-regulatory-instrument-of-very-large-online-platforms-acting-as-gatekeepers

17 Montero, J. (2019). Asymmetric regulation for competition in European railways? *Competition and Regulation in Network Industries*, 20(2), 184–201.

18 Article 4(3) of Directive 97/33/EC, of June 30, 1997 on interconnection in Telecommunications with regard to ensuring universal service and interoperability through application of the principles of Open Network Provision, OJ 1997 L 199/32: "3. An organization shall be presumed to have significant market power when it has a share of more than 25 percent of a particular telecommunications market in the geographical area in a Member State within which it is authorized to operate."

19 Article 14(2) of Directive 2002/21/EC, of March 7, 2002 on a common regulatory framework for electronic communications networks and services (Framework Directive), OJ 2002 L 108/33, of 24.4.2002: "An undertaking shall be deemed to have significant market power if, either individually or jointly with others, it enjoys a position equivalent to dominance, that is to say a position of economic strength affording it the power to behave to an appreciable extent independently of competitors, customers and ultimately consumers."

20 This is the preferred approach in Alexiadis, P. & de Streel, A. (2020). *Designing an EU Intervention Standard for Digital Platforms*, EUI Working Papers, RSCAS 2020/14, retrieved

from https://cadmus.eui.eu/bitstream/handle/1814/66307/RSCAS%202020_14.pdf?sequence=1 &isAllowed=y

21 Furman et al. (2019). *Unlocking Digital Competition*. Report of the Digital Competition Expert Panel, p. 55.

22 ARCEP (2019). *Plateformes numériques structurantes. Eléments de réflexion relatif à leur caractérisation*, December 2019.

23 *Stigler Committee on Digital Platforms*, Final Report, September 2019, p. 32, available at https:// research.chicagobooth.edu/stigler/media/news/committee-on-digital-platforms-final-report

24 Alexiadis, P. & de Streel, A. (2020). *Designing an EU Intervention Standard for Digital Platforms*, EUI Working Papers, RSCAS 2020/14.

25 See CtrlShift (2018). *Data Mobility: The Personal Data Portability Growth opportunity for the UK Economy*, report for the Department for Digital, Culture, Media & Sport, retrieved from https://assets.publishing.service.gov.uk/government/uploads/system/uploads/attachment_data/ file/755219/Data_Mobility_report.pdf

26 Art. 20 Regulation (EU) 2016/679 of the European Parliament and of the Council of 27 April 2016 on the protection of natural persons with regard to the processing of personal data and on the free movement of such data, and repealing Directive 95/46/E, OJ 2016 L 119/1, 4.5.2016.

27 Decision of the Spanish Comisión Nacional de la Competencia y los Mercados, *Just Eat/La Nevera Roja*, 31.3.2016.

28 Decision of the Competition and Consumer Commission of Singapore *Grab/Uber*, of 13.4.2018.

29 ISO/IEC 2382:2015 Information technology–Vocabulary, available at www.iso.org/obp/ui/ #iso:std:iso-iec:2382:ed-1:v1:en

30 See story in Seidenberg, I & McMurray, S. (2018). *Verizon Untethered. An Insider's Story of Innovation and Disruption*.Post Hill Press, New York, 23–24.

31 Robertson, A. (2018, Dec 18). *Mark Zuckerberg Personally Approved Cutting off Vine's Friend-finding Feature. Verge*, retrieved from www.theverge.com/2018/12/5/18127202/ mark-zuckerberg-facebook-vine-friends-api-block-parliament-documents.

32 Stigler Committee on Digital Platforms, Final Report, September 2019, p. 16, available at https:// research.chicagobooth.edu/stigler/media/news/committee-on-digital-platforms-final-report

33 Art. 61(2) Directive (EU) 2018/1972 of the European Parliament and of the Council of 11 December 2018 establishing the European Electronic Communications Code (Recast), OJ L 321, 17.12.2018, pp. 36–214.

34 *Stigler Committee on Digital Platforms*, Final Report, September 2019, p.118.

35 Furman, J. et al. (2019). *Unlocking Digital Competition*. Report of the Digital Competition Expert Panel, p. 74.

36 Chernow, R. (1998). *Titan. The Life of John D. Rockefeller*. Vintage Books, New York.

37 Temin, P. (1987). *The Fall of the Bell System*. Cambridge University Press, New York.

38 Baker, J. (2019). *The Digital Paradigm. Restoring a Competitive Economy*. Harvard University Press, Cambridge, MA.

39 Wu, T. (2018). *The Curse of Bigness. Antitrust in the New Gilded Age*. Columbia Global Reports, New York.

40 Wu, T. (2018, Sep. 28). The case for breaking up Facebook and Instagram, *Washington Post*, retrieved from www.washingtonpost.com/outlook/2018/09/28/case-breaking-up-facebook-instagram.

41 Warren, E. (2019, Mar 8). *Here's How We Can Break Up Big Tech, Medium*, retrieved from https://medium.com/@teamwarren/heres-how-we-can-break-up-big-tech-9ad9e0da324c

42 OECD (2016). *Structural Separation in Regulated Industries: Report on Implementing the OECD Recommendation*, retrieved from www.oecd.org/daf/competition/Structural-separation-in-regulated-industries-2016report-en.pdf

43 Bundeskartellamt Decision February 6, 2019 in case B6-22/16, English translation available at www.bundeskartellamt.de/SharedDocs/Entscheidung/EN/Entscheidungen/Missbrauchsaufsicht/ 2019/B6-22-16.pdf?__blob=publicationFile&v=5

44 Khan, L. (2019). The separation of platforms and commerce. *Columbia Law Review*, 119, 973, retrieved from https://papers.ssrn.com/id=3180174

45 Layne-Farrar, A., Padilla, A. J., & Schmalensee, R. (2007). Pricing patents for licensing in standard-setting organizations: Making sense of FRAND commitments. *Antitrust Law Journal*, 74, 671.

46 Commission Decision of April 29, 2014, Case AT.39939 *Samsung–Enforcement of UMTS Standard Essential Patents*.

47 Judgement of the Court of Justice of the European Union of 16.7.2015, C-170/13, *Huawey v. ZTE*, ECLI:EU:C:2015:477.

48 Nikolic, I. & Heim, M. (2019). A FRAND regime for dominant digital platforms. *Journal of Intellectual Property, Information Technology and E-Commerce Law*, 18, 38–55.

49 Art. 61(5) of Directive 2018/1972, establishing the European Electronic Communications Code.

50 Ibid., Article 70(2).

51 This seems to be the approach in consideration by the UK authorities, as recommended in the Furman Report.

Chapter 24

Regulating Platform Ecosystems

Platform regulation requires still a wider perspective. As platforms create ecosystems around themselves, regulation should not only focus on the platform itself, or on the underlying intermediated companies, but also on the entire ecosystem. This is particularly relevant in the traditional network industries, as they have always been subject to regulation for the protection and promotion of the general interest. Platformization will not change this.

However, new institutions and new techniques will have to be developed to ensure the effectiveness of platform regulation. Traditional antitrust and sector-specific regulators at a national level might not be in the position to regulate platforms, other than in very large countries (US, China, and Russia). Platform regulation requires more transparency and control over the working of algorithms. Europe is currently in the lead in terms of developing such institutions and techniques.

24.1 Promoting Network Effects

Platform regulation should take into account the entire ecosystem around platforms, something we could call the digital ecosystem. Public authorities should not impede and should instead protect and promote the creation of network effects by platforms in multi-sided markets. Scale is necessary for the creation of network effects. Public authorities should protect and promote investments to build scale in all sides of multi-sided markets. Unnecessary regulatory restrictions to pool supply and demand should be eliminated. Furthermore, public authorities should not restrain, but instead provide the necessary freedom for platforms to find the right balance in the distribution of the value derived from network effects across the ecosystem.

On the supply side, the provision of goods and services is subject to all kinds of regulatory obligations. These obligations were defined in the industrial era and do not take into consideration the new organization model led by platforms. Regulation has to be adapted to the new reality.

Platforms empower non-professional individuals to provide goods and services previously provided only by professional providers. Platforms are at the center of the so-called "sharing economy." Consumers of services such as transportation, accommodation, etc. contract with their peers, with non-professional service providers.

In the industrial era, service providers required scale, which was necessary to reduce the average cost of goods and services. Mass-production, massive advertising, and sophisticated distribution systems, all coordinated inside a large hierarchical corporation, were the source of efficiency. Corporate law, labor law, and tax law were tailored to this industrial production model.

In the digital era, scale is built by platforms as they coordinate multi-sided markets. The service providers do not have to build scale in order to reduce transaction costs. Small service providers can obtain the benefits of scale by using platforms pooling together large pools of consumers. Even non-professional service providers can now join the market and compete with well-established corporations. Individuals can generate audiovisual content and monetize it, provide transport services through Uber, accommodation services through Airbnb, etc.

However, non-professional and small service providers face substantial regulatory obstacles to enter the market, as market access conditions were defined for the industrial era. Regulatory requirements were often imposed in order to protect consumers against information asymmetries and, more generally, risks related to poor services. Nevertheless, platforms reduce information asymmetries, often in a more satisfactory way than traditional regulation. For example, platforms provide a lot of information about accommodation, including pictures, evaluations by other users, etc., which clearly improves the guarantees provided by traditional regulation such as hotel ratings in the form of stars, minimum quality requirements, etc.

Particularly relevant are restrictions in the form of licenses or authorizations when they are limited in number according to the traditional demand of the service, as is the case with taxi services. Platforms transform the industries in which they operate. Network effects reduce costs, which leads to price reductions, which attract new demand. As a consequence, traditional quantitative restrictions make it impossible to meet all the new demand. These type of restrictions impede the creation of new network effects. Coming back to the example of Uber, regulatory restrictions in the supply of taxis and private hire vehicles in most European cities make it impossible to provide UberPooL services. European citizens do not benefit from the low prices that their American and Asians counterparts enjoy.

Platforms are not limited to the sharing economy and the coordination of non-professional service providers. Google's long-lasting CEO, Eric Schmidt – one of the individuals who best understands platforms in multi-sided markets – thought that platforms had the potential to disrupt any industry.[1] This is also the case of industries that rely on long-established network effects, such as telecommunications, transport, or energy. Platforms can create even larger network effects, particularly by pooling together the existing service providers and their networks, to provide access to more sophisticated and efficient services.

It is legitimate to ask whether public authorities should promote the construction of network effects by forcing traditional players to join a platform and make their services available through the platform. Such a measure would certainly facilitate growth, as platforms would not have to invest to attract suppliers. This is not a rhetorical question. Mobility-as-a-service platforms are calling for the introduction of such an obligation. Net neutrality obligations imposed on telecommunications carriers have a similar effect in their relation with communications platforms.

Platforms claim that traditional players active in the network industries often have exclusive or special rights, or at least a position of market power. By refusing to work with platforms, these players would be breaching antitrust rules,[2] as they would be refusing to deal with no objective reason, or refusing access to an essential facility. Furthermore, as common carriers or services of general interest, they might be often under an obligation to provide their services to everyone, also to platforms trying to intermediate their services.

As these legal bases might not be always clear, platforms are lobbying for the adoption of legislation imposing upon traditional players the obligation to work with platforms. The best example is the Transport Law adopted in Finland, which imposes upon the traditional transport services providers (railways, bus services, etc.) the obligation to make their services and their data available to platforms.

While we believe that public authorities should promote the construction of network effects by platforms, it might be a step too far to impose upon traditional players a general obligation to deal with platforms. On one hand, it would not be sufficient to define the obligation to work with platforms. Once the obligation would be defined, and as the actors would not always reach an agreement, public authorities would have to define the conditions for the provision of the service. Coming back to transport, should platforms have access to services at the same conditions as the final users? Should platforms benefit of discounts granted to final users, such as monthly passes, etc., which are often publicly subsidized? Should platforms benefit of have even better terms than final users, as wholesalers often do when they contract larger volumes?

On the other hand, the relationship between service providers and platforms is highly strategic. Service providers are reluctant to work with platforms, as they are aware of the risk of becoming platformed; that is, of seeing their service commoditized by a platform that coordinates the multi-sided market and acquires market power, so as to be able to set the contracting conditions. The balance of power between platforms and the underlying service providers will be one of the most relevant evolutions in the platform economy. Platforms have the potential to become dominant, as has happened in other industries (newspapers, music, etc.). Voluntary contracting under market conditions seems better fitted to determine the contracting conditions and the balance of power between platforms and traditional service providers. Only under exceptional circumstances, when absolute refusals to deal by dominant companies take place, should public intervention be recommended.

On the demand side, obstacles to growing scale can derive from fragmentation. Fragmentation can derive from differences in language or consumption habits, which is why platforms tend to originate in large national markets such as the US and China. Even if European consumers are as sophisticated and as eager to use platforms as their American and Asian counterparts, it is more difficult for European platforms to grow scale on the demand side, as they have to work across languages, consumption habits, and regulations. Local regulations are another obstacle to aggregate demand across borders and the obstacle is particularly relevant during the period of construction of the customer base of a new platform.

In any case, fragmentation is a challenge, but also an opportunity, as platforms can build seamless customer experiences across borders, reducing the effects of fragmentation on the supply side. Airbnb is a good example of how a global seamless service for consumers can be built over the most fragmented of supplies: individual homes across the world, owned by individuals of different nationalities, different languages, subject to different legislations, etc. It is part of the business of a platform to identify how to overcome fragmentation in supply for the benefit of demand.

It is not sufficient to promote the aggregation of supply and demand. Platforms must also be given the necessary freedom to experiment and identify the right balance among all the interests in the ecosystem, particularly when supply is highly fragmented and subject to different conditions and regulations.

Platforms have the ability to create colossal value, but the key to success is how value is distributed across the ecosystem. Platforms must identify the key parameters driving the use of the platform for each side of the market and be in the position to meet the expectations of all parties. It is not sufficient to embark on one side in the platform, such as demand attracted by low prices. It is also necessary to attract supply and to grant some benefit to suppliers so as to make their participation in the platform attractive. Good access to demand might be sufficient sometimes, but might be insufficient on other occasions. Supply tends to be the weakest side, in that the platform must firstly invest to attract supply and, once the platform is more mature, share a significant part of the value created by the platform with the underlying service providers.

To further complicate matters, the balance in the distribution of the value created by the platform is dynamic. It changes over time as the platform matures and, even after that point, it might change as markets evolve and new circumstances (such as new competition) might make it necessary to redefine the balance. Platforms are constantly redefining their proposition. They are constantly testing with some users or at a small scale in one town or country, the conditions of the provision of the service, to identify how to keep all sides of the multi-sided market engaged and active in the platform.

Over-regulation kills network effects, not only when regulation limits supply or fragments demand, but also when over-regulation is imposed upon the intermediation service itself. Public authorities might be tempted to intervene too early and too much in the process of construction and curation of the multi-sided market by digital platforms.

Over-regulation is particularly dangerous at the initial phase of the creation and growth of a platform. A new platform with a small number of users does not create significant network effects. Platforms must invest to attract users. If regulatory obstacles are introduced at this stage, it will take longer to reach the minimum scale to create network effects. Regulatory obligations might even make it impossible to grow and reach such critical mass.

We are at the early stages of developing the new industrial organization model. Platforms are still only mature in the primary data industries (search engines, social networks, communications and alike). Platforms intermediating physical goods and services are still in their infancy, so it is too early to systematically regulate all platforms.

Even when platforms become mature, they need room to adapt their terms and conditions to the unstable and evolving circumstances of multi-sided markets. Regulation that ossifies the conditions provided to supply and demand might reduce the competitiveness of the platform. Just as network effects create virtuous cycles when they grow, they can turn into a vicious cycle if regulation introduces rigidity and reduces the number of active members in the platform.

24.2 Platforms and the General Interest

The regulation of digital platforms and digital ecosystems must take the general interest into consideration. This was understood by the business leaders who ran infrastructures in the late nineteenth and early twentieth centuries. The great economic and political power derived from concentration in the management of networks was only accepted as it became controlled by regulators.

Throughout this book we have analyzed how the traditional network industries were monopolized in the US by consolidating the pioneer providers of telecommunications, transport, and electricity services. Market leaders such as Theodore Vail (the president of AT&T), supported by financiers such as JP Morgan, understood that network effects required large volumes of customers and that the larger the network effects, the greater the value created. They understood that a single provider would exhaust all network effects, creating maximum value. They also understood that such a value would benefit not only their organizations, but society as a whole.

The most prescient of the network industry leaders, such as Vail, also understood that market power and monopolization would only be accepted by the public if subject to close scrutiny and control by the public authorities. Monopolization was opposed by citizens suffering the abuse of the new monopolies. Private monopolies such as AT&T, General Electric, and the railroad trusts were only allowed because independent agencies controlled the rates of the provision of the service and discrimination was outlawed. Building such a regulatory framework (economic theory, legal base, specific statutes, institutions, etc.) took more than 30 years.

In Europe, the trend was the same, but even more blatant. Monopolies were also created, although they were not under the regulatory control of the state, but directly owned by public authorities. The largest network effects were exhausted, and abuse was excluded, at least in theory, thanks to public management.

Today, individuals leading the largest platforms in the US do not seem to be as inclined to accept and even promote regulation as a counterbalance to their market power. Maybe they have not felt the immediate threat of forced divestiture, as was the case for Vail and Morgan a century ago, after the divestiture of Rockefeller's Standard Oil in 1911.

However, platformization is not changing the fact that all these industries are affecting the general interest. Platformization of media, telecoms, transport, and energy is transforming the organization, the structure, and the operations of all these industries. Such innovation does not change the underlying fact that these services affect the community at large, not only the platforms, the underlying service providers, and the direct users of the service. The Supreme Court made the following declaration in 1876, in one of the first cases upholding the regulation of the network industries (*Munn v. Illinois*): "Property does become clothed with a public interest when used in a manner to make it of public consequence, and affect the community at large. When, therefore, one devotes his property to a use in which the public has an interest, he, in effect, grants to the public an interest in that use, and must submit to be controlled by the public for the common good, to the extent of the interest he has thus created. He may withdraw his grant by discontinuing the use; but, so long as he maintains the use, he must submit to the control."[3] Ever since then, this has been a fundamental principle in regulation in the US and is a well-established principle in Europe.

Some of the general interest obligations traditionally imposed upon service providers will have to be shifted to digital platforms. Traditional providers of media, telecom, transport, and energy services were and remain subject to regulation to protect the general interest. As digital platforms take a leadership position in the coordination of these systems, substituting the coordinating role of the service providers, some of the public service obligations will have to be shifted from the service providers to the digital platforms.

Pluralism has always been a general interest objective in media regulation. Pluralism takes different forms; it can be ensured by a minimum number of media outposts competing with each other, but also through "internal pluralism" in a specific media outpost, ensuring different voices have room in it. This is particularly relevant when media is highly concentrated, as was the case in the early days of television broadcasting. This general interest objective has never disappeared. It might have become less relevant as the number of broadcaster multiplied over recent decades. As digital platforms become the new gatekeepers, curating news pieces to be displayed to the audience, new and imaginative forms of supervision over the editorial role of the platforms will have to be developed.

Guaranteeing universal access to the most basic telecom, transport, and energy services has traditionally been one of the roles of regulatory authorities in these industries. While the industries might be in transformation, the need to ensure universal access to these services is not going to change. As digital platforms become more relevant as system organizers, the role of increasingly fragmented, commoditized, and platformed service providers as guarantors of the universality of the service will have to be transformed. Digital platforms will have to be increasingly seen as being guarantors of such universal access.

These are just two basic examples of the need to introduce the general interest into the algorithms increasingly governing media, telecom, transport, and energy services. Algorithms with the power to organize these industries cannot merely be designed to maximize the benefit of the platform against the interest of the underlying service providers and the community as a whole.

Algorithms must ensure a balanced distribution of the new value created by network effects in multi-sided markets and, in particular, distribute to the ecosystem the funds to sustain the underlying infrastructure. The role of platforms as superintermediaries, and particularly the market power of certain platforms, could easily pose a threat to such a balanced distribution. This is particularly important when platforms do not displace traditional service providers, but instead rely on the provision of services by such traditional service providers. In the case of infrastructures, massive investment is necessary to build and maintain the underlying infrastructure and the services provided over them. If platforms extract value from these industries instead of distributing the efficiency gains back to infrastructure managers, the ecosystem will not be sustainable in the long run.[4] This is why platform regulation is particularly necessary in infrastructure industries.

24.3 Institutional Framework

It is still early to identify the optimum institutional form to regulate digital platforms. Independent sector-specific agencies were the preferred institutional form for regulating the traditional infrastructure networks. Sector-specific agencies in Europe are taking a leading role in regulating platforms active in their jurisdictions. For instance, the Italian telecommunications and media regulator AGCOM is active in the regulation of fake news in online platforms. Transport and energy regulatory authorities might also take a leading role in the regulation of online platforms in the industries under their control.

Remember that platforms are expected to be active in any industry and not only in the regulated industries in which a sector-specific agency has been established. Therefore, action by existing sector-specific regulatory agencies cannot be the universal solution.

In some jurisdictions, the idea of a digital regulatory authority has been proposed. This is the case of the House of Lords of the United Kingdom. In March 2019 the House of Lords published the report "Regulating in a digital world,"[5] proposing the establishment of a "Digital Authority" that would build knowledge and "scan the horizon" coordinating regulators across different sectors. This coordinating role seems more sensible than an overarching regulator with powers across all digital markets. The Furman Report confirmed this approach with a proposal to create a digital markets unit, without specifying whether it should be a new institution or a department in an existing institutions, such as the antitrust authority, the telecoms regulator, or the data protection agency.

The geographic scope of platforms is a further challenge. Infrastructure network industries are closely connected to the territory in which they are deployed and operate, typically the nation-state. It is not surprising that regulation emerged at the local level, as it was mostly connected to the right to occupy public space for the deployment of the infrastructure. The leadership role in regulating infrastructure networks evolved from the local level to a wider level as local infrastructures were connected into regional and national networks. States have been the natural level for the regulation of infrastructure networks.

Digital platforms are virtual and disconnected from the territory and they operate on a large scale, often globally. They do not need to occupy public space subject to a license issued by a municipality. Most platforms operate across jurisdictions, so that nation-states have limited tools to discipline platforms. Only the largest states, such as the US, China, and Russia, are in a position to effectively regulate platforms.

The European Union is particularly suited for the regulation of digital platforms. The European Union is large enough in terms of population and wealth to be a credible counterpart to the global platforms. It is already specialized in the regulation of trade across borders and is a sophisticated organization with decades of experience in the application of competition law and sector-specific regulation in the different network industries. It is no coincidence, therefore, that the European Union is leading on platform regulation.

24.4 Transparency in Algorithms

Behavioral obligations in digital multi-sided markets often require the modification of the machine learning algorithms[6] that characterize the functioning of the platforms. However, such obligations face a relevant technical challenge, as it is difficult to enforce them due to the lack of transparency around the algorithms.

One of the main difficulties of regulating digital platforms is the lack of transparency in the intermediation services they provide. Most of the intermediation activity is automated thanks to algorithms. Individuals and organizations affected by algorithms have little visibility regarding how algorithms reach the decisions that affect them. This was underlined in 1998 in the seminal paper by Google's founders "Since it is very difficult even for experts to evaluate search engines, search engine bias is particularly insidious."[7] However, some kind of transparency is necessary in order to make regulation effective.

The massive (algorithmic) network effects created by platforms rely on machine learning algorithms to automate the intermediation between all the parties interacting through the platform. This is how large pools of users can interact without congestion. Algorithms increasingly rely not on specific code with instructions, but on machine learning techniques. Machine learning has been described as "a family of techniques that allow computers to learn directly from examples, data and experience, finding rules or patterns that a human programmer did not specifically identify."[8]

Machine learning algorithms can be opaque, even for the individuals that have created them. These algorithms do not fully rely on specific instructions from the programmers. On the contrary, they rely on the rules and patterns identified by the algorithms themselves. However, algorithms can hide all kinds of mistakes, biases, and unlawful purposes, which is why they have been colorfully called "weapons of math destruction."[9]

Accountability in the operation of algorithms is necessary in order to make regulation effective. Most behavioral obligations that can be imposed upon platforms, such as the ones defined so far, have to be executed by the algorithms. This is the case of obligations regarding fairness, objectivity, non-discrimination, or interoperability. Regulation will only be effective if regulators and market players are in a position to ascertain that algorithms are respecting the obligations defined in the regulatory framework.

During the industrial era, market players and regulators could ascertain whether regulated companies were fulfilling their contractual and regulatory obligations. Mass market goods and services were mass-produced and sold at homogeneous conditions. Pricing and other conditions would often be advertised in the media. Even the conditions for the provision of bespoke goods and services could be traced in contracts, invoices, etc.

In the digital era, however, it is more difficult to identify the contracting conditions. Conditions can be personalized by machine learning algorithms. It is difficult to ascertain the conditions under which a good or a service is being provided, as each user might be getting different and personalized conditions. Under these conditions, it is also difficult to identify whether the algorithm running a platform is matching demand following the principles of objectivity and non-discrimination potentially imposed by regulation. It is difficult for market players and is even difficult for the regulatory authorities.

Transparency has been identified by various public authorities around the world as a key enabler to introduce accountability in the operation of algorithms. The report "Algorithms in decision-making"[10] by the House of Commons of the United Kingdom is particularly interesting.

The House of Commons has debated how to make algorithms transparent. It might be possible to make the inner working of the algorithms transparent, but different difficulties and negative side effects are also identified. An alternative would be to provide only an explanation of how the algorithm works and the results it produces.

Full transparency is usually requested in traditional legal and regulatory procedures. However, full transparency poses some relevant challenges when dealing with algorithms. Firstly, the inner working of machine learning algorithms is highly complex. A full team of data experts will usually be required in order to analyze and understand the inner working of a machine learning algorithm used by a large online platform. A regular user would certainly not be in a position to understand the information disclosed by the platforms, if it were obliged to do so. Complexity might just create mistrust in users when provided with information they are not in the position to process.[11] It has been noted that even the team that designed the algorithm might not be in a position to fully understand all the implications of its own algorithm.

Furthermore, by disclosing the algorithm, the most sophisticated players could play around it in their favor, making it useless. The most common example is how Google must constantly review its search algorithm to prevent professional spammers gaming the algorithm and making it useless. For instance, fraudsters could second-guess the algorithm to avoid security measures and make their goods and service available for final users.

Finally, disclosing the full information about the algorithm could damage intellectual property rights over the algorithm itself, confidential information, and even the personal data upon which the algorithm learns.

The alternative is to limit transparency to a right to explanation. The platform would be obliged to explain the key metrics relating to an algorithm operation, providing information on the outcome of the operation and how to modify the outcome in the future. Data visualization tools could help citizens better understand the explanations on how algorithms work. It has been reported that the Defense Advanced Research Projects Agency (DARPA) in the US is investing in different projects to have a better understanding of the conclusions reached by deep learning algorithms and how to display the corresponding information.[12]

The "right to explanation" is referred in recital 71 of the EU General Data Protection Regulation. The data subject has the right "to obtain an explanation of the decision reached" based solely on automated processing, which produces legal effects. However, such a right is not formally defined in Article 22 or any other provision of the Regulation. It is a first step, but limited in scope and uncertain in its legal value.

EU Regulation 2019/1150 imposes on intermediaries and search engines the obligation to set out in their terms and conditions the main parameters determining ranking, as well as the reasons for the relative importance of those main parameters as opposed to other parameters.[13]

The two alternatives are not mutually exclusive. Citizens affected by the decision of the algorithm running a platform might be satisfied by the right to explanation, as long as the explanation empowers them to challenge the decision by the algorithm. Public authorities, on the contrary, can pool the resources to have the full understanding of the inner working of the algorithm. Concerns such as confidentiality, intellectual property rights, and personal data are not of concern for investigations by public authorities, or at least no more than in any other type of investigation.

A middle ground can be identified in the form of audits by expert entities. Auditors can gather the technical know-how, but should have no full access to data and information compromising confidentiality, intellectual property rights, and personal data. A precedent can be identified in the EU Regulation on Computerized Reservation Systems in aviation, as it empowers the Commission to request external audits on the intermediation activity of the platforms.[14]

A further proposal has been to introduce experimentation for the definition of obligation to be imposed on digital platforms.[15] Platforms are constantly testing changes to their algorithms. They test changes in a specific country or with a specific set of users. If the results are positive, the changes are then exported to the entire customer base. Remedies imposed upon platforms should be subject to the same testing. Regulators could test different alternatives and, based on the results, opt for one of them.

Codes of conduct, certification of algorithms by third parties, and ethics boards could be other instruments to control algorithms,[16] as they could all help private entities and public authorities make platform regulation effective. Algorithm transparency will certainly be one of the main topics for discussion in platform regulation in the coming years.

24.5 The European Model

European authorities are leading the regulation of online platforms. Both at a national level and at the European Union level, different initiatives are maturing, from the application of antitrust rules to the definition of the first examples of sector-specific regulation.

It is perhaps no coincidence that France is leading and putting pressure on the regulation of platforms. It was in France where the first structured economic analysis of platforms in multi-sided markets was made back in 2003 by Nobel Laureate Jean Tirole. At the same time, France, like the rest of Europe, has not been able to develop large native platforms. The large US platforms dominate the European market: Google has a market share of more than 90 percent in search, Facebook is the leading social network, Airbnb and Uber dominate the market, and so on. Not having produced such platforms by itself, Europe definitely has the ambition to regulate the US platforms.

The French Digital Council (*Conseil National du Numérique*) is an advisory body of the French Government set up in 2011. In its first report in 2013, the Council identified the central role of platforms in the digital transformation. It initiated the first organized, systematic analysis of online platforms for the introduction of regulation. In 2013 a working group of ten members held four workshops and, subsequently, a series of meetings were organized with economic and legal experts to develop a proposal for the regulation of online platforms. The result was an opinion published in 2014 under the title "Platform Neutrality. Building an open and sustainable digital environment."[17]

The French Digital Council made various proposals to achieve a more open internet. Some proposals were related to data sharing as well as the transparency in algorithms. Nevertheless, the most ambitious proposal was to adopt a platform neutrality regulation; that is, to extend to platforms the principles defined for telecom operators under the net neutrality principles. The position of the Council was further refined in a report published in 2015.[18]

Parallel exercises were developed at national levels. The French and the German competition authorities published a joint report on competition law and data in May 2016, with particular attention devoted to platforms. They identified the relevance of algorithmic network effects: "The relevance of data as a strategic input and the opportunities for foreclosure depend in part on the volume levels: (i) at which a firm can reap the economic benefits of data; (ii) beyond which these benefits decline or cease to exist altogether. These levels will vary, depending on the type and purpose of the data."[19]

After the diagnosis, implementation moved from France and the member-states to the European Union. Platform regulation cannot take place at a national level. Most states are too small to impose effective obligations on global platforms. However, the European Union comprises a significant portion of the global economy and is in a position to effectively influence the strategy of the global platforms.

The European Union launched the Digital Agenda for Europe in May 2010 to complete the so-called digital single market.[20] However, this agenda did not include a reference to the regulation of online platforms. It was only after the work of the French Digital Council in 2014 that the European Union introduced platform regulation in the digital single market strategy.[21]

On one hand, the European Commission reinvigorated some pending antitrust cases involving online platforms. A complaint was filed against Google by some European companies for abuse of dominant position in the form of discrimination in search results. A parallel complaint against Google was filed in 2014 for abusive practices related to the mobile operating system Android.

In June 2017, the European Commission imposed upon Google a fine of €2.42 billion for abuse of dominant position in the search market by giving illegal advantage to another Google service, its comparison shopping service.[22] Google Shopping allows consumers to compare products and prices and find deals from digital retailers of all types, including digital shops of manufacturers, platforms (such as Amazon and eBay), as well as other resellers. The Commission underlined the relevance of direct, indirect, and algorithmic network effects. The Commission found that Google had systematically given prominent placement to its own comparison shopping service and had demoted rival comparison shopping services in its search results.[23]

In July 2018 the Commission fined Google €4.34 billion in the Google Android Case.[24] The Commission concluded that Google committed an abuse in the markets for general internet search services, licensable smart mobile operating systems, and app stores for the Android mobile operating system to reinforce its dominant position in such markets. Firstly, it imposed an illegal tying of Google's search and browser apps. Secondly, it granted illegal payments conditional on exclusive pre-installation of Google Search. Finally, Google imposed an illegal obstruction of development and distribution of competing Android operating systems.

In March 2019 the Commission also fined Google €1.49 billion for abusing its market dominance by imposing a number of restrictive clauses in contracts with third-party websites, which prevented Google's rivals from placing their search adverts on these websites.[25] Google first imposed an exclusive supply obligation, which prevented competitors from placing any search adverts on the most commercially significant websites. Google then introduced what it called its "relaxed exclusivity" strategy, aimed at reserving for its own search adverts the most valuable positions and controlling the performance of competing adverts.

These cases are particularly remarkable because similar practices were common in the US, but the US authorities had decided not to file actions against Google. It seems clear that a divergence exists across the Atlantic in the application of antitrust rules to digital platforms.

The European Parliament went even further, voting in favor of a non-enforceable declaration to "unbundle search engines from commercial services." Even if Google was not specifically mentioned, the divestiture of Google was in the mind of the members of Parliament.

On the other hand, the European Commission initiated the procedure to adopt legislation on platforms. In 2016 the Commission launched a consultation on the regulatory environment for platforms.[26] As a result, the Commission proposed the adoption of a regulation on fairness in platform-to-business relations, which in 2019 became Regulation 2019/1150 for the promotion of fairness and transparency for business users of online intermediation services.[27]

The new Commission, appointed by the end of 2019, had ambitious plans for the regulation of the digital platforms. Commissioner Vestager, in charge not only of competition but also of digital affairs, asked for a report for the evolution of the digital regulation.[28] The report set the main lines for the regulation of the digital platforms. Competition law would always be fundamental to blocking anticompetitive strategies and it would be reviewed so that it was adapted to the new digital markets. However, it was concluded that competition law would have to be complemented with ex-ante regulation. The regulatory measures under consideration are inspired by the pro-competition asymmetric regulation enacted for the liberalization of the traditional network industries, particularly telecommunications. Digital platforms with market power would be subject to regulatory obligations with the objective to reduce barriers to entry in the form of data portability and the promotion of multi-homing, but in particular by eliminating the competitive advantage of network effects. Platform interoperability would share direct and indirect network effects among all players. Data sharing would share algorithmic network effects. But it still remains to be confirmed whether some kind of FRAND regime will be imposed upon platforms with market power to ensure fairness.

Notes

1 Schmidt, E. & Rosenberg, J. (2014). *How Google Works*. John Murray (Publishers), London, p. 81.
2 Valdiani Vicari & Associati (2019). *Study on Market Access and Competition issues Related to MaaS*, June 2019, retrieved from https://maas-alliance.eu/wp-content/uploads/sites/7/2019/07/Market-access-and-competition-issues-related-to-MaaS_final_040719.pdf
3 *Munn v. Illinois*, 94 U.S. 113, (1876).
4 Finger, M., Bert, N., Kupfer, D., Montero, J. J. & Wolek, M. (2017). *Infrastructure funding challenges in the sharing economy*, Research for the TRAN Committee of the European Parliament, retrieved from www.europarl.europa.eu/RegData/etudes/STUD/2017/601970/IPOL_STU(2017)601970_EN.pdf
5 House of Lords (2019). *Regulating in a digital world*, Select Committee on Communications, 2nd Report of Session 2017–19.
6 Christian, B. & Griffiths, T. (2016). *Algorithms to Live By. The Computer Science of Human Decisions*. William Collins, London.
7 They even argued for the need to keep the search engine out of the commercial realm: "But we believe the issue of advertising causes enough mixed incentives that it is crucial to have a competitive search engine that is transparent and in the academic realm." In Brin, S. & Page, L. (1998). *The Anatomy of Large-Scale Hypertextual Web Search Engine*, retrieved from https://storage.googleapis.com/pub-tools-public-publication-data/pdf/334.pdf.
8 Rieke, A., Bogen, M., & Robinson, D. (2017). *Public Scrutiny of Automated Decisions: Early Lessons and Emerging Methods*, An Upturn and Omidyar Network Report, p. 9, retrieved from www.omidyar.com/sites/default/files/file_archive/Public%20Scrutiny%20of%20Automated%20Decisions.pdf
9 O'Neil, C. (2016). *Weapons of Math Destruction: How big data increases inequality and threatens democracy*. Broadway Books, New York.
10 House of Commons (2018). *Algorithms in Decision Making*. Fourth Report of Session 2017–2019, Science and Technology Committee,
11 Hosanagar, K. & Jair, V. (2018). We Need Transparency in Algorithms, But Too Much Can Backfire. *Harvard Business Review*, retrieved from https://hbr.org/2018/07/we-need-transparency-in-algorithms-but-too-much-can-backfire
12 Knight, W. (2017). The US Military Wants Its Autonomous Machines to Explain Themselves, *MIT Technology Review*, retrieved from www.technologyreview.com/s/603795/the-us-military-wants-its-autonomous-machines-to-explain-themselves/
13 Article 5 Regulation (EU) 2019/1150 of the European Parliament and of the Council of June 20, 2019 on promoting fairness and transparency for business users of online intermediation services, OJ 2019 L 186/57, 11.7.2019.

14 Article 14 Regulation (EC) No 80/2009 of the European Parliament and of the Council of January 14, 2009 on a Code of Conduct for computerized reservation systems and repealing Council Regulation (EEC) No 2299/89, OJ 2009 L 35/47, 4.2.2009.

15 Feasey, R. & Krämer, J. (2019). *Implementing Effective Remedies for Anti-competitive Intermediation Bias on Vertically Integrated Platforms*, Report for CERRE.

16 Koene, A., Clifton, C., Hatada, Y., Webb, H., & Richardson, R. (2019). *A Governance Framework for Algorithmic Accountability and Transparency*, European Parliament Research Service, retrieved from www.europarl.europa.eu/thinktank/en/document.html?reference=EPRS_ STU(2019)624262

17 Opinion no. 2014–2 of the French Digital Council on platform neutrality, May 2014, available at https://ec.europa.eu/futurium/en/system/files/ged/platformneutrality_va.pdf

18 Conseil National du Numérique. (2015). *Ambition numérique: Pour une politique française et européenne de la transition numérique.* Report, retrieved from www.entreprises.gouv.fr/ files/files/directions_services/secteurs-professionnels/economie-numerique/CNNum--rapport-ambition-numerique.pdf

19 Autorité de la Concurrence & Bundeskartellamt (2016). *Competition Law and Data*, p. 54, retrieved from www.bundeskartellamt.de/SharedDocs/Publikation/DE/Berichte/Big%20Data%20 Papier.pdf?__blob=publicationFile&v=2

20 European Commission. *A Digital Agenda for Europe*, COM/2010/0245, retrieved from https:// eur-lex.europa.eu/legal-content/EN/ALL/?uri=CELEX:52010DC0245R(01)

21 European Commission (2015), *Digital Single Market Strategy for Europe*, COM/2015/0192, retrieved from https://eur-lex.europa.eu/legal-content/EN/TXT/?qid=1447773803386&uri=CE LEX:52015DC0192

22 European Commission Decision, Case AT.39740, *Google Search (Shopping)*, 27.6.2017.

23 European Commission Decision, Case AT.39740, *Google Search (Shopping)*, 27.6.2017.

24 European Commission Decision, Case AT.40099, *Google Android*, 18.7.2018.

25 European Commission Decision, Case AT.40411, *Google Adsense*, 20.3.2019.

26 European Commission (2016b). Public consultation on the regulatory environment for platforms, online intermediaries and the collaborative economy. Synopsis Report. Available at   https://ec.europa.eu/digital-single-market/en/news/results-public-consultation-regulatory-environment-platforms-online-intermediaries-data-and

27 Regulation (EU) 2019/1150 of the European Parliament and of the Council of June 20, 2019 on promoting fairness and transparency for business users of online intermediation services, OJ 2019 L 186/57, 11.7.2019.

28 Crémer, J., de Montjoye, Y.-A., & Schweitzer, H. (2019). *Competition Policy for the Digital Era*, Report to the European Commission, retrieved from https://op.europa.eu/es/publication-detail/-/ publication/21dc175c-7b76–11e9–9f05–01aa75ed71a1

Afterword

We are on the verge of a profound transformation of the network industries. As infrastructures are increasingly being digitalized and as all assets of a physical network is soon to be connected through 5G wireless technologies, new ways of coordinating the assets that constitute each network industry will arise, creating unprecedented and larger network effects. The Internet of Things (IoT) will be different from the internet that had connected mainframe computers, PCs, and smartphones.

Originally, the internet that connected mainframe computers had little impact on the infrastructures and the physical world more generally. It was only in the 1990s, when PCs became popular and interconnected, that the impact of the internet on communications services started to be felt. Postal services were substituted by email, and telephony and media services were disrupted by new digital platforms. When smartphones were launched in 2007, the internet became personal and ubiquitous, generating new digital platforms with a wider impact upon the physical world. Transport started to be disrupted and the first impact on the energy industry was felt.

As the IoT spreads, and sensors, smart meters and so on are installed and connected through 5G telecom networks, the opportunities to build new network effects on top of the existing assets will multiply. The physical world, and infrastructures and network industries in particular, are headed towards a profound disruption.

The IoT is multiplying the number of interconnected elements. Not only humans making use of PCs and smartphones will be interconnected; basically any asset will be able to interact: goods to be transported, vehicles, roads, railways, electric appliances, and so on. The number of interconnected elements will multiply from a few billion (the number of human beings) to trillions, enabling new complementarities, new forms of coordination, and unprecedented network effects.

The IoT will not only transform the scale, but also the type of interactions. The transition from the PC to the smartphone has enabled the interaction of assets by proxy. Vehicles can now form a network as drivers have their own smartphone and both the driver and the vehicles can be geolocalized. Even more importantly, as individuals have their own smartphones, they can sign on to the service provided by a platform located in Silicon Valley. The platform is in a position to create network effects with little or no presence on the territory. Digital platforms have grown global, beyond the San Francisco suburbs.

Furthermore, digital platforms have become centralized and hierarchical, capturing a large share of the value created by the different network effects, whether they are direct, indirect, or algorithmic. The most vivid example is the gatekeeper role of the mobile smart-phone operating systems as developed by Apple and Google. The gatekeeper is taking 30 percent of the revenue made by the apps installed in the smartphone, but has the power to self-preference its own apps, not to approve new app features, and even the power to exclude the app from the app store and, consequently, from the market altogether. Exclusion can be the result of a commercial decision or the result of a governmental order, as it has been the case in China with a large number of apps, from Facebook to the *New York Times*.

The IoT challenges the current balance of power around digital platforms, as is par-ticularly clear in the case of infrastructures. Digitalizing infrastructures require a large investment that can only be made by the infrastructure managers. Even more importantly, the infrastructure manager will have the control not only over the infrastructure, but over the sensors, meters, and the like attached to the infrastructure, as well as over the data generated by such devices. Platforms based in Silicon Valley will not be able to rely on the smartphones acquired by users. They will have to negotiate with the infrastructure man-agers to have access to data and to control the interactions of the assets, or alternatively, to deploy their own data-generating devices. This is Tesla's strategy behind its construction of batteries, solar roofs and electric vehicles. Only as far as digital platforms have access to data will they be in a position to platformize the infrastructure managers and the services provided over the infrastructures. Therefore, the regulation of data sharing will be one of the key debates in the following years.

We have already learnt that the battle for the hegemony in the data layer is not only limited to access to data. The control of the infrastructure supporting the data layer is equally important. This can be best illustrated by the operating systems. Apple is powerful not because it has access to the data generated by their smartphones, but because it is the gatekeeper of the apps to be installed in the smartphone, the apps that create the network effects making use of the data.

What form will the operating systems supporting the Internet of Things take? Microsoft's operating system for PCs did not set out to become a gatekeeper with the power to approve any software to be installed on the PC. Apple and Google, by contrast, actively sought to become gatekeepers to the smartphone. It is not clear at this stage who will be the winner in the battle to develop the operating system of the new generation of the internet. However, companies managing infrastructures and physical assets more generally should learn not to support operating systems that lock them into a centralized system with a gatekeeper.

There is growing awareness of the risks of network effects being created and monopolized by digital platforms. An alternative would be a distributed network as imagined by Baran back in 1964 for the internet itself (see Chapter 6). Such a distributed networks could be built upon the collaboration of the intermediated parties. This is how Amadeus, the avi-ation global distribution system, was born in 1987 from the collaboration of the European airlines (Chapter 12). Less successful attempts for collaboration can be identified between taxi drivers to build an alternative to Uber, for instance. Regulation could be used to incen-tivize such collaboration. The recent regulation of payment systems in Europe can be an example of enforced collaboration to build network effects.

The role of digital platforms as gatekeepers raises questions that go beyond compe-tition in the market. China has blocked most of the US-based platforms that provide communications services: Google, YouTube, Facebook, Instagram, WhatsApp, Twitter, Snapchat, and so on. Access to these platforms was restricted by the so-called "Great Firewall of China," a series of different technical and legislative measures to censor content in the internet. It includes measures such as orders to the app stores managed by Apple and Google not to support specific apps.

In parallel, copycat platforms have been developed in China: Baidu for searches, Alibaba for e-commerce, Tencent's WeChat for instant messaging and social networks, and other platforms imitating Uber, Tinder, Twitter, and so on. China has both the power to block US platforms and the scale to develop their own platforms with massive mostly domestic network effects. In practice, China has decoupled from the global internet, while actively pursuing commercial ventures abroad: Tencent acquired minority stakes in the Swedish music platform Spotify, the US communications platform Snapchat, the US gaming platform Epic Games, and even Tesla. The Chinese platform TikTok became popular worldwide.

Russia has followed a similar strategy. US-based platforms have not been prohibited, but national alternatives have been developed and the pre-installation of such apps in all smartphones has been mandated: Yandex is the most popular search engine, VK is the most popular social network and FaceApp is a popular instant messaging platform.

The US has reacted to this asymmetry, with actions taken against Chinese platforms TikTok and WeChat. Even more relevantly, the US has prohibited the use of Huawei hardware for the development of 5G wireless networks. Huawei is becoming the leader in 5G technology amid accusations of cybersecurity risks, but also of disproportionate state aid from Chinese authorities. Huawei has an active strategy to digitalize infrastructure, in China and around the world, selling equipment to digitalize ports, airports, postal systems, etc.

It is commonly said that the US innovates, China imitates, and Europe regulates. While some European platforms are leaders in their markets, such as Amadeus, particularly in business-to-business platforms, the mass-market has been more of a challenge for European platforms.

However, the regulation of digital platforms by EU authorities should not be interpreted as a protectionist trade measure to favor the local industry, as it is sometimes criticized by the US. On the contrary, the weakness of the local industry has reduced regulatory capture in such a way that authorities have more room to build the tools to regulate the platforms.

Actually, the US actions against Chinese technology companies demonstrate how strategic data-generating platforms can get, as well as the gatekeeper role of the largest US platforms, particularly the app stores. Actions against Huawei show how relevant the risk of relying on a short list of gatekeepers for the functioning of the most basic infrastructures in a society is, whether they are communications, transport, or energy infrastructures.

The rise of the new network industries shows that digitalization has the power to transform the traditional network industries and to enable new digital players to platformize them. Regulating the new digital industries becomes necessary, particularly when they become active in general interest industries such as communications, transport, and energy.

References

Abeyratne, R. (2020). *Aviation in the Digital Age*. Springer, Cham.

ADEME/6T Bureau de recherche (2015). *Enquête auprès des utilisateurs du covoiturage longue distance*, September 2015.

AGCM, AGCOM, & Garante (2019). *Indagine cognoscitiva sui big data*. Report, retrieved from https://www.garanteprivacy.it/documents/10160/0/Indagine+conoscitiva+sui+Big+Data.pdf/58490808-c024-bf04-7e4e-e953b3d38a9a?version=1.0

Agrawal, A., Gans, J., & Goldfarb, A. (2018). *Prediction Machines. The Simple Economics of Artificial Intelligence*. Harvard Business Review Press, Boston, MA.

Alexiadis, P. & de Streel, A. (2020). *Designing an EU Intervention Standard for Digital Platforms*. EUI Working Papers, RSCAS 2020/14, retrieved from https://cadmus.eui.eu/bitstream/handle/1814/66307/RSCAS%202020_14.pdf?sequence=1&isAllowed=y

Alter. A. (2017). *Irresistible. The Rise of Addictive Technology and the Business of Keeping Us Hooked*. Penguin, New York.

Ammirati, S. (2016). *The Science of Growth. How Facebook Beat Friendster and How Nine Other Startups Left the Rest in the Dust*. St. Martin's Press, New York.

Andrews, D., Nicoletti G., & Timiliotis, C. (2018). *Digital technology diffusion: A matter of capabilities, incentives or both?* OECD Economics Department Working Papers, No. 1476, OECD Publishing, Paris, retrieved from http://dx.doi.org/10.1787/7c542c16-en

ARCEP (2019). *Plateformes numériques structurantes. Eléments de réflexion relatif à leur caractérisation*, December 2019.

Argenton, C. & Prüfer, J. (2012). Search engine competition with network externalities. *Journal of Competition Law & Economics*, 8(1), 73–105.

Arthur D. Little (1978). *The Impact of Electronic Communications Systems on First Class Mail Volume in 1980–1990*, Cambridge, MA, April.

Asimov, I. (1982). *On Numbers*. Random House, New York.

Auletta, K. (2009). *Google. The End of the World as We Know It*. Penguin Books, New York.

Auletta, K. (2018). *Frenemies. The Epic Disruption of the Ad Business (and Everything Else)*.Penguin Press, New York.

Australian Competition & Consumer Commission (2019). *Digital Platforms Enquiry*, retrieved from https://www.accc.gov.au/system/files/Digital%20platforms%20inquiry%20-%20final%20report.pdf

Autorité de la Concurrence & Bundeskartellamt (2016). *Competition Law and Data*. retrieved from https://www.bundeskartellamt.de/SharedDocs/Publikation/DE/Berichte/Big%20Data%20Papier.pdf?__blob=publicationFile&v=2

Autoriteit Consument & Markt (2019). *Market Study into Mobile App Stores*, Report, retrieved from https://www.acm.nl/sites/default/files/documents/market-study-into-mobile-app-stores.pdf

Baak, J. & Sioshansi, F. (2019). Integrated Energy Services, Load Aggregation, and Intelligent Storage. In F.Sioshansi (ed.), *Consumer, Prosumer, Prosumager. How Service Innovations Will Disrupt the Utility Business Model*, pp. 53–73, Academic Press and Elsevier, London.

Baker, J. (2019). *The Digital Paradigm. Restoring a Competitive Economy*. Harvard University Press, Cambridge, MA.

Baldwin, C. & Woodard, C. J. (2009). The Architecture of Platforms: A Unified View. In A. Gawer (ed.), *Platforms, Markets and Innovation*. Edward Elgar, Cheltenham, UK and Northampton, MA.

Baran, P. (1964). *On Distributed Communications: I. Introduction to Distributed Communications Networks*. RAND Corporation, Santa Monica, CA, retrieved from https://www.rand.org/content/dam/rand/pubs/research_memoranda/2006/RM3420.pdf

Bárd, P. & Bayer, J. (2016). *A Comparative Analysis of Media Freedom and Pluralism in the EU Member States*. Study for the LIBE Committee of the European Parliament, retrieved from https://www.europarl.europa.eu/thinktank/en/document.html?reference=IPOL_STU%282016%29571376

Berlin, L. (2014). The First Venture Capital Firm in Silicon Valley: Draper, Gaither & Anderson. In *Making of the American Century. Essays on the Political Culture of Twentieth Century America*, Oxford University Press, New York, pp. 155–170.

Berlin, L. (2017). *Trouble Makers. How a Generation of Silicon Valley Upstarts Invented the Future*. Simon & Schuster, London.

Berschadsky, A. (2000). RIIA v Napster: A window onto the future of copyright law in the internet age. *The John Marshall Journal of Information Technology & Privacy Law*, 18(3), 755.

Blanco, F.G.B. & Chen, H. (2014). The implementation of building information modelling in the United Kingdom by the transport industry. *Procedia–Social and Behavioral Sciences*, 138(2014), 510–520.

Blum, A. (2012). *Tubes. A Journey to the Center of the Internet*. HarperCollins, New York.

Bostoen, F. (2019). *Regulating Online Platforms: Lessons from 100 Years of Telecommunications Regulation*. Harvard Law School, Boston.

Botsman, R. & Rogers, R. (2010). *What's Mine Is Yours: How Collaborative Consumption Is Changing the Way We Live*. HarperCollins, New York.

Brafman, O. & Beckstrom, R. A. (2006). *The Starfish and the Spider. The Unstoppable Power of Leaderless Organizations*. Penguin, New York.

Brautigan, R. (1968). *The Pill versus the Springhill Mine Disaster*. Dell Publishing, New York.

Brin, S. & Page, L. (1998). *The Anatomy of Large-Scale Hypertextual Web Search Engine*, retrieved from https://storage.googleapis.com/pub-tools-public-publication-data/pdf/334.pdf.

Briscoe, B., Odlyzko, A., & Tilly, B. (2006). Metcalfe's law is wrong-communications networks increase in value as they add members-but by how much? *IEEE Spectrum*, 43(7), 34–39.

Brogi, E. et al. (2020). *Monitoring Media Pluralism in the Digital Era: Application of the Media Pluralism Monitor 2020 in the European Union, Albania & Turkey: Policy Report*, Centre for Media Pluralism and Media Freedom, Florence, retrieved from https://cadmus.eui.eu//handle/1814/67828

Bronson, P. (1999). *The Nudist on the Late Shift and Other True Tales of Silicon Valley*. Random House, New York.

Brown, M., Woodhouse, S., & Sioshansi, F. (2019). Digitalization of Energy. In *Consumer, Prosumer, Prosumager. How Service Innovations Will Disrupt the Utility Business Model*, Elsevier, London.

Bundeskartellamt (2016). *The Market Power of Platforms and Networks*, working paper, available at https://www.bundeskartellamt.de/SharedDocs/Publikation/EN/Berichte/Think-Tank-Bericht-Zusammenfassung.pdf?__blob=publicationFile&v=4

Bundesnetzagentur (2017). *Digital Transformation in the Network Sectors: Recent Developments and Regulatory Challenges*, Report, retrieved from https://www.bundesnetzagentur.de/SharedDocs/Downloads/DE/Sachgebiete/Telekommunikation/Unternehmen_Institutionen/Digitalisierung/Grundsatzpapier/KurzfassungDigitalisierungEN.pdf?__blob=publicationFile&v=1

Burgess, J. & Green. J. (2018). *YouTube*. Polity Press, Cambridge, 2nd edition.

Caillaud, B. & Jullien, B. (2003). Chicken andegg: Competition among intermediation service providers. *RAND Journal of Economics*, 309, 328.

Cantelon. P. L. (1993). *The History of MCI: 1968–1988. The Early Years*. MCI, Washington DC.

Carr, N. G. (2008). *The Big Switch: Rewiring the World, from Edison to Google*. W. W. Norton, New York.

CGDD (2016). Covoiturage longue distance: état des lieux et potentiel de croissance, *Etudes & Documents du Commissariat Général au Développement Durable*, no. 146, May 2016.

Chan, N. D. & Shaheen, S. A. (2012). Ridesharing in North America: Past, present and future. *Transport Review*, 32(1), 93–112.

Chernow, R. (1998). *Titan. The Life of John D. Rockefeller SR*. Vintage Books, New York.

Choudary, S. P. (2015). *Platform Scale*. Platform Thinking Labs Pte, Ltd.

Christensen, C. M. (1997). *The Innovator's Dilemma. When New Technologies Cause Great Firms to Fail*. Harvard Business Review Press, Cambridge MA.

Christian, B. & Griffiths, T. (2016). *Algorithms to Live By. The Computer Science of Human Decisions*. William Collins, London.

Clark, D. (2016). *Alibaba, The House that Jack Ma Built*. HarperCollins, New York.

Clewlow, R. & Mishra, G. S. (2017). *Disruptive Transportation. The Adoption, Utilization, and Impacts of Ride-Hailing in the United States*, Institute of Transportation Studies, University of California Davis, Research Report UCD-ITS-RR-17–07.

Coase, R. (1937). The nature of the firm. *Economica*, 4(16), 386–405.

Competition & Markets Authority (2020). *Online Platforms and Digital Advertising*, Report, retrieved from https://assets.publishing.service.gov.uk/media/5efc57ed3a6f4023d242ed56/Final_report_1_July_2020_.pdf

Competition Authorities Working Group on Digital Economy (2019). *BRICS in the Digital Economy. Competition Law in Practice*, retrieved from http://www.cade.gov.br/acesso-a-informacao/publicacoes-institucionais/brics_report.pdf

Conseil National du Numérique (2015). *Ambition numérique: Pour une politique française et européenne de la transition numérique*. Report, retrieved from https://www.entreprises.gouv.fr/files/files/directions_services/secteurs-professionnels/economie-numerique/CNNum--rapport-ambition-numerique.pdf

Costa, P., Montero, J. J., & Roson, R. (2018). *The Impact of Disruptive Technologies on Infrastructure Networks*, 11th Meeting of the Network of Economic Regulators, retrieved from http://www.dirittoepoliticadeitrasporti.it/wp-content/uploads/2018/12/OECD-The-impact-of-disruptive-technologies-on-infrastructure-networks-2-Nov-2018.pdf

Couture, T, Cader, C., & Blechinger, P. (2019). Off-Grid Prosumers: Electrifying the Next Billion with PAYGO Solar. In F. Sioshansi (ed.), *Consumer, Prosumer, Prosumager. How Service Innovations Will Disrupt the Utility Business Model*, pp. 311–329, Academic Press and Elsevier, London.

Crémer, J., de Montjoye, Y.-A., & Schweitzer, H. (2019). *Competition Policy for the Digital Era*, Report to the European Commission, retrieved from https://op.europa.eu/es/publication-detail/-/publication/21dc175c-7b76-11e9-9f05-01aa75ed71a1

CtrlShift (2018). *Data Mobility: The Personal Data Portability Growth Opportunity for the UK Economy*, report for the Department for Digital, Culture, Media & Sport, retrieved from https://assets.publishing.service.gov.uk/government/uploads/system/uploads/attachment_data/file/755219/Data_Mobility_report.pdf

Cusumano, M.A., Gawer, A., & Yoffie, D.B. (2019). *The Business of Platforms. Strategy in the Age of Digital Competition, Innovation, and Power*. Harper Business, New York.

Daneshkhah, A., Stocks, N.G., & Jeffrey, P. (2017). Probabilistic sensitivity analysis of optimised preventive maintenance strategies for deteriorating infrastructure assets. Reliability. *Engineering & System Safety*, 163, 33–45.

Davidson, J. et al. (2010). The YouTube Video Recommendation System, *Proceedings of the Fourth ACM Conference on Recommender Systems*. ACM, 2010, 293–296.

Domingos, P. (2015). *The Master Algorithm. How the Quest for the Ultimate Learning Machine Will Remake Our World*. Penguin Books, New York.

Economiedes, N. (1996). The Economics of Networks. *International Journal of Industrial Organization*, 14(2), 673–699.

Eli, J. W. (2001). *Railroads and American Law*. University Press of Kansas, Lawrence.

Evans, D. S. (2011). *Platform Economics: Essays on Multi-sided Businesses*. CPI.

Evans, D. S. & Schmalensee, R. (2005). *Paying with Plastic. The Digital Revolution in Buying and Borrowing*. MIT Press, Cambridge, MA, 2nd ed.

Evans, D. S. & Schmalensee, R. (2016). *Matchmakers. The New Economics of Multi-sided Platforms.* Harvard Business Review Press, Boston.

Evans, D. S. & Schmalensee, R. (2019). *Antitrust Analysis of Platform Markets. Why the Supreme Court Got it Right in American Expre*ss. Competition Policy International, Boston.Expert Group for the Observatory on the Online Platform Economy (2020).*Work Stream on Differentiated Treatment*, Progress Report, available at https://platformobservatory.eu/app/uploads/2020/07/ProgressReport_Workstream_on_Differentiated_treatment_2020.pdf

FCC (1939). *Investigation of the Telephone Industry in the United States.* US Government Printing Office, Washington DC.

FCC (1966). *Notice of Inquiry in the Matter of Regulatory and Policy Problems Presented by the Interdependence of Computer and Communications Services and Facilities.* Docket F.C.C. No. 16979, November 9, 1966.

Feasey, R. & Krämer, J. (2019). *Implementing Effective Remedies for Anti-competitive Intermediation Bias on Vertically Integrated Platforms.* Report for CERRE, retrieved from https://www.cerre.eu/sites/cerre/files/cerre_intermediationbiasremedies_report.pdf

Finger, M. (2019). *Network Industries. A Research Overview.* Routledge, New York.

Finger, M. & Montero, J.J. (2020). *Handbook on Railway Regulation: Concepts and Practice.* Edward Elgar, Cheltenham.

Finger, M., Bert, N., Kupfer, D., Montero, J. J., & Wolek, M. (2017). *Infrastructure Funding Challenges in the Sharing Economy*, Research for the TRAN Committee of the European Parliament, retrieved from http://www.europarl.europa.eu/RegData/etudes/STUD/2017/601970/IPOL_STU(2017)601970_EN.pdf

Finger, M., Bukovc, B., & Burhan, M. (eds.) (2014). *Postal Services in the Digital Age.* IOS Press, Amsterdam.Fisher, C. (1987). The Revolution in Rural Telephony. *Journal of Social History, 21*, 6.

Freeberg, E. (2014). *The Age of Edison. Electric Light and the Invention of Modern America.* Penguin Books, New York.

Frischmann, B. (2012). *Infrastructure. The Social Value of Shared Resources.* Oxford University Press, Oxford.

Fuentes, R., Hunt, L. C., Lopez-Ruiz, H., & Manzano, B. (2020). The "iPhone effect": The impact of dual technological disruptions on electrification. *Competition and Regulation in Network Industries, 21*(2), 110–123.

Furman, J. et al. (2019). *Unlocking Digital Competition.* Report of the Digital Competition Expert Panel, retrieved from https://assets.publishing.service.gov.uk/government/uploads/system/uploads/attachment_data/file/785547/unlocking_digital_competition_furman_review_web.pdf

Gabel, R. (1969). The early competitive era in telephone communication, 1893–1920. *Law and Contemporary Problems*, Spring, 343–346.

Gallagher, L. (2017). *The Airbnb Story.* Houghton Mifflin Harcourt, Boston.

Galloway, S. (2017). *The Four. The Hidden DNA of Amazon, Facebook and Google.* Portfolio/Penguin, New York.

Gans, J. (2016). *The Disruption Dilemma.* The MIT Press, Cambridge MA.

GART/UTP (2014). *Covoiturage et transports collectives: concurrence ou complementarite sur les deplacements longue distance?* July 2014.

Gawer, A. & Cusumano, M. A. (2002). *Platform Leadership.* Harvard Business School Publishing, Boston, MA.

Geradin, D. & Katsifis, D. (2018). *An EU Competition Law Analysis of Online Display Advertising in the Programmatic Age* (December 12, 2018), retrieved from SSRN: https://ssrn.com/abstract=3299931

Gerardin, D. & Katsifis, D. (2020). *The Antitrust Case against the Apple App Store*, retrieved from https://papers.ssrn.com/sol3/papers.cfm?abstract_id=3583029

Gillespie, T. (2018). *Custodians of the Internet. Platforms, Content Moderation, and the Hidden Decision that Shape Social Media.*Yale University Press, New Haven, CT.

Gitlin, M. (2011). *eBay. The Company and Its Founders*, ABDO Publishing Company, Edina MI.

Goldman, E. (2010). The Regulation of Reputational Information. In *The Next Digital Decade. Essays on the Future of the Internet*, TechFreedom, Washington.

Hafner, K. & Lyon, M. (1996). *Where the Wizards Stay Up Late. The Origins of the Internet*. Simon & Schuster Paperbacks, New York.

Hagiu, A. & Jullien, B. (2011). Why Do Intermediaries Divert Search? *The RAND Journal of Economics*, 42(2), 337–362.

Hall, J.D., Palsson, C., & Price, J. (2017). *Is Uber a Substitute or Complement for Public Transit?* Working Paper 585, University of Toronto, Department of Economics.

Harris, B. J. (2014). *Console Wars. Sega, Nintendo and the Battle that Defined a Generation*. HarperCollins, New York.

Heikkilä, S. (2014). *Mobility as a Service: A Proposal for Action for the Public Administration*, Case Helsinki, Master's Thesis, Aalto University.

Holmberg, P., Collado, M., Sarasini, S., & Williander, M. (2016). *Mobility as a Service: Describing the Framework*, Viktoria Swedish ICT AB, January 15, 2016.

Hosanagar, K. & Jair, V. (2018). We Need Transparency in Algorithms, But Too Much Can Backfire. *Harvard Business Review*, retrieved from https://hbr.org/2018/07/we-need-transparency-in-algorithms-but-too-much-can-backfire

House of Lords (2019). *Regulating in a Digital World*, Select Committee on Communications, 2nd Report of Session 2017–19.

Hovenkamp, E. (2018). Platform Antitrust. *Journal of Competition Law*, 2019, 49.

Hughes, T. P. (1979). The Electrification of America: The System Builders. *Technology and Culture*, 20 (1), 124–161.

Hughes, T. P. (1983). *Networks of Power. Electrification in Western Society, 1880–1930*. The Johns Hopkins University Press, Baltimore.

Hughes, T. P. (1987). The Evolution of Large Technological Systems, *The Social Construction of Technological Systems. New Directions in the Sociology and History of Technology*. MIT Press, Boston, 2nd ed. 2012.

Iansiti, M. & Lakhani, K. R. (2020). *Competing in the Age of AI. Strategy and Leadership when Algorithms and Networks Run the World*, Harvard University Press. Cambridge, MA.

Isaac, M. (2019). *Superpumped. The Battle for Uber*. W.W. Norton & Company, New York.

Isaacson, W. (2011). *Steve Jobs*. Abacus, London.

Japan Fair Trade Commission (2019). *Report Regarding Trade Practices on Digital Platforms*, retrieved from https://www.jftc.go.jp/en/pressreleases/yearly-2019/October/191031Report.pdf

Jardine, A. K., Lin, D., & Banjevic, D. (2006). A review on machinery diagnostics and prognostics implementing condition-based maintenance. *Mechanical Systems and Signal Processing*, 20(7), 1483–1510.

Kalra, N. & Paddock, S. (2016). Driving to safety: How many miles of driving would it take to demonstrate autonomous vehicle reliability? *Transportation Research Part A: Policy and Practice*, 94: 182–193.

Kamargianni, M., Matyas, M., Li, W. & Schäfer, A. (2015). *Feasibility Study for "Mobility as a Service" Concept in London*. UCL Energy Institute for the Department for Transport.

Kaplan, D. A. (2000). *The Silicon Boys and Their Valley of Dreams*. HarperCollins, New York.

Khan, L. M. (2016). Amazon's antitrust paradox, *Yale Law Journal*, 126, 710.

Khan, L. M. (2019). The separation of platforms and commerce. *Columbia Law Review*, 119, 973, retrieved from https://papers.ssrn.com/id=3180174

Kirkpatrick, D. (2010). *The Facebook Effect*. Virgin Books, New York.

Knieps, G. (2017). Internet of Things, future networks, and the economics of virtual networks. *Competition and Regulation in Network Industries*, 18(3–4), 240–255.

Knieps, G. (2019). Network economics of operator platforms. *Networks Industry Quarterly*, 21(3), 17.

Knight, W. (2017). The US military wants its autonomous machines to explain themselves, *MIT Technology Review*, retrieved from https://www.technologyreview.com/s/603795/the-us-military-wants-its-autonomous-machines-to-explain-themselves/

Koene, A., Clifton, C., Hatada, Y., Webb, H., & Richardson, R. (2019). *A Governance Framework for Algorithmic Accountability and Transparency*, European Parliament Research Service, retrieved from https://www.europarl.europa.eu/thinktank/en/document.html?reference=EPRS_STU(2019)624262

Krämer, J., Schnurr, D., & de Streel, A. (2017). *Internet Platforms and Non-Discrimination*, CERRE Project Report, retrieved from https://www.cerre.eu/sites/cerre/files/171205_CERRE_PlatformNonDiscrimination_FinalReport.pdf

Krattenmaker, T. (1996). The Telecommunications Act of 1996. *Federal Communications Law Journal*, 29, 1.

Lashinsky, A. (2017). *Wild Ride. Inside Uber's Quest for World Domination*. Penguin, New York.

Layne-Farrar, A., Padilla, A. J. & Schmalensee, R. (2007). Pricing Patents for Licensing in Standard-Setting Organizations: Making Sense of FRAND Commitments. *Antitrust Law Journal*, 74, 671.

Levison, M. (2016). *The Box. How the Shipping Container Made the World Smaller and the World Economy Bigger*. Princeton University Press, Princeton and Oxford.

Levy, S. (2020). *Facebook. The Inside Story*. Penguin, New York.

Libert, B., Beck, M., & Wind, J. (2016). *The Network Imperative. How to Survive and Grow in the Age of Digital Business Models*. Harvard Business Review Press, Boston, MA.

Licklider, J. C. R. (1960). *Man-Computer Symbiosis*. IRE Transactions on Human Factors in Electronics, Volume: HFE-1, Issue: 1.

Lynskey, O. (2017). *Regulating 'Platform Power'*. LSE Law, Society and Economy Working Papers 1/2017, retrieved from http://eprints.lse.ac.uk/73404/1/WPS2017-01_Lynskey.pdf

MAIF (2009). *Usages et attitudes des utilisateurs du site Internet covoiturage.fr*, December 2009, retrieved from https://www.maif.fr/content/pdf/particuliers/auto-moto/covoiturage/maif-etude-covoiturage-12-2009.pdf. More altruistic reasons were also considered: protection of the environment (12 percent) or providing a service (4 percent).

McAfee, A. & Brynjolsson, E. (2017). *Machine, Platform Crowd: Harnessing Our Digital Future*. W. W. Norton & Company, New York.

McKinsey & Co. (2016). *Huge Value Pool Shifts Ahead–How Rolling Stock Manufacturers Can Lay Track for Profitable Growth*.Report, September 2016.

McLuhan, M. (1964). *Understanding Media. The Extensions of Man*. MIT Press, Cambridge, MA edition, 1994.

Montero, J. (2019). Asymmetric regulation for competition in European railways? *Competition and Regulation in Network Industries*, 20(2), 184–201.

Montero, J. (2019). Regulating Transport Platforms: The Case of Carpooling in Europe. M. Finger & M. Audouin (eds.), *The Governance of Smart Transportation Systems*. Springer, Cham, 13–35.

Montero, J. (2020). The Digitalization Dilemma in the Railways Industry. In M. Finger & J. Montero (eds.), *Handbook on Railway Regulation. Concepts and Practice*. Edward Elgar, Cheltenham, UK. and Northampton, MA.

Montero, J. & Finger, M. (2017). "Platformed! Network industries and the new digital paradigm," *Competition and Regulation in Network Industries*, 18(3–4), 217–239.

Montero, J. & Finger, M. (2021). *The Modern Guide to the Digitalization of Infrastructures*. Edward Elgar, Cheltenham.

Monti, G. & Augenhofer, S. (1998). *Consumer Choice and Fair Competition on the Digital Single Market in the Areas of Air Transportation and Accommodation*, Research for the IMCO Committee of the European Parliament, retrieved from https://www.europarl.europa.eu/RegData/etudes/STUD/2018/626082/IPOL_STU(2018)626082_EN.pdf

Mueller, M. (1997). *Universal Service. Competition, Interconnection and Monopoly in the Making of the American Telephone System*. MIT/AEI Press, Boston.

Munger, M. C. (2015). Coase and the "Sharing Economy." In *Forever Contemporary: The Economics of Ronald Coase*. Institute of Economic Affairs, London, 187–208.

Nidermeyer, E. (2019). *Ludicrous. The Unvarnished Story of Tesla Motors*. BenBella Books, Dallas (TX).

Nikolic, I. & Heim, M. (2019). A FRAND regime for dominant digital platforms. *Journal of Intellectual Property, Information Technology and E-Commerce Law*, 18, 38–55.

O'Mara, M. (2019). *The Code*. Penguin Press, New York.

O'Neil, C. (2016). *Weapons of Math Destruction: How Big Data Increases Inequality and Threatens Democracy*. Broadway Books, New York.

OECD (2009). *Two-Sided Markets*, Policy Roundtables, retrieved from https://www.oecd.org/daf/competition/44445730.pdf

OECD (2016). *Structural Separation in Regulated Industries: Report on Implementing the OECD Recommendation*, retrieved from https://www.oecd.org/daf/competition/Structural-separation-in-regulated-industries-2016report-en.pdf

OECD (2018). *Rethinking Antitrust Tools for Multi-sided Platforms*, retrieved from https://www.oecd.org/daf/competition/Rethinking-antitrust-tools-for-multi-sided-platforms-2018.pdf

OECD (2019). *Digital Dividend: Policies to Harness the Productivity Potential of Digital Technologies*. Economic Policy Paper No. 26, OECD Publishing, Paris. Retrieved from https://doi.org/10.1787/273176bc-en

Paine, A. B. (1921). *In One Man's Life*. Forgotten Books, 2012 ed.

Parcu, P. L. (2020). New digital threats to media pluralism in the information age. *Competition and Regulation in Network Industries*, 21(2), 91–109.

Parker, G. G., Van Alstyne, M. W., & Choudary, S. P. (2016). *Platform Revolution*. W. W. Norton & Company, New York.

Pascuale, F. (2015). *The Black Box Society. The Secret Algorithms that Control Money and Information.*Harvard University Press, Cambridge, MA.

Penenberg, A. L. (2009). *Viral Loop. The Power of Pass-it-on*. Hyperion Books, New York.

Picard, R. G. (2001). Strategic responses to free distribution daily newspapers. *The International Journal on Media Management*, 3(3), 167–172.

Randolph, M. (2019). *That Will Never Work. The Birth of Netflix and the Amazing Life of an Idea*. Endeavour, London.

Ravich, T. M. (2004). Deregulation of the Airline Computer Reservation Systems (CRS) Industry. *Journal of Air Law and Commerce*, 69, 387.

Rill, J., Wilson, C., & Bauers, S. (2000). The Amadeus Global Travel Distribution Case. In S. J. Evenett, A. Lehman, and B. Steil (eds.), *Antitrust Goes Global, What Future for Transatlantic Cooperation?*Brookings Institution Press, Washington DC.

Rochet, J.C. & Tirole, J. (2003). Platform competition in two-sided markets, *Journal of the European Economic Association*, 1(4), 990–1029.

Ryan, J. (2010). A *History of the Internet and the Digital Future*. Reaktion Books, London.

Schmidt, E. & Rosenberg, J. (2014). *How Google Works*. John Murray, London.

Schumpeter, J. A. (1942). *Capitalism, Socialism and Democracy*. Routledge, London.

Schwab, K. (2016). *The Fourth Industrial Revolution*. Penguin, New York

Seidenberg, I. & McMurray, S. (2018). *Verizon Untethered. An Insider's Story of Innovation and Disruption.*Post Hill Press, New York.

Shaller, B. (2017). *Empty Seats, Full Streets. Fixing Manhattan's Traffic Problem*, Schaller Consulting, retrieved from http://schallerconsult.com/rideservices/emptyseats.pdf

Shy, O. (2011). A short survey of network economics. *Review of Industrial Organization*, 38, 119–149.

Simon, P. (2011). *The Age of Platform*. Motion Publishing, Henderson, NE.

Sismanidou, A., Palacios, M., & Tafur, J. (2008). New developments in Global Distribution Systems (GDSs) for the airline industry: first-mover mechanisms that enabled incumbent firms to maintain a leading position, *IIInternational Conference on Industrial Engineering and Industrial Management*, pp. 727–734.

Soni, J. (2021). *The Founders: The Story of PayPal and the Entrepreneurs Who Shaped Silicon Valley*. Simon & Schuster, New York.

Srnicek, N. (2017). *Platform Capitalism*. Polity Press, Cambridge.

Standage, T. (1998). *The Victorian Internet. The Remarkable Story of the Telegraph and the Nineteenth Century's On-line Pioneers*. Bloomsbury, New York.

Starr, P. (2005). *The Creation of Media: Political Origins of Modern Communications*. Basic Books, New York.

Stasi, M. L. (2020). *Ensuring Pluralism in Social Media Markets: Some Suggestions*, Robert Schuman Centre for Advanced Studies Research Paper No. RSCAS 2020/05, retrieved from https://papers.ssrn.com/sol3/papers.cfm?abstract_id=3531794

Stigler Committee on Digital Platforms, Final Report, September 2019, retrieved from https://research.chicagobooth.edu/stigler/media/news/committee-on-digital-platforms-final-report

Stone, B. (2013). *The Everything Store. Jeff Bezos and the Age of Amazon.* Corgi Books, London.

Stone, B. (2017). *The Upstarts. How Uber, Airbnb and the Killer Companies of the New Silicon Valley are Changing the World.* Transworld Publishers, London.

Stross, R. (2007). *The Wizard of Menlo Park.* Three Rivers Press, New York.

Strouse, J. (2000). *Morgan. American Financier.* Harper Perennial, New York.

Succar, B. (2009). Building information modelling framework: A research and delivery foundation for industry stakeholders. *Automation in Construction*, 18(3), 357–375.

SUMC (2016). *Shared Mobility and the Transformation of Public Transit*, Research Analysis for the American Public Transportation Association. TCRP Report 188 Pre-Publication Draft–Subject to Revision, retrieved from www.tcrponline.org/PDFDocuments/tcrp_rpt_188.pdf

Sundararajan, A. (2016). *The Sharing Economy: The End of Employment and the Rise of Crowd-Based Capitalism.* MIT Press, Cambridge, MA.

Temin, P. (1987). *The Fall of the Bell System.* Cambridge University Press, New York.

Thiel, P. (2014). *Zero to One. Notes on Startups, of How to Build the Future.* Virgin Books, New York.

Tiwana, A. (2014). *Platform Ecosystems.* Elsevier, Waltham, MA.

Tseng, P. H., Lin, D. Y., & Chien, S. (2014). Investigating the impact of highway electronic toll collection to the external cost: A case study in Taiwan. *Technological Forecasting and Social Change*, 86, 265–272.

UITP (2016). Public transport at the heart of the integrated urban mobility solution, Policy Brief, April 2016, retrieved from http://www.uitp.org/sites/default/files/cck-focus-papers-files/Public%20transport%20at%20the%20heart%20of%20the%20integrated%20urban%20mobility%20solution.pdf (08–02-2017).

Valdiani Vicari & Associati (2019). *Study on Market Access and Competition issues Related to MaaS*, June 2019, retrieved from https://maas-alliance.eu/wp-content/uploads/sites/7/2019/07/Market-access-and-competition-issues-related-to-MaaS_final_040719.pdf

van Damme, E. E. C., Filistrucchi, L., Geradin, D., Keunen, S., Klein, T. J., Michielsen, T. O., & Wileur, J. (2010). *Mergers in Two-Sided Markets–A Report to the NMA.* Netherlands Competition Authority, pp. 1–183.

Van Dijck, J., Poell, T., & De Waal, M. (2018). *The Platform Society. Public Values in a Connective World.* Oxford University Press, Oxford.

Vance, A. (2015). *Elon Musk. How the Billionaire CEO of SpaceX and Tesla is Shaping Our Future.* Virgin Books, London.

Wappelhorst, S., Hall, D., Nicholas, M., & Lutsey, N. (2020). *Analyzing Policies to Grow the Electricity Vehicle Market in European Cities*, White Paper of the International Council on Clean Transportation, retrieved from https://theicct.org/sites/default/files/publications/EV_city_policies_white_paper_fv_20200224.pdf

Weinberger, S. (2017). *The Imagineers of War. The Untold Story of DARPA, the Pentagon Agency That Changed the World.* Vintage Books, New York.

Weyl, G. & White, A. (2014). Let the Right "One" Win: Policy Lessons from the New Economics of Platforms, *Competition Policy International*, 12(2), 29–51.

Wood, A. et al. (1977). *USPS and the Communications Revolution: Impacts, Options, and Issues.* Final Report to the Commission on Postal Service, prepared by the Program of Policy Studies in Science and Technology, The George Washington University, Washington, D. C., March 5, 1977.

Wu, T. (2003). Network neutrality, broadband discrimination. *Journal of Telecommunications and High Technology Law*, 2, 141, retrieved from SSRN: https://ssrn.com/abstract=388863.

Wu, T. (2010). *The Master Switch. The Rise and Fall of Information Empires.* Alfred A. Knopf, New York.

Wu, T. (2016). *The Attention Merchants. From the Daily Newspaper to the Social Media, How Our Time and Attention is Harvested and Sold*. Atlantic Books, London.

Wu, T. (2018). *The Curse of Bigness. Antitrust in the New Gilded Age*. Columbia Global Reports, New York.

Wylia, C. (2019). *Mindf*ck. Inside Cambridge Analytica's Plot to Break the World*. Profile Books, London.

Zhou, R., Khemmarat, S., & Gao, L. (2010). The Impact of YouTube Recommendation System on Video Views, *Proceedings of the 10th ACM SIGCOMM Conference on Internet Measurement*, 404–410.

Zuboff, S. (2019). *The Age of Surveillance Capitalism. The Fight for a Human Future at the New Frontier of Power*. Profile Books, London.

Index

For Product Safety Concerns and Information please contact our EU
representative GPSR@taylorandfrancis.com
Taylor & Francis Verlag GmbH, Kaufingerstraße 24, 80331 München, Germany

www.ingramcontent.com/pod-product-compliance
Lightning Source LLC
Chambersburg PA
CBHW081141180526
45170CB00006B/1887

9 780367 693053